McGraw Hill
Illustrative Mathematics®
Algebra 1

Mc Graw Hill

Cover Credit: Anna Bliokh/iStockphoto/Getty Images

mheducation.com/prek-12

Send all inquiries to:
McGraw Hill
8787 Orion Place
Columbus, OH 43240

ISBN: 978-0-07-693043-2
MHID: 0-07-693043-2

Illustrative Mathematics, Algebra 1
Student Edition, Volume 1

Printed in the United States of America.

9 10 11 12 13 MER 28   27 26 25 24 23

Contents in Brief

Welcome to

McGraw Hill
Illustrative
Mathematics®

Illustrative Mathematics is designed to provide opportunities for you to learn math by doing math. By engaging in problem-solving activities, you will make and test your own conclusions and try out different strategies as you learn math.

Let's get started!

Unit 1

One-variable Statistics

Unit 2

Linear Equations, Inequalities, and Systems

Syda Productions/Shutterstock

Systems of Linear Inequalities in Two Variables

Unit 3

Two-variable Statistics

sauce7/123RF

Unit 4

Functions

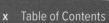

Functions and Their Representations

Analyzing and Creating Graphs of Functions

A Closer Look at Inputs and Outputs

Inverse Functions

Putting it All Together

Phoenixns/Shutterstock

Unit 5

Introduction to Exponential Functions

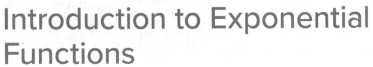

Unit 6

Introduction to Quadratic Functions

Unit 7

Quadratic Equations

MIND AND I/Shutterstock

Vertex Form Revisited

Putting It All Together

One-variable Statistics

Gardeners can use data gathered from the soil to determine how much fertilizer needs to be applied. You will learn more about analyzing data in this unit.

Kingarion/Shutterstock

Topics

- Getting to Know You
- Distribution Shapes
- How to Use Spreadsheets
- Manipulating Data
- Analyzing Data

Unit 1

One-variable Statistics

Lesson 1-1

Getting to Know You

NAME _____ DATE _____ PERIOD _____

Learning Goal Let's work together to collect data and explore statistical questions.

Warm Up

1.1 Which One Doesn't Belong?: Types of Data

Which one doesn't belong?

Question A: How many potato chips are in this bag of chips?

Question B: What is the typical number of chips in a bag of chips?

Question C: What type of chips are these?

Question D: What type of chips do students in this class prefer?

Activity

1.2 Representing Data About You and Your Classmates

Your teacher will assign you a set of 3 questions.

- Write another question of your own that will require data collected from the class to answer.

- For each of the 4 questions, write a survey question that will help you collect data from the class that can be analyzed to answer the questions.

- Ask the 4 survey questions to 15 classmates and record their responses to collect data.

- After collecting the data return to your group.

1. What is the question of your own that will require data collected from the class to answer?

2. What are the 4 survey questions you will ask your classmates?

3. Summarize the data for each question in a sentence or two and share the results with your group.

4. With your group, decide what the responses for questions numbered 1 have in common. Then do the same for questions numbered 2 and 3.

5. Does the question you wrote fit best with the questions numbered 1, 2, or 3? Explain your reasoning.

NAME _____ DATE _____ PERIOD _____

Responder's Name	Question 1 Response	Question 2 Response	Question 3 Response	My Question Response

Responder's Name	Question 1 Response	Question 2 Response	Question 3 Response	My Question Response

Are you ready for more?

1. Find a news article that uses numerical data to discuss a statistical question.

2. Find a news article that uses categorical data to discuss a statistical question.

NAME _____ DATE _____ PERIOD _____

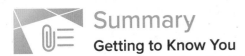

Summary
Getting to Know You

Statistics is about using data to solve problems or make decisions. There are two types of data:

- **Numerical data** are expressed using a number. For example, to answer the question "How tall are the students in this class?" you would measure the height of each student which would result in numerical data.

- **Categorical data** are expressed using characteristics. For example, to answer the question "What brand of phones do people use?" you would survey several people and their answers result in categorical data.

The question that you ask determines the type of data that you collect and whether or not there is *variability* in the data collected. In earlier grades, you learned that there is variability in a data set if all of the values in the data set are not the same. These are examples of **statistical questions** because they are answered by collecting data that has variability:

- "What is the average class size at this school?" would produce numerical data with some variability.

- "What are the favorite colors of students in this class?" would produce categorical data with some variability.

These are examples of **non-statistical questions** because they are answered by collecting data that does not vary:

- "How many students are on the roster for this class?" would produce numerical data that does not vary. There is only one value in the data set, so there is no variability.

- "What color is this marker?" would produce categorical data that does not vary. There is only one value in the data set, so there is no variability.

Glossary

categorical data
non-statistical
numerical data
statistical question

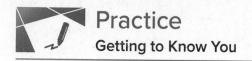
1. Write a survey question for which you would expect to collect numerical data.

2. Write a survey question for which you would expect to collect categorical data.

3. Select **all** the statistical questions.

   A. What is the typical amount of rainfall for the month of June in the Galapagos Islands?

   B. How much did it rain yesterday at the Mexico City International Airport?

   C. Why do you like to listen to music?

   D. How many songs does the class usually listen to each day?

   E. How many songs did you listen to today?

   F. What is the capital of Canada?

   G. How long does it typically take for 2nd graders to walk a lap around the track?

Lesson 1-2

Data Representations

NAME _____ DATE _____ PERIOD _____

Learning Goal Let's represent and analyze data using dot plots, histograms, and box plots.

Warm Up

2.1 Notice and Wonder: Battery Life

The dot plot, histogram, and box plot summarize the hours of battery life for 26 cell phones constantly streaming video. What do you notice? What do you wonder?

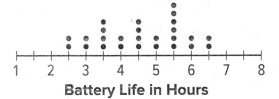

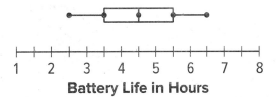

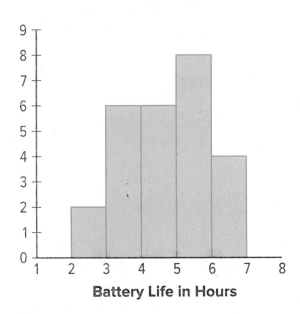

Activity

2.2 Tomato Plants: Histogram

A histogram can be used to represent the distribution of numerical data.

1. The data represent the number of days it takes for different tomato plants to produce tomatoes. Use the information to complete the frequency table.

47	52	53	55	57
60	61	62	63	65
65	65	65	68	70
72	72	75	75	75
76	77	78	80	81
82	85	88	89	90

Days to Produce Fruit	Frequency
40–50	
50–60	
60–70	
70–80	
80–90	
90–100	

2. Use the set of axes and the information in your table to create a histogram.

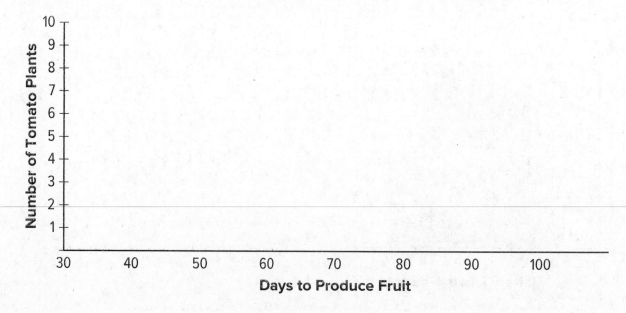

NAME _____ DATE _____ PERIOD _____

3. The histogram you created has intervals of width 10 (like 40–50 and 50–60). Use the set of axes and data to create another histogram with an interval of width 5. How does this histogram differ from the other one?

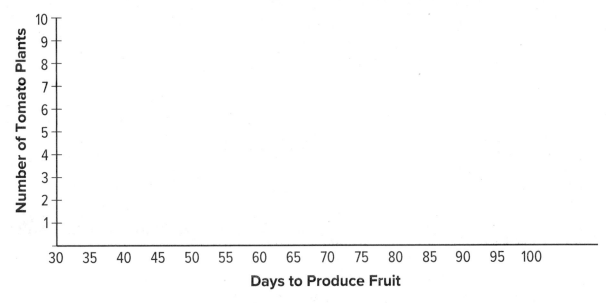

Days to Produce Fruit

Are you ready for more?

It often takes some playing around with the interval lengths to figure out which gives the best sense of the shape of the distribution.

1. What might be a problem with using interval lengths that are too large?

2. What might be a problem with using interval lengths that are too small?

3. What other considerations might go into choosing the length of an interval?

Activity

2.3 Tomato Plants: Box Plot

A box plot can also be used to represent the distribution of numerical data.

Minimum	Q1	Median	Q3	Maximum

1. Using the same data as the previous activity for tomato plants, find the median and add it to the table. What does the median represent for these data?

2. Find the median of the least 15 values to split the data into the first and second quarters. This value is called the first quartile. Add this value to the table under Q1. What does this value mean in this situation?

3. Find the value (the third quartile) that splits the data into the third and fourth quarters and add it to the table under Q3. Add the minimum and maximum values to the table.

4. Use the **five-number summary** to create a box plot that represents the number of days it takes for these tomato plants to produce tomatoes.

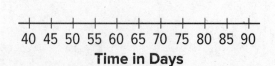

40 45 50 55 60 65 70 75 80 85 90
Time in Days

NAME _____ DATE _____ PERIOD _____

Summary
Data Representations

The table shows a list of the number of minutes people could intensely focus on a task before needing a break. 50 people of different ages are represented. In a situation like this, it is helpful to represent the data graphically to better notice any patterns or other interesting features in the data. A dot plot can be used to see the shape and **distribution** of the data.

19	7	1	16	20	3	7	19	9	13
3	9	18	13	20	8	3	14	13	2
8	5	17	7	18	17	8	8	7	6
2	20	7	7	10	7	6	19	3	18
8	19	7	13	20	14	6	3	19	4

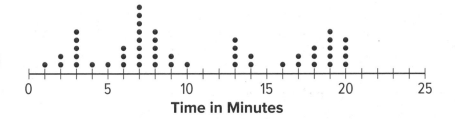

Time in Minutes

There were quite a few people that lost focus at around 3, 7, 13, and 19 minutes and nobody lost focus at 11, 12, or 15 minutes. Dot plots are useful when the data set is not too large and shows all of the individual values in the data set. In this example, a dot plot can easily show all the data. If the data set is very large (more than 100 values, for example) or if there are many different values that are not exactly the same, it may be hard to see all of the dots on a dot plot.

A histogram is another representation that shows the shape and distribution of the same data.

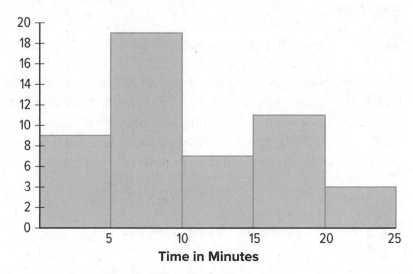

Most people lost focus between 5 and 10 minutes or between 15 and 20 minutes, while only 4 of the 50 people got distracted between 20 and 25 minutes. When creating histograms, each interval includes the number at the lower end of the interval but not the upper end. For example, the tallest bar displays values that are greater than or equal to 5 minutes but less than 10 minutes. In a histogram, values that are in an interval are grouped together. Although the individual values get lost with the grouping, a histogram can still show the shape of the distribution.

Here is a box plot that represents the same data.

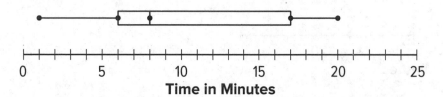

Box plots are created using the **five-number summary**. For a set of data, the five-number summary consists of these five statistics: the minimum value, the first quartile, the median, the third quartile, and the maximum value. These values split the data into four sections each representing approximately one-fourth of the data. The median of this data is indicated at 8 minutes and about 25% of the data falls in the short second quarter of the data between 6 and 8 minutes. Similarly, approximately one-fourth of the data is between 8 and 17 minutes. Like the histogram, the box plot does not show individual data values, but other features such as quartiles, range, and median are seen more easily. Dot plots, histograms, and box plots provide 3 different ways to look at the shape and distribution while highlighting different aspects of the data.

Glossary

distribution

five-number summary

NAME _____ DATE _____ PERIOD _____

Practice
Data Representations

1. The dot plot displays the number of bushes in the yards for houses in a neighborhood. What is the median?

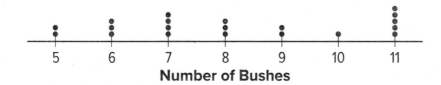

Number of Bushes

2. The data set represents the shoe sizes of 19 students in a fifth-grade physical education class.

   4, 5, 5, 5, 6, 6, 6, 6, 7, 7, 7, 7, 7.5, 7.5, 8, 8, 8.5, 8.5, 9

   Create a box plot to represent the distribution of the data.

3. The data set represents the number of pages in the last book read by each of 20 students over the summer.

   163, 170, 171, 173, 175, 205, 220, 220, 220, 253, 267, 281, 305, 305, 305, 355, 371, 388, 402, 431

   Create a histogram to represent the distribution of the data.

4. Each set of data was collected from surveys to answer statistical questions. Select **all** of the data sets that represent numerical data. (Lesson 1-1)

   (A.) {1, 1.2, 1.4, 1.4, 1.5, 1.6, 1.8, 1.9, 2, 2, 2.1, 2.5}

   (B.) {Red, Red, Yellow, Yellow, Blue, Blue, Blue}

   (C.) {45, 60, 60, 70, 75, 80, 85, 90, 90, 100, 100, 100}

   (D.) {-7, -5, -3, -1, -1, -1, 0}

   (E.) {98.2, 98.4, 98.4, 98.6, 98.6, 98.6, 98.6, 98.7, 98.8, 98.8}

   (F.) {Yes, Yes, Yes, Yes, Maybe, Maybe, No, No, No}

   (G.) {A, A, A, B, B, B, C, C, C}

5. Is "What is the typical distance a moped can be driven on a single tank of gas?" a statistical question? Explain your reasoning. (Lesson 1-1)

Lesson 1-3

A Gallery of Data

NAME _____ DATE _____ PERIOD _____

Learning Goal Let's make, compare, and interpret data displays.

 ## Warm Up
3.1 Notice and Wonder: Dot Plots

The dot plots represent the distribution of the amount of tips, in dollars, left at 2 different restaurants on the same night.

What do you notice? What do you wonder?

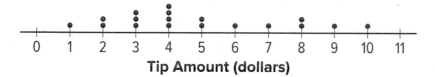

Tip Amount (dollars)

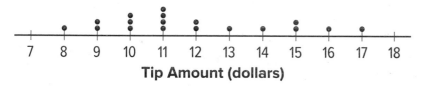

Tip Amount (dollars)

Activity

3.2 Data Displays

Your teacher will assign your group a statistical question. As a group:

1. Create a dot plot, histogram, and box plot to display the distribution of the data.

2. Write 3 comments that interpret the data.

As you visit each display, write a sentence or two summarizing the information in the display.

Are you ready for more?

Choose one of the more interesting questions you or a classmate asked and collect data from a larger group, such as more students from the school. Create a data display and compare results from the data collected in class.

NAME _____ DATE _____ PERIOD _____

Summary
A Gallery of Data

We can represent a distribution of data in several different forms, including lists, dot plots, histograms, and box plots. A list displays all of the values in a data set and can be organized in different ways. This list shows the pH for 30 different water samples.

5.9	7.6	7.5	8.2	7.6	8.6
8.1	7.9	6.1	6.3	6.9	7.1
8.4	6.5	7.2	6.8	7.3	8.1
5.8	7.5	7.1	8.4	8.0	7.2
7.4	6.5	6.8	7.0	7.4	7.6

Here is the same list organized in order from least to greatest.

5.8	5.9	6.1	6.3	6.5	6.5
6.8	6.8	6.9	7.0	7.1	7.1
7.2	7.2	7.3	7.4	7.4	7.5
7.5	7.6	7.6	7.6	7.9	8.0
8.1	8.1	8.2	8.4	8.4	8.6

With the list organized, you can more easily:

- interpret the data
- calculate the values of the five-number summary
- estimate or calculate the mean
- create a dot plot, box plot, or histogram

Here is a dot plot and histogram representing the distribution of the data in the list.

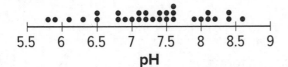

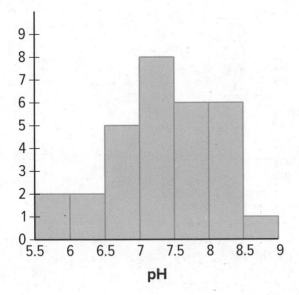

A dot plot is created by putting a dot for each value above the position on a number line. For the pH dot plot, there are 2 water samples with a pH of 6.5 and 1 water sample with a pH of 7. A histogram is made by counting the number of values from the data set in a certain interval and drawing a bar over that interval at a height that matches the count. In the pH histogram, there are 5 water samples that have a pH between 6.5 and 7 (including 6.5, but not 7). Here is a box plot representing the distribution of the same data as the dot plot and histogram.

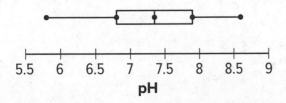

To create a box plot, you need to find the minimum, first quartile, median, third quartile, and maximum values for the data set. These 5 values are sometimes called the *five-number summary*. Drawing a vertical mark and then connecting the pieces as in the example creates the box plot. For the pH box plot, we can see that the minimum is about 5.8, the median is about 7.4, and the third quartile is around 7.9.

NAME _____ DATE _____ PERIOD _____

Practice
A Gallery of Data

1. The box plot represents the distribution of speeds, in miles per hour, of 100 cars as they passed through a busy intersection.

   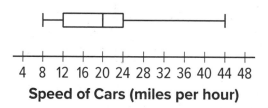

   Speed of Cars (miles per hour)

   a. What is the smallest value in the data set? Interpret this value in the situation.

   b. What is the largest value in the data set? Interpret this value in the situation.

   c. What is the median? Interpret this value in the situation.

   d. What is the first quartile (Q1)? Interpret this value in the situation.

   e. What is the third quartile (Q3)? Interpret this value in the situation.

2. The data set represents the number of eggs produced by a small group of chickens each day for ten days: 7, 7, 7, 7, 7, 8, 8, 8, 8, 9. Select **all** the values that could represent the typical number of eggs produced in a day.

   (A.) 7.5 eggs

   (B.) 7.6 eggs

   (C.) 7.7 eggs

   (D.) 8 eggs

   (E.) 9 eggs

3. The dot plot displays the lengths of pencils (in inches) used by students in a class. What is the mean? (Lesson 1-2)

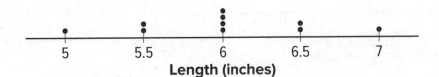

Length (inches)

4. The histogram represents ages of 40 people at a store that sells children's clothes. Which interval contains the median? (Lesson 1-2)

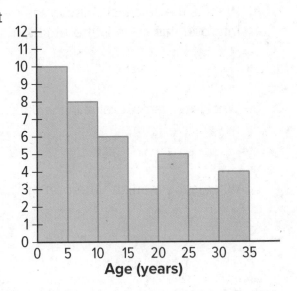

Age (years)

A. The interval from 0 to 5 years.

B. The interval from 5 to 10 years.

C. The interval from 10 to 15 years.

D. The interval from 15 to 20 years.

NAME _____ DATE _____ PERIOD _____

5. The data set represents the responses, in degrees Fahrenheit, collected to answer the question "How hot is the sidewalk during the school day?".
(Lesson 1-2)

92, 95, 95, 95, 98, 100, 100, 100, 103, 105, 105, 111, 112, 115, 115, 116, 117, 117, 118, 119, 119, 119, 119, 119, 119

a. Create a dot plot to represent the distribution of the data.

b. Create a histogram to represent the distribution of the data.

c. Which display gives you a better overall understanding of the data? Explain your reasoning.

6. Is "What is the area of the floor in this classroom?" a statistical question? Explain your reasoning. (Lesson 1-1)

Lesson 1-4

The Shape of Distributions

NAME _____ DATE _____ PERIOD _____

Learning Goal Let's explore data and describe distributions.

Warm Up
4.1 Which One Doesn't Belong: Distribution Shape

Which one doesn't belong?

A.

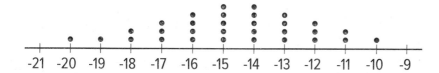

B.

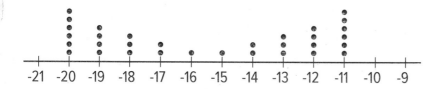

C.

D.

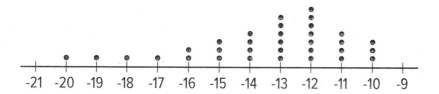

Activity

4.2 Matching Distributions

Take turns with your partner matching 2 different data displays that represent the distribution of the same set of data.

1. For each set that you find, explain to your partner how you know it's a match.

2. For each set that your partner finds, listen carefully to their explanation. If you disagree, discuss your thinking and work to reach an agreement.

3. When finished with all ten matches, describe the shape of each distribution.

NAME _____ DATE _____ PERIOD _____

Activity

4.3 Where Did The Distribution Come From?

Your teacher will assign you some of the matched distributions. Using the information provided in the data displays, make an educated guess about the survey question that produced this data. Be prepared to share your reasoning.

Are you ready for more?

This distribution shows the length in inches of fish caught and released from a nearby lake.

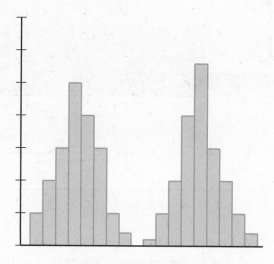

Length of Fish (inches)

1. Describe the shape of the distribution.

2. Make an educated guess about what could cause the distribution to have this shape.

We can describe the shape of distributions as *symmetric, skewed, bell-shaped, bimodal,* or *uniform*. Here is a dot plot, histogram, and box plot representing the distribution of the same data set. This data set has a symmetric distribution.

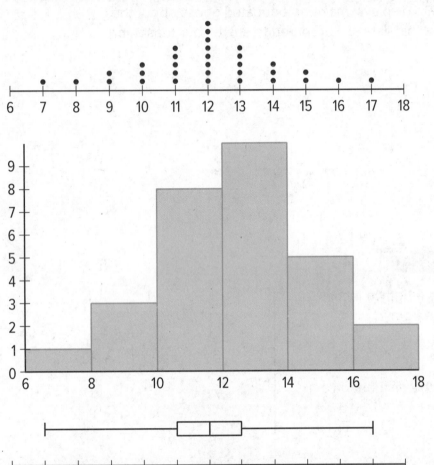

In a **symmetric distribution**, the mean is equal to the median and there is a vertical line of symmetry in the center of the data display. The histogram and the box plot both group data together. Since histograms and box plots do not display each data value individually, they do not provide information about the shape of the distribution to the same level of detail that a dot plot does. This distribution, in particular, can also be called bell-shaped. A **bell-shaped distribution** has a dot plot that takes the form of a bell with most of the data clustered near the center and fewer points farther from the center. This makes the measure of center a very good description of the data as a whole. Bell-shaped distributions are always symmetric or close to it.

NAME _____ DATE _____ PERIOD _____

Here is a dot plot, histogram, and box plot representing a skewed distribution.

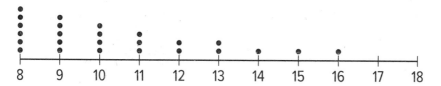

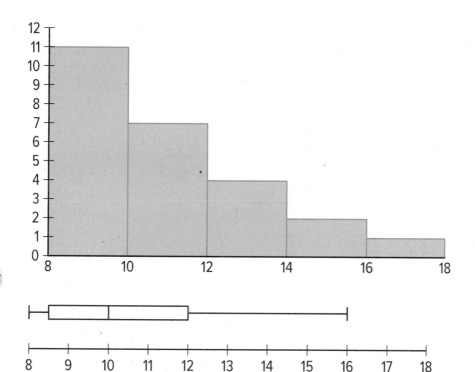

In a **skewed distribution**, one side of the distribution has more values farther from the bulk of the data than the other side. This results in the mean and median not being equal. In this skewed distribution, the data is skewed to the right because most of the data is near the 8 to 10 interval, but there are many points to the right. The mean is greater than the median. The large data values to the right cause the mean to shift in that direction while the median remains with the bulk of the data, so the mean is greater than the median for distributions that are skewed to the right. In a data set that is skewed to the left, a similar effect happens but to the other side. Again, the dot plot provides a greater level of detail about the shape of the distribution than either the histogram or the box plot.

A **uniform distribution** has the data values evenly distributed throughout the range of the data. This causes the distribution to look like a rectangle.

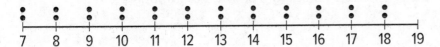

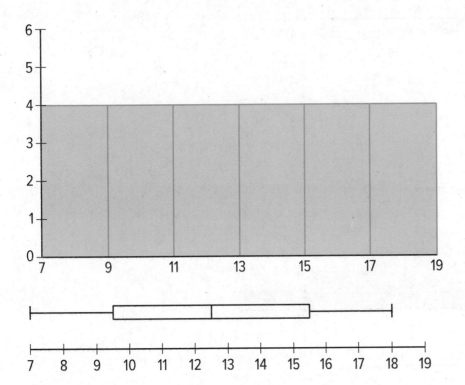

In a uniform distribution the mean is equal to the median since a uniform distribution is also a symmetric distribution. The box plot does not provide enough information to describe the shape of the distribution as uniform, though the even length of each quarter does suggest that the distribution may be approximately symmetric.

NAME _____ DATE _____ PERIOD _____

A **bimodal distribution** has two very common data values seen in a dot plot or histogram as distinct peaks.

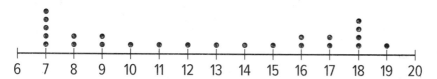

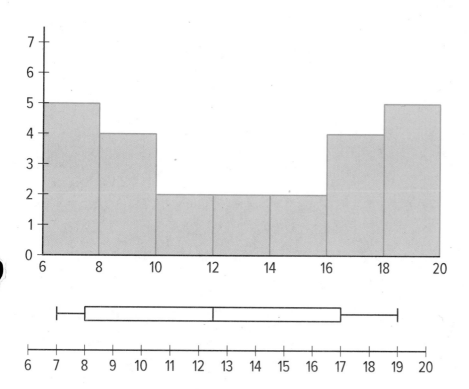

Sometimes, a bimodal distribution has most of the data clustered in the middle of the distribution. In these cases the center of the distribution does not describe the data very well. Bimodal distributions are not always symmetric. For example, the peaks may not be equally spaced from the middle of the distribution or other data values may disrupt the symmetry.

Glossary
bell-shaped distribution
bimodal distribution
skewed distribution
symmetric
uniform distribution

1. Which of the dot plots shows a symmetric distribution?

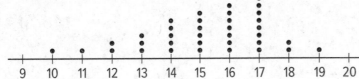

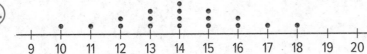

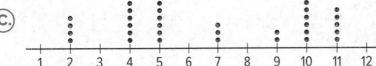

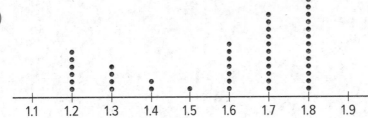

NAME _____ DATE _____ PERIOD _____

2. Which of the dot plots shows a skewed distribution?

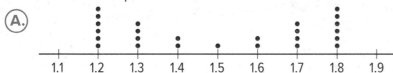

(A.)

| 1.1 | 1.2 | 1.3 | 1.4 | 1.5 | 1.6 | 1.7 | 1.8 | 1.9 |

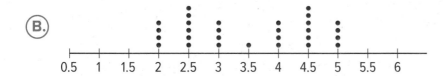

(B.)

| 0.5 | 1 | 1.5 | 2 | 2.5 | 3 | 3.5 | 4 | 4.5 | 5 | 5.5 | 6 |

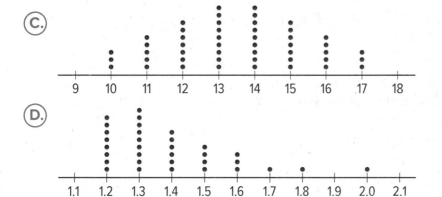

(C.)

| 9 | 10 | 11 | 12 | 13 | 14 | 15 | 16 | 17 | 18 |

(D.)

| 1.1 | 1.2 | 1.3 | 1.4 | 1.5 | 1.6 | 1.7 | 1.8 | 1.9 | 2.0 | 2.1 |

3. Create a dot plot showing a uniform distribution.

4. The data represent the number of ounces of water that 25 students drank before donating blood: 8, 8, 8, 16, 16, 16, 32, 32, 32, 32, 32, 32, 64, 64, 64, 64, 64, 64, 64, 80, 80, 80, 80, 88, 88, 88. **(Lesson 1-3)**

   a. Create a dot plot for the data.

b. Create a box plot for the data.

c. What information about the data is provided by the box plot that is not provided by the dot plot?

d. What information about the data is provided by the dot plot that is not provided by the box plot?

e. It was recommended that students drink 48 or more ounces of water. How could you use a histogram to easily display the number of students who drank the recommended amount?

5. The box plot represents the distribution of the number of points scored by a cross country team at 12 meets. **(Lesson 1-2)**

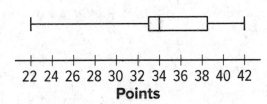

a. If possible, find the mean. If not possible, explain why not.

b. If possible, find the median. If not possible, explain why not.

c. Did the cross country team ever score 30 points at a meet?

Lesson 1-5

Calculating Measures of Center and Variability

NAME _____ DATE _____ PERIOD _____

Learning Goal Let's calculate measures of center and measures of variability and know which are most appropriate for the data.

Warm Up
5.1 Calculating Centers

Decide if each situation is *true* or *false*. Explain your reasoning.

1. The mean can be found by adding all the numbers in a data set and dividing by the number of numbers in the data set.

2. The mean of the data in the dot plot is 4.

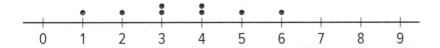

3. The median of the data set is 9 for the data: 4, 5, 9, 1, 10.

4. The median of the data in the dot plot is 3.5.

The heart rates of eight high school students are listed in beats per minute:

72   75   81   76   76   77   79   78

1. What is the interquartile range?

2. How many values in the data set are:

    a. less than Q1?

    b. between Q1 and the median?

    c. between the median and Q3?

    d. greater than Q3?

3. A pod of dolphins contains 800 dolphins of various ages and lengths. The median length of dolphins in this pod is 5.8 feet. What information does this tell you about the length of dolphins in this pod?

4. The same vocabulary test with 50 questions is given to 600 students from fifth to tenth grades and the number of correct responses is collected for each student in this group. The interquartile range is 40 correct responses. What information does this tell you about the number of correct responses for students taking this test?

NAME _____ DATE _____ PERIOD _____

Activity
5.3 Heartbeats: Part 2

1. Calculate the mean absolute deviation (MAD) using the same data from the previous activity by finding the average distance from each data value to the mean. You may find it helpful to organize your work by completing the table provided.

Data Values	Mean	Deviation from Mean (Data Value – Mean)	Absolute Deviation |Deviation|
72			
75			
81			
76			
76			
77			
79			
78			

MAD:

2. For another data set, all of the values are either 3 beats per minute above the mean or 3 beats per minute below the mean. Is that enough information to find the MAD for this data set? If so, find the MAD. If not, what other information is needed? Explain your reasoning.

3. Several pennies are placed along a meter stick and the position in centimeters of each penny is recorded. The mean position is the 50 centimeter mark and the MAD is 10 centimeters. What information does this tell you about the position of the pennies along the meter stick?

Suppose there are 6 pennies on a meter stick so that the mean position is the 50 centimeter mark and the MAD is 10 centimeters.

1. Find possible locations for the 6 pennies.

2. Find a different set of possible locations for the 6 pennies.

NAME _____ DATE _____ PERIOD _____

Summary
Calculating Measures of Center and Variability

The *mean absolute deviation*, or MAD, and the *interquartile range*, or IQR, are measures of variability. Measures of variability tell you how much the values in a data set tend to differ from one another. A greater measure of variability means that the data is more spread out while a smaller measure of variability means that the data is more consistent and close to the measure of center.

To calculate the MAD of a data set:

1. Find the mean of the values in the data set.

2. Find the distance between each data value and the mean on the number line.

   $|\text{data value} - \text{mean}|$

3. Find the mean of the distances. This value is the MAD.

To calculate the IQR, subtract the value of the first quartile from the value of the third quartile. Recall that the first and third quartiles are included in the five-number summary.

1. The data set represents the number of errors on a typing test.

   5   6   8   8   9   9   10   10   10   12

   a. What is the median? Interpret this value in the situation.

   b. What is the IQR?

2. The data set represents the heights, in centimeters, of ten model bridges made for an engineering competition.

   13   14   14   16   16   16   16   18   18   19

   a. What is the mean?

   b. What is the MAD?

3. Describe the shape of the distribution shown in the dot plot. The dot plot displays the golf scores from a golf tournament. **(Lesson 1-4)**

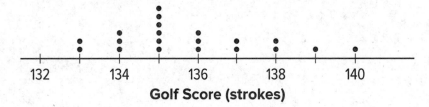

NAME _____ DATE _____ PERIOD _____

4. The dot plot shows the weight, in grams, of several different rocks.

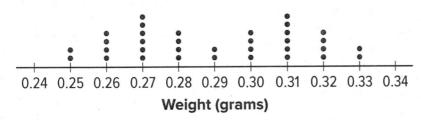

Weight (grams)

Select **all** the terms that describe the shape of the distribution. **(Lesson 1-4)**

(A.) bell-shaped

(B.) bimodal

(C.) skewed

(D.) symmetric

(E.) uniform

5. The dot plot represents the distribution of wages earned during a one-week period by 12 college students. **(Lesson 1-3)**

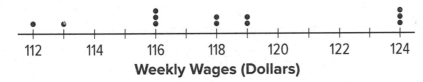

Weekly Wages (Dollars)

a. What is the mean? Interpret this value based on the situation.

b. What is the median? Interpret this value based on the situation.

c. Would a box plot of the same data have allowed you to find both the mean and the median?

6. The box plot displays the temperature of saunas in degrees Fahrenheit. What is the median? (Lesson 1-2)

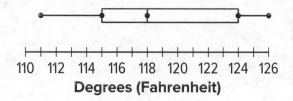

Lesson 1-6

Mystery Computations

NAME _____ DATE _____ PERIOD _____

Learning Goal Let's explore spreadsheets.

 ## Warm Up
6.1 Make 24

Your teacher will give you 4 numbers. Use these numbers, along with mathematical operations like addition and multiplication, to make 24.

Activity

6.2 Mystery Operations

Navigate to the spreadsheet in the digital version of the materials.

Input different numbers in column A, and try to predict what will happen in column B. (Do not change anything in column B.)

1. How is the number in cell B2 related to all or some of the numbers in cells A2, A3, A4, and A5?

2. How is the number in cell B3 related to all or some of the numbers in cells A2, A3, A4, and A5?

3. How is the number in cell B4 related to all or some of the numbers in cells A2, A3, A4, and A5?

4. How is the number in cell B5 related to all or some of the numbers in cells A2, A3, A4, and A5?

Are you ready for more?

Check your conjectures by entering different kinds of numbers in column A, for example: fractions, decimals, very large numbers.

NAME _____ DATE _____ PERIOD _____

Activity
6.3 More Spreadsheets!

Navigate to the spreadsheet in the digital version of the materials.

1. Change the spreadsheet so that B2 contains = A2 + A4. To edit the formula in B2, you may have to click it twice.

2. Change the numbers in A2 through A5. Make sure that your new formula does what it is supposed to do by doing a mental calculation and checking the result in B2.

3. Change the contents of B3 so that B3 does something different.

4. Before trading with a partner, make sure your new formula is not visible by clicking in a different cell.

5. Trade with your partner.

6. Change the numbers in Column A to try and figure out your partner's new rule.

Spreadsheets are useful mathematical and statistical tools. Here is an example of a spreadsheet. Each cell in the spreadsheet can be named with its column and row. For example, cell B2 contains the value 99. Cell A4 contains the value -17. Cell D1 is selected.

	A	B	C	D	E
1					
2		99			
3					
4	-17				
5				0.25	
6					
7					

It is possible for the value in a cell to depend on the value in other cells. Let's type the formula = B2 − D5 into cell D1.

	A	B	C	D	E
1				=B2-D5	
2		99			
3					
4	-17				
5				0.25	
6					
7					

When we press enter, D1 will display the result of subtracting the number in cell D5 from the number in cell B2.

	A	B	C	D	E
1				98.75	
2		99			
3					
4	-17				
5				0.25	
6					
7					

If we type new numbers into B2 or D5, the number in D1 will automatically change.

	A	B	C	D	E
1				79	
2		99			
3					
4	-17				
5				20	
6					
7					

NAME _____ DATE _____ PERIOD _____

Practice
Mystery Computations

1. What could be the formula used to compute the value shown in cell B3?

	A	B
1	change these	what happens here?
2	7	20
3	0	350
4	13	0
5	50	69
6	-1	

   Ⓐ = B3 * B4

   Ⓑ = A2 + A5

   Ⓒ = A2 * A5

   Ⓓ = Sum(A2:A6)

2. Refer to the spreadsheet above. Select **all** the formulas that could be used to calculate the value in cell B4.

   Ⓐ = Product(A2:A6)

   Ⓑ = Sum(A2:A6)

   Ⓒ = A2 + A3

   Ⓓ = A2 * A3

   Ⓔ = A3 * A4 * A5

   Ⓕ = A3 + A4 + A5

3. What number will appear in cell B2 when the user presses Enter?

	A	B
1	change these	what happens here?
2	10	=Sum(A3:A5)
3	5	
4	0	
5	-7	

4. The formula in cell B2 is = Product(A2:A5). Describe a way to change the contents of column A so that the value in cell B2 becomes -70.

	A	B
1	change these	what happens here?
2	10	0
3	5	
4	0	
5	-7	

5. The dot plot displays the number of books read by students during the semester. (Lesson 1-5)

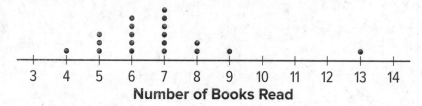

Number of Books Read

a. Which measure of center would you use given the shape of the distribution in the dot plot? Explain your reasoning.

b. Which measure of variability would you use? Explain your reasoning.

6. The dot plot displays the number of families living in different blocks of a town. (Lesson 1-5)

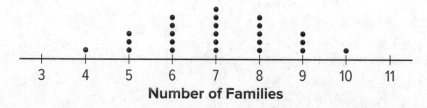

Number of Families

a. Which measure of center would you use, given the shape of the distribution in the dot plot? Explain your reasoning.

b. Which measure of variability would you use? Explain your reasoning.

Lesson 1-7

Spreadsheet Computations

NAME _____ DATE _____ PERIOD _____

Learning Goal Let's use spreadsheets as calculators.

 Warm Up
7.1 Dust Off Those Cobwebs

1. A person walks 4 miles per hour for 2.5 hours. How far do they walk?

2. A rectangle has an area of 24 square centimeters. What could be its length and width?

3. What is the area of this triangle?

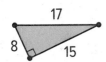

Activity

7.2 A Spreadsheet Is a Calculator

Use a spreadsheet to compute each of the following. Type each computation in a new cell, instead of erasing a previous computation.

1. $2 + 7$
2. $2 - 7$
3. $7 \cdot 2$
4. 7^2
5. $7 \div 2$
6. $\frac{1}{7}$ of 91
7. $0.1 \cdot 2 + 3$
8. $0.1(2 + 3)$
9. $13 \div \frac{1}{7}$
10. The average of 2, 7, 8, and 11

Activity

7.3 Use the Contents of a Cell in a Calculation

1. Type any number in cell A1, and another number in cell A2. Then in cell A3, type $= A1 + A2$. What happens?

2. In cell A4, compute the product of the numbers in A1 and A2.

3. In cell A5, compute the number in A1 raised to the power of the number in A2.

4. Now, type a new number in cell A1. What happens?

5. Type a new number in cell A2. What happens?

6. Use nearby cells to label the contents of each cell. For example in cell B3, type "the sum of A1 and A2." (This is a good habit to get into. It will remind you and anyone else using the spreadsheet what each cell means.)

NAME _____ DATE _____ PERIOD _____

Activity

7.4 Solve Some Problems

For each problem:

- Estimate the answer before calculating anything.

- Use the spreadsheet to calculate the answer.

- Write down the answer and the formula you used in the spreadsheet to calculate it.

1. The speed limit on a highway is 110 kilometers per hour. How much time does it take a car to travel 132 kilometers at this speed?

2. In a right triangle, the lengths of the sides that make a right angle are 98.7 cm and 24.6 cm. What is the area of the triangle?

3. A recipe for fruit punch uses 2 cups of seltzer water, $\frac{1}{4}$ cup of pineapple juice, and $\frac{2}{3}$ cup of cranberry juice. How many cups of fruit punch are in 5 batches of this recipe?

4. Check in with a partner and resolve any discrepancies with your answer to the last question. Next, type 2, $\frac{1}{4}$, $\frac{2}{3}$, and 5 in separate cells. (You may find it helpful to label cells next to them with the meaning of each number.) In a blank cell, type a formula for the total amount of fruit punch that uses the values in the other four cells. Now you should be able to easily figure out:

   a. How much in 7.25 batches?

   b. How much in 5 batches if you change the recipe to 1.5 cups of seltzer water per batch?

   c. Change the ratio of the ingredients in the fruit punch so that you would like the flavor. How many total cups are in $\frac{1}{2}$ batch?

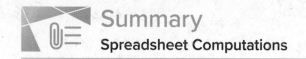

Summary
Spreadsheet Computations

A spreadsheet can be thought of as a type of calculator. For example, in a cell, you could type $= 2 + 3$, and then the sum of 5 is displayed in the cell. You can also perform operations on the values in other cells. For example, if you type a number in A1 and a number in A2, and then in A3 type $= A1 + A2$, then A3 will display the sum of the values in cells A1 and A2.

Familiarize yourself with how your spreadsheet software works on your device.

- On some spreadsheet programs, an $=$ symbol must be typed before the expression in the cell. (On others, it does not matter if your expression begins with $=$.)

- Know how to "submit" the expression so the computation takes place. If your device has a keyboard, it's likely the enter key. On a touchscreen device, you may have to tap a check mark.

- Learn symbols to use for various operations, and how to find them on your keyboard. Here are the symbols used for some typical operations.

Symbol	Operation
+	Add
–	Subtract or a negative number (this symbol does double duty in most spreadsheets)
*	Multiply
/	Divide
a/b	Fraction $\frac{a}{b}$
^	Exponent
.	Decimal point
()	To tell it what to compute first (often needed around fractions)

NAME _____ DATE _____ PERIOD _____

Practice
Spreadsheet Computations

1. Write a formula you could type into a spreadsheet to compute the value of each expression.

   a. $(19.2) \cdot 73$

   b. 1.1^5

   c. $2.34 \div 5$

   d. $\dfrac{91}{7}$

2. A long-distance runner jogs at a constant speed of 7 miles per hour for 45 minutes. Which spreadsheet formula would give the distance she traveled?

   (A.) $= 7 * 45$

   (B.) $= 7 / 45$

   (C.) $= 7 * (3 / 4)$

   (D.) $= 7 / (3 / 4)$

3. In a right triangle, the lengths of the sides that make a right angle are 3.4 meters and 5.6 meters. Select **all** the spreadsheet formulas that would give the area of this triangle.

   (A.) $= 3.4 * 5.6$

   (B.) $= 3.4 * 5.6 * 2$

   (C.) $= 3.4 * 5.6 / 2$

   (D.) $= 3.4 * 5.6 * (1/2)$

   (E.) $= (3.4 * 5.6) / 2$

4. This spreadsheet should compute the total ounces of sparkling grape juice based on the number of batches, ounces of grape juice in a single batch, and ounces of sparkling water in a single batch.

	A	B
1	number of batches	4
2	ounces of grape juice in 1 batch	3
3	ounces of sparkling water in 1 batch	7
4	total ounces	

   a. Write a formula for cell B4 that uses the values in cells B1, B2, and B3, to compute the total ounces of sparkling grape juice.

   b. How would the output of the formula change if the value in cell B1 was changed to 10?

   c. What would change about the sparkling grape juice if the value in B3 was changed to 10?

5. The dot plot and the box plot represent the same distribution of data. (Lesson 1-5)

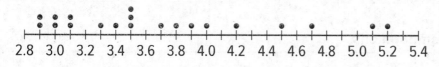

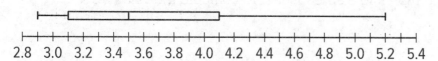

a. How does the median change when the highest value, 5.2, is removed?

b. How does the IQR change when the highest value, 5.2, is removed?

6. Describe the shape of the distribution shown in the histogram which displays the light output, in lumens, of various light sources. (Lesson 1-4)

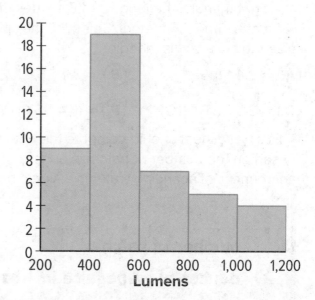

7. The dot plot represents the distribution of the number of goals scored by a soccer team in 10 games. (Lesson 1-2)

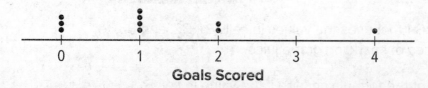

a. If possible, find the mean. If not possible, explain why not.

b. If possible, find the median. If not possible, explain why not.

c. Did the soccer team ever score exactly 3 goals in one of the games?

Lesson 1-8

Spreadsheet Shortcuts

NAME _____ DATE _____ PERIOD _____

Learning Goal Let's explore recursive formulas in spreadsheets.

 ## Warm Up
8.1 Tables of Equivalent Ratios

Here is a table of equivalent ratios:

1. Complete the table with the missing values.

2. Explain what it means to say that the pairs of numbers are equivalent ratios.

a	b
3	15
10	50
6	30
1	
	80

 ## Activity
8.2 The Birthday Trick

Navigate to the spreadsheet in the digital version of the materials.

1. In cell B4, we want to enter = B1 * 5 to multiply the month by 5. Enter this, but when you are about to type B1, instead, click on cell B1. This shortcut can be used any time: click on a cell instead of typing its address.

2. Practice this technique as you program each cell in B5 through B10 to perform the right computation.

3. When you are finished, does cell B10 show a number that contains the month and day of your birthday? If not, troubleshoot your computations.

4. Try changing the month and day in cells B1 and B2. The rest of the computations should automatically update. If not, troubleshoot your computations.

Why does this trick work? Try using *m* for the month and *d* for the day, and writing the entire computation as an algebraic expression. Can you see why the resulting number contains the month and day?

NAME _____ DATE _____ PERIOD _____

Activity

8.3 Using Spreadsheet Patterns

Navigate to the spreadsheet in the digital version of the materials.

The spreadsheet contains a table of equivalent ratios.

1. Use spreadsheet calculations to continue the pattern in columns A and B, down to row 5. Pause for discussion.

2. Click on cell A5. See the tiny blue square in the bottom right corner of the cell? Click it and drag it down for several cells and let go.

3. Repeat this, starting with cell B5.

Sometimes you want to create a list of numbers based on a rule. For example, let's say that the cost of a gym membership is $25 sign-up fee followed by monthly dues of $35. We may want to know how much the membership will cost over the course of 6 months. We could use a spreadsheet and set it up this way:

	A	B	C
1	sign-up fee	25	
2	total cost after 1 month	=B1+35	
3	total cost after 2 months		
4	total cost after 3 months		
5	total cost after 4 months		
6	total cost after 5 months		
7	total cost after 6 months		
8			
9			

Which results in:

	A	B	C
1	sign-up fee	25	
2	total cost after 1 month	60	
3	total cost after 2 months		
4	total cost after 3 months		
5	total cost after 4 months		
6	total cost after 5 months		
7	total cost after 6 months		
8			
9			

See the little square on the lower-right corner of cell B2? If we click and drag that down, it will keep adding 35 to the value above to find the value in the next row. Drag it down far enough, and we can see the total cost after 6 months.

	A	B	C
1	sign-up fee	25	
2	total cost after 1 month	60	
3	total cost after 2 months	95	
4	total cost after 3 months	130	
5	total cost after 4 months	165	
6	total cost after 5 months	200	
7	total cost after 6 months	235	
8			
9			

Any time you need to repeat a mathematical operation several times, continuing a pattern by dragging in a spreadsheet might be a good choice.

NAME _____ DATE _____ PERIOD _____

Practice
Spreadsheet Shortcut

1. *Technology required.* Open a blank spreadsheet. Use "fill down" to recreate this table of equivalent ratios. You should not need to type anything in rows 3–10.

	A	B
1	3	7
2	6	14
3	9	21
4	12	28
5	15	35
6	18	42
7	21	49
8	24	56
9	27	63
10	30	70

2. A list of numbers is made with the pattern: Start with 11, and subtract 4 to find the next number. Here is the beginning of the list: 11, 7, 3, . . .

    Explain how you could use "fill down" in a spreadsheet to find the tenth number in this list. (You do *not* need to actually find this number.)

3. Here is a spreadsheet showing the computations for a different version of the birthday trick:

    Explain what formulas you would enter in cells B4 through B8 so that cell B8 shows a number representing the month and day. (In this example, cell B8 should show 704.) If you have access to a spreadsheet, try your formulas with a month and day to see whether it works.

	A	B
1	month	7
2	day	4
3		
4	multiply month by 50	
5	add 30	
6	multiply by 2	
7	add the day	
8	subtract 60	
9		
10		

4. Write a formula you could type into a spreadsheet to compute the value of each expression. **(Lesson 1-7)**

   a. $\frac{2}{5}$ of 35

   b. $25 \div \frac{5}{3}$

   c. $\left(\frac{1}{11}\right)^{4}$

   d. The average of 0, 3, and 17

5. The data set represents the number of cars in a town given a speeding ticket each day for 10 days. **(Lesson 1-5)**

   | 2 | 4 | 5 | 5 | 7 | 7 | 8 | 8 | 8 | 12 |

   a. What is the median? Interpret this value in the situation.

   b. What is the IQR?

6. The data set represents the most recent sale price, in thousands of dollars, of ten homes on a street. **(Lesson 1-5)**

   | 85 | 91 | 93 | 99 | 99 | 99 | 102 | 108 | 110 | 115 |

   a. What is the mean?

   b. What is the MAD?

Lesson 1-9

Technological Graphing

NAME _____ DATE _____ PERIOD _____

Learning Goal Let's use technology to represent data.

Warm Up
9.1 It Begins With Data

Open a spreadsheet window and enter the data so that each value is in its own cell in column A.

1. How many values are in the spreadsheet? Explain your reasoning.

2. If you entered the data in the order that the values are listed, the number 7 is in the cell at position A1 and the number 5 is in cell A5. List all of the cells that contain the number 13.

3. In cell C1 type the word "Sum", in C2 type "Mean", and in C3 type "Median". You may wish to double-click or drag the vertical line between columns C and D to allow the entire words to be seen.

Using this data, we can calculate a few **statistics** and look at the data.

- Next to the word Sum, in cell D1, type =Sum(A1:A20)
- Next to the word Mean, in cell D2, type =Mean(A1:A20)
- Next to the word Median, in cell D3, type =Median(A1:A20)

4. What are the values for each of the statistics?

5. Change the value in A1 to 8. How does that change the statistics?

6. What value can be put into A1 to change the mean to 10.05 and the median to 9?

	A
1	7
2	8
3	4
4	13
5	5
6	15
7	14
8	8
9	12
10	2
11	8
12	13
13	12
14	13
15	6
16	1
17	9
18	4
19	9
20	15

Activity

9.2 Finding Spreadsheet Statistics

We can also use technology to create data displays. Navigate to the digital version of the materials.

- In Column A of the tool, enter the data you ended with in the warm up.

- Select Histogram as the graph type.

- Change the settings to show 7 classes.

1. Click the Data Summary button to see many of the statistics. What statistics do you recognize?

2. Change the settings so that 5 classes of data are shown. What does this do to the histogram?

3. Change the graph type to look at a box plot of the data. When looking at the box plot, notice there is an x near the box plot. This represents a data point that is considered an outlier. Click on the settings button and uncheck the box labeled Outliers to remove this point from the box plot. What changes? Why might you want to show outliers? Why might you want to include or exclude outliers?

NAME _____ DATE _____ PERIOD _____

Activity
9.3 Making Digital Displays

Use the data you collected from the numerical, statistical question from a previous lesson. Use technology to create a dot plot, boxplot, and histogram for your data. Then find the mean, median, and interquartile range for the data.

A stem and leaf plot is a table where each data point is indicated by writing the first digit(s) on the left (the stem) and the last digit(s) on the right (the leaves). Each stem is written only once and shared by all data points with the same first digit(s). For example, the values 31, 32, and 45 might be represented like:

3	1	2
4	5	

Key: 3 | 1 means 31

A class took an exam and earned the scores:

86, 73, 85, 86, 72, 94, 88, 98, 87, 86, 85, 93, 75, 64, 82, 95, 99, 76, 84, 68

1. Use technology to create a stem and leaf plot for this data set.

2. How can we see the shape of the distribution from this plot?

3. What information can we see from a stem and leaf plot that we cannot see from a histogram?

4. What do we have more control of in a histogram than in a stem and leaf plot?

NAME _____ DATE _____ PERIOD _____

Summary
Technological Graphing

Data displays (like histograms or box plots) are very useful for quickly understanding a large amount of information, but often take a long time to construct accurately using pencil and paper. Technology can help create these displays as well as calculate useful *statistics* much faster than doing the same tasks by hand. Especially with very large data sets (in some experiments, millions of pieces of data are collected), technology is essential for putting the information into forms that are more easily understood.

A **statistic** is a quantity that is calculated from sample data as a measure of a distribution. *Mean* and *median* are examples of statistics that are measures of center. *Mean absolute deviation (MAD)* and *interquartile range (IQR)* are examples of statistics that are measures of variability. Although the interpretation must still be done by people, using the tools available can improve the accuracy and speed of doing computations and creating graphs.

Glossary
_____

statistic

Practice
Technological Graphing

1. *Technology required.* The data represent the average customer ratings for several items sold online.

0.5	1	1.2	1.3	2.1	2.1	2.1	2.3
2.5	2.6	3.5	3.6	3.7	4	4.1	4.1
4.2	4.2	4.5	4.7	4.8			

   a. Use technology to create a histogram for the data with intervals
      0–1, 1–2, and so on.

   b. Describe the shape of the distribution.

   c. Which interval has the highest frequency?

NAME _____ DATE _____ PERIOD _____

2. *Technology required.* The data represent the amount of corn, in bushels per acre, harvested from different locations.

133, 133, 134, 134, 134, 135, 135, 135, 135, 135, 135, 136, 136, 136, 137, 137, 138, 138, 139, 140

a. Use technology to create a dot plot and a box plot.

b. What is the shape of the distribution?

c. Compare the information displayed by the dot plot and box plot.

3. Refer to the histogram. **(Lesson 1-4)**

   a. Describe the shape of the distribution.

   b. How many values are represented by the histogram?

   c. Write a statistical question that could have produced the data set summarized in the histogram.

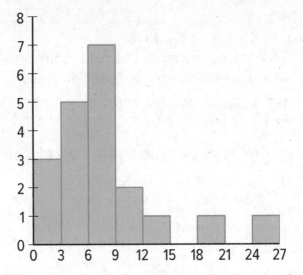

4. The dot plot represents the distribution of satisfaction ratings for a landscaping company on a scale of 1 to 10. Twenty-five customers were surveyed. On average, what was the satisfaction rating of the landscaping company? **(Lesson 1-3)**

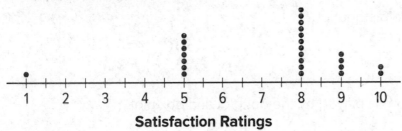

Satisfaction Ratings

Lesson 1-10

The Effect of Extremes

NAME _____ DATE _____ PERIOD _____

Learning Goal Let's see how statistics change with the data.

Warm Up
10.1 Battle Royale

Several video games are based on a genre called "Battle Royale" in which 100 players are on an island and they fight until only 1 player remains and is crowned the winner. This type of game can often be played in solo mode as individuals or in team mode in groups of 2.

1. What information would you use to determine the top players in each mode (solo and team)? Explain your reasoning.

2. One person claims that the best solo players play game A. Another person claims that game B has better solo players. How could you display data to help inform their discussion? Explain your reasoning.

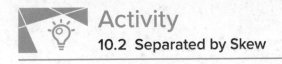

Activity

10.2 Separated by Skew

1. Use technology to create a dot plot that represents the distribution of the data, then describe the shape of the distribution.

6	7	8	8	9	9	9	10	10
10	10	11	11	11	12	12	13	14

2. Find the mean and median of the data.

3. Find the mean and median of the data with 2 additional values included as described.

   a. Add 2 values to the original data set that are greater than 14.

   b. Add 2 values to the original data set that are less than 6.

   c. Add 1 value that is greater than 14 and 1 value that is less than 6 to the original data set.

   d. Add the two values, 50 and 100, to the original data set.

4. Change the values so that the distribution fits the description given to you by your teacher, then find the mean and median.

NAME _____ DATE _____ PERIOD _____

5. Find another group that created a distribution with a different description. Explain your work and listen to their explanation, then compare your measures of center.

 Activity

10.3 Plots Matching Measures

Create a possible dot plot with at least 10 values for each of the conditions listed. Each dot plot must have at least 3 values that are different.

1. a distribution that has both mean and median of 10

2. a distribution that has both mean and median of -15

3. a distribution that has a median of 2.5 and a mean greater than the median

4. a distribution that has a median of 5 and a median greater than the mean

The mean and the median are by far the most common measures of center for numerical data. There are other measures of center, though, that are sometimes used. For each measure of center, list some possible advantages and disadvantages. Be sure to consider how it is affected by extremes.

1. *Interquartile mean:* The mean of only those points between the first quartile and the third quartile.

2. *Midhinge:* The mean of the first quartile and the third quartile.

3. *Midrange:* The mean of the minimum and maximum value.

4. *Trimean:* The mean of the first quartile, the median, the median again, and the third quartile. So we are averaging four numbers as the median is counted twice.

NAME _____ DATE _____ PERIOD _____

Summary
The Effect of Extremes

Is it better to use the mean or median to describe the center of a data set?

The mean gives equal importance to each value when finding the center. The mean usually represents the typical values well when the data has a symmetric distribution. On the other hand, the mean can be greatly affected by changes to even a single value.

The median tells you the middle value in the data set, so changes to a single value usually do not affect the median much. So, the median is more appropriate for data that is not very symmetric.

We can look at the distribution of a data set and draw conclusions about the mean and the median.

Here is a dot plot showing the amount of time a dart takes to hit a target in seconds. The data produces a symmetric distribution.

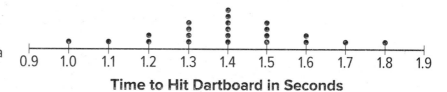

Time to Hit Dartboard in Seconds

When a distribution is symmetric, the median and mean are both found in the middle of the distribution. Since the median is the middle value (or mean of the two middle values) of a data set, you can use the symmetry around the center of a symmetric distribution to find it easily. For the mean, you need to know that the sum of the distances from the mean to the values greater than the mean is equal to the sum of the distances from the mean to the values less than the mean. Using the symmetry of the symmetric distribution you can see that there are four values 0.1 second above the mean, two values 0.2 seconds above the mean, one value 0.3 seconds above the mean, and one value 0.4 seconds above the mean. Likewise, you can see that there are the same number of values that are the same distances below the mean.

Here is a dot plot using the same data, but with two of the values changed, resulting in a skewed distribution.

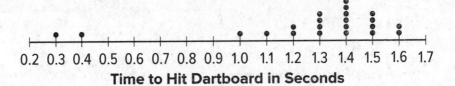

Time to Hit Dartboard in Seconds

When you have a skewed distribution, the distribution is not symmetric, so you are not able to use the symmetry to find the median and the mean. The median is still 1.4 seconds since it is still the middle value. The mean, on the other hand, is now about 1.273 seconds. The mean is less than the median because the lower values (0.3 and 0.4) result in a smaller value for the mean.

The median is usually more resistant to extreme values than the mean. For this reason, the median is the preferred measure of center when a distribution is skewed or if there are extreme values. When using the median, you would also use the IQR as the preferred measure of variability. In a more symmetric distribution, the mean is the preferred measure of center and the MAD is the preferred measure of variability.

NAME _____ DATE _____ PERIOD _____

Practice
The Effect of Extremes

1. Select **all** the distribution shapes for which it is most often appropriate to use the mean.

   (A.) bell-shaped   (D.) symmetric

   (B.) bimodal   (E.) uniform

   (C.) skewed

2. For which distribution shape is it usually appropriate to use the median when summarizing the data?

   (A.) bell-shaped   (C.) symmetric

   (B.) skewed   (D.) uniform

3. The number of writing instruments in some teachers' desks is displayed in the dot plot. Which is greater, the mean or the median? Explain your reasoning using the shape of the distribution.

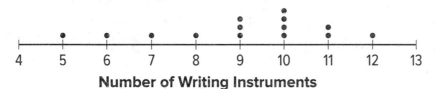

Number of Writing Instruments

4. A student has these scores on their assignments. The teacher is considering dropping a lowest score. What effect does eliminating the lowest value, 0, from the data set have on the mean and median? **(Lesson 1-9)**

   0, 40, 60, 70, 75, 80, 85, 95, 95, 100

5. Refer to the data: 2, 2, 4, 4, 5, 5, 6, 7, 9, 15. (Lesson 1-9)

   a. What is the five-number summary for the data?

   b. When the maximum, 15, is removed from the data set, what is the five-number summary?

6. The box plot summarizes the test scores for 100 students:

   Which term best describes the shape of the distribution? (Lesson 1-4)

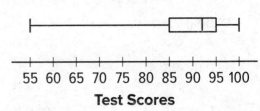
Test Scores

   (A.) bell-shaped

   (B.) uniform

   (C.) skewed

   (D.) symmetric

7. The histogram represents the distribution of lengths, in inches, of 25 catfish caught in a lake. (Lesson 1-2)

   a. If possible, find the mean. If not possible, explain why not.

   b. If possible, find the median. If not possible, explain why not.

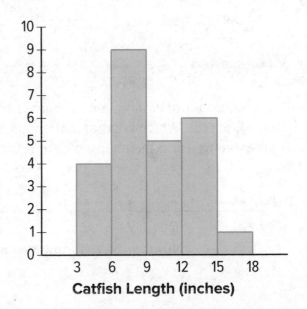
Catfish Length (inches)

   c. Were any of the fish caught 12 inches long?

   d. Were any of the fish caught 19 inches long?

Lesson 1-11

Comparing and Contrasting Data Distributions

NAME _____ DATE _____ PERIOD _____

Learning Goal Let's investigate variability using data displays and summary statistics.

Warm Up
11.1 Math Talk: Mean

Evaluate the mean of each data set mentally.

27, 30, 33

61, 71, 81, 91, 101

0, 100, 100, 100, 100

0, 5, 6, 7, 12

Activity

11.2 Describing Data Distributions

1. Your teacher will give you a set of cards. Take turns with your partner to match a data display with a written statement.

   a. For each match that you find, explain to your partner how you know it's a match.

   b. For each match that your partner finds, listen carefully to their explanation. If you disagree, discuss your thinking and work to reach an agreement.

2. After matching, determine if the mean or median is more appropriate for describing the center of the data set based on the distribution shape. Discuss your reasoning with your partner. If it is not given, calculate (if possible) or estimate the appropriate measure of center. Be prepared to explain your reasoning.

NAME _____ DATE _____ PERIOD _____

Activity

11.3 Visual Variability and Statistics

Each box plot summarizes the number of miles driven each day for 30 days in each month. The box plots represent, in order, the months of August, September, October, November, and December.

1. The five box plots have the same median. Explain why the median is more appropriate for describing the center of the data set than the mean for these distributions.

2. Arrange the box plots in order of least variability to greatest variability. Check with another group to see if they agree.

A.

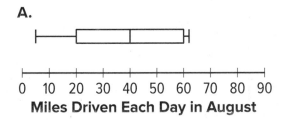

Miles Driven Each Day in August

B.

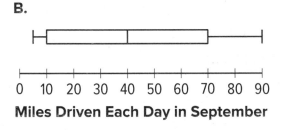

Miles Driven Each Day in September

C.

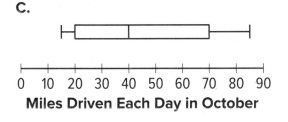

Miles Driven Each Day in October

D.

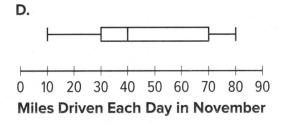

Miles Driven Each Day in November

E.

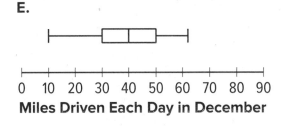

Miles Driven Each Day in December

3. The five dot plots have the same mean. Explain why the mean is more appropriate for describing the center of the data set than the median.

4. Arrange the dot plots in order of least variability to greatest variability. Check with another group to see if they agree.

A.

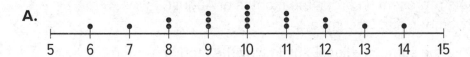

B.

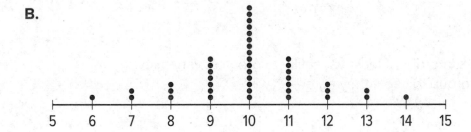

C.

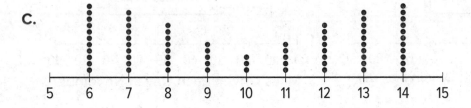

D.

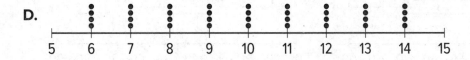

E.

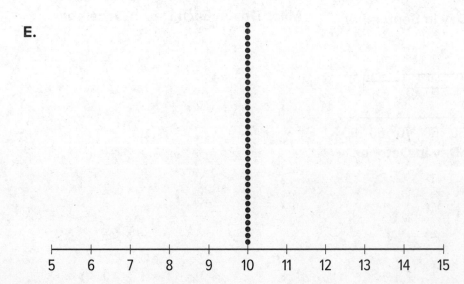

NAME _____ DATE _____ PERIOD _____

Are you ready for more?

1. These two box plots have the same median and the same IQR. How could we compare the variability of the two distributions?

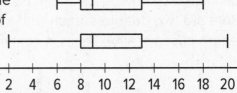

2. These two dot plots have the same mean and the same MAD. How could we compare the variability of the two distributions?

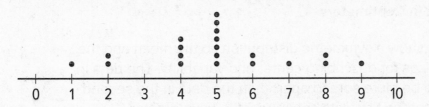

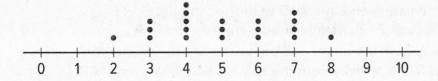

The mean absolute deviation, or MAD, is a measure of variability that is calculated by finding the mean distance from the mean of all the data points. Here are two dot plots, each with a mean of 15 centimeters, displaying the length of sea scallop shells in centimeters.

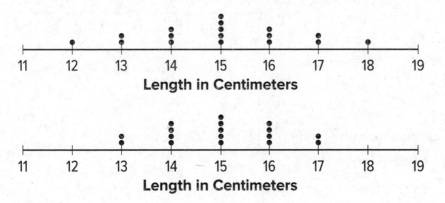

Notice that both dot plots show a symmetric distribution so the mean and the MAD are appropriate choices for describing center and variability. The data in the first dot plot appear to be more spread apart than the data in the second dot plot, so you can say that the first data set appears to have greater variability than the second data set. This is confirmed by the MAD. The MAD of the first data set is 1.18 centimeters and the MAD of the second data set is approximately 0.94 cm. This means that the values in the first data set are, on average, about 1.18 cm away from the mean and the values in the second data set are, on average, about 0.94 cm away from the mean. The greater the MAD of the data, the greater the variability of the data.

NAME _____ DATE _____ PERIOD _____

The interquartile range, IQR, is a measure of variability that is calculated by subtracting the value for the first quartile, Q1, from the value for the third quartile, Q3. These two box plots represent the distributions of the lengths in centimeters of a different group of sea scallop shells, each with a median of 15 centimeters.

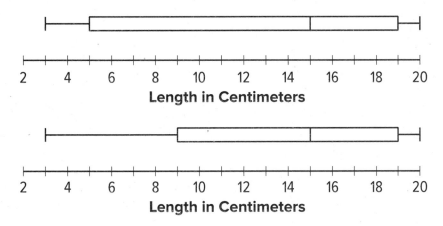

Notice that neither of the box plots have a symmetric distribution. The median and the IQR are appropriate choices for describing center and variability for these data sets. The middle half of the data displayed in the first box plot appear to be more spread apart, or show greater variability, than the middle half of the data displayed in the second box plot. The IQR of the first distribution is 14 cm and 10 cm for the second data set. The IQR measures the difference between the median of the second half of the data, Q3, and the median of the first half, Q1, of the data, so it is not impacted by the minimum or the maximum value in the data set. It is a measure of the spread of middle 50% of the data.

The MAD is calculated using every value in the data while the IQR is calculated using only the values for Q1 and Q3.

Practice

Comparing and Contrasting Data Distributions

1. In science class, Clare and Lin estimate the mass of eight different objects that actually weigh 2,000 grams each. Some summary statistics:

Clare	Lin
mean: 2,000 grams	mean: 2,000 grams
MAD: 275 grams	MAD: 225 grams
median: 2,000 grams	median: 1,950 grams
IQR: 500 grams	IQR: 350 grams

Which student was better at estimating the mass of the objects? Explain your reasoning.

2. Four amateur miniature golfers attempt to finish 100 holes under par several times. Each round of 100, the number of holes they successfully complete under par is recorded. Due to the presence of extreme values, box plots were determined to be the best representation for the data. List the four box plots in order of variability from least to greatest.

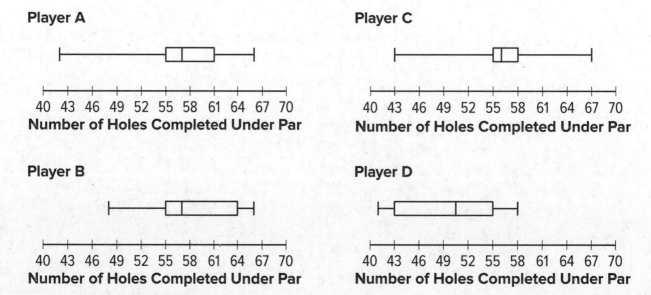

Player A

40 43 46 49 52 55 58 61 64 67 70
Number of Holes Completed Under Par

Player C

40 43 46 49 52 55 58 61 64 67 70
Number of Holes Completed Under Par

Player B

40 43 46 49 52 55 58 61 64 67 70
Number of Holes Completed Under Par

Player D

40 43 46 49 52 55 58 61 64 67 70
Number of Holes Completed Under Par

NAME _____ DATE _____ PERIOD _____

3. A reporter counts the number of times a politician talks about jobs in their campaign speeches. What is the MAD of the data represented in the dot plot?

Number of Mentions of "Jobs"

(A.) 1.1 mentions

(B.) 2 mentions

(C.) 2.5 mentions

(D.) 5.5 mentions

4. Select **all** the distribution shapes for which the median *could be* much less than the mean. (Lesson 1-10)

(A.) symmetric

(B.) bell-shaped

(C.) skewed left

(D.) skewed right

(E.) bimodal

5. Refer to the data: 0, 2, 2, 4, 5, 5, 5, 5, 7, 11. (Lesson 1-9)

   a. What is the five-number summary for the data?

   b. When the minimum, 0, is removed from the data set, what is the five-number summary?

6. What effect does eliminating the highest value, 180, from the data set have on the mean and median? (Lesson 1-9)

25, 50, 50, 60, 70, 85, 85, 90, 90, 180

7. The histogram represents the distribution of the number of seconds it took for each of 50 students to find the answer to a trivia question using the internet. Which interval contains the median? (Lesson 1-3)

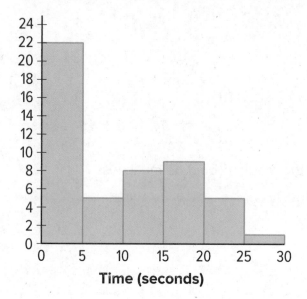

Time (seconds)

(A.) 0 to 5 seconds

(B.) 5 to 10 seconds

(C.) 10 to 15 seconds

(D.) 15 to 20 seconds

Lesson 1-12

Standard Deviation

NAME _____ DATE _____ PERIOD _____

Learning Goal Let's learn about standard deviation, another measure of variability.

 Warm Up

12.1 Notice and Wonder: Measuring Variability

What do you notice? What do you wonder?

mean: 10, MAD: 1.56, standard deviation: 2

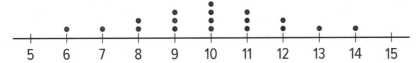

mean: 10, MAD: 2.22, standard deviation: 2.58

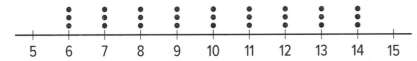

mean: 10, MAD: 2.68, standard deviation: 2.92

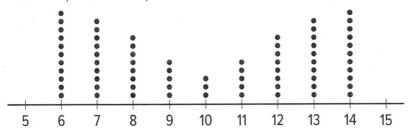

mean: 10, MAD: 1.12, standard deviation: 1.61

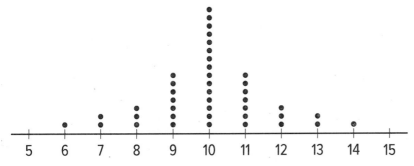

mean: 10, MAD: 2.06, standard deviation: 2.34

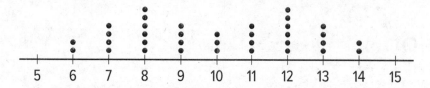

mean: 10, MAD: 0, standard deviation: 0

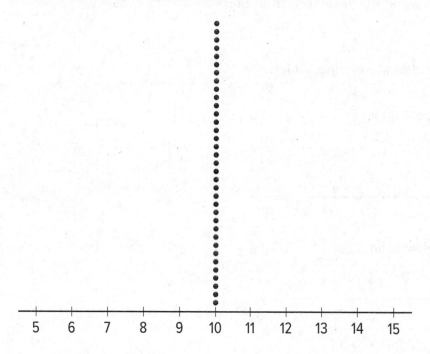

NAME _____ DATE _____ PERIOD _____

Activity

12.2 Investigating Standard Deviation

Use technology to find the mean and the standard deviation for the data in the dot plots.

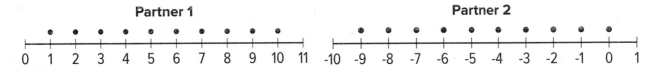

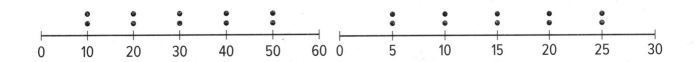

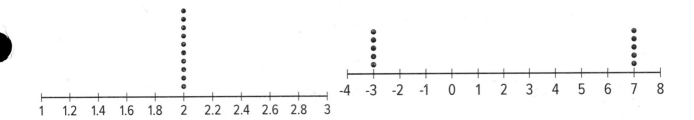

1. What do you notice about the mean and standard deviation you and your partner found for the three dot plots?

2. Invent some data that fits the conditions. Be prepared to share your data set and reasoning for choice of values.

Partner 1
Conditions:

- 10 numbers with a standard deviation equal to the standard deviation of your first dot plot with a mean of 6.

- 10 numbers with a standard deviation three times greater than the data in the first row.

- 10 different numbers with a standard deviation as close to 2 as you can get in 1 minute.

Partner 2
Conditions:

- 10 numbers with a standard deviation equal to the standard deviation of your first dot plot with a mean of 12.

- 10 numbers with a standard deviation four times greater than the data in the first row.

- 10 different numbers with a standard deviation as close to 2 as you can get in 1 minute.

NAME _____ DATE _____ PERIOD _____

Activity

12.3 Investigating Variability

Begin with the data:

1, 2, 3, 4, 5, 6, 7, 8, 9, 10, 11, 12, 13, 14, 15, 16, 17, 18, 19, 20

1. Use technology to find the mean, standard deviation, median, and interquartile range.

2. How do the standard deviation and mean change when you remove the greatest value from the data set? How do they change if you add a value to the data set that is twice the greatest value?

3. What do you predict will happen to the standard deviation and mean when you remove the least value from the data set? Check to see if your prediction was correct.

4. What happens to the standard deviation and mean when you add a value to the data set equal to the mean? Add a second value equal to the mean. What happens?

5. Add, change, and remove values from the data set to answer the question: What appears to change more easily, the standard deviation or the interquartile range? Explain your reasoning.

How is the standard deviation calculated? We have seen that the standard deviation behaves a lot like the mean absolute deviation and that is because the key idea behind both is the same.

1. Using the original data set, calculate the deviation of each point from the mean by subtracting the mean from each data point.

2. If we just tried to take a mean of those deviations what would we get?

3. There are two common ways to turn negative values into more useful positive values: take the absolute value or square the value. To find the MAD we find the absolute value of each deviation, then find the mean of those numbers. To find the standard deviation we square each of the deviations, then find the mean of those numbers. Finally, we take the square root of that mean. Compute the MAD and the standard deviation of the original data set.

Summary
Standard Deviation

We can describe the variability of a distribution using the **standard deviation**. The standard deviation is a measure of variability that is calculated using a method that is similar to the one used to calculate the MAD, or mean absolute deviation.

A deeper understanding of the importance of standard deviation as a measure of variability will come with a deeper study of statistics. For now, know that standard deviation is mathematically important and will be used as the appropriate measure of variability when mean is an appropriate measure of center.

Like the MAD, the standard deviation is large when the data set is more spread out, and the standard deviation is small when the variability is small. The intuition you gained about MAD will also work for the standard deviation.

Glossary

standard deviation

NAME _____ DATE _____ PERIOD _____

Practice
Standard Deviation

1. The shoe size for all the pairs of shoes in a person's closet are recorded.

   7   7   7   7   7   7   7   7   7   7

   a. What is the mean?

   b. What is the standard deviation?

2. Which of these best estimates the standard deviation of points in a card game?

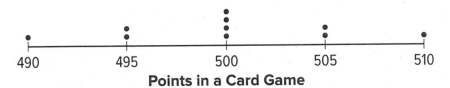

Points in a Card Game

   Ⓐ 5 points

   Ⓑ 20 points

   Ⓒ 50 points

   Ⓓ 500 points

3. Refer to the data set.

   1   2   3   3   4   4   4   4   5   5   6   7

   a. What happens to the mean and standard deviation of the data set when the 7 is changed to a 70?

   b. For the data set with the value of 70, why would the median be a better choice for the measure of center than the mean?

4. The mean of data set A is 43.5 and the MAD is 3.7. The mean of data set B is 12.8 and the MAD is 4.1. **(Lesson 1-11)**

   a. Which data set shows greater variability? Explain your reasoning.

   b. What differences would you expect to see when comparing the dot plots of the two data sets?

5. What is the IQR? **(Lesson 1-11)**

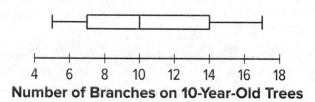

Number of Branches on 10-Year-Old Trees

A. 5 branches

B. 7 branches

C. 10 branches

D. 12 branches

6. Select **all** the distribution shapes for which the mean and median *must be* approximately the same. **(Lesson 1-10)**

   A. bell-shaped

   B. bimodal

   C. skewed

   D. symmetric

   E. uniform

7. The data represent the number of cans collected by different classes for a service project. **(Lesson 1-9)**

   12   14   22   14   18   23   42   13   9   19   22   14

   a. Find the mean.

   b. Find the median.

   c. Eliminate the greatest value, 42, from the data set. Explain how the measures of center change.

Lesson 1-13

More Standard Deviation

NAME _____ DATE _____ PERIOD _____

Learning Goal Let's continue to interpret standard deviation.

 ## Warm Up
13.1 Math Talk: Outlier Math

Evaluate mentally.

$0.5 \cdot 30$

$1.5 \cdot 30$

$100 - 1.5 \cdot 30$

$100 - 1.5 \cdot 18$

Activity

13.2 Info Gap: African and Asian Elephants

Your teacher will give you either a problem card or a data card. Do not show or read your card to your partner.

If your teacher gives you the problem card:

1. Silently read your card and think about what information you need to answer the question.

2. Ask your partner for the specific information that you need.

3. Explain to your partner how you are using the information to solve the problem.

4. When you have enough information, share the problem card with your partner, and solve the problem independently.

5. Read the data card, and discuss your reasoning.

If your teacher gives you the data card:

1. Silently read the information on your card.

2. Ask your partner "What specific information do you need?" and wait for your partner to ask for information. Only give information that is on your card. (Do not figure out anything for your partner!)

3. Before telling your partner the information, ask "Why do you need to know (that piece of information)?"

4. Read the problem card, and solve the problem independently.

5. Share the data card, and discuss your reasoning.

NAME _____ DATE _____ PERIOD _____

Activity

13.3 Interpreting Measures of Center and Variability

For each situation, you are given two graphs of data, a measure of center for each, and a measure of variability for each.

    a. Interpret the measure of center in terms of the situation.

    b. Interpret the measure of variability in terms of the situation.

    c. Compare the two data sets.

1. The heights of the 40 trees in each of two forests are collected.

    mean: 44.8 feet, standard deviation: 4.72 feet

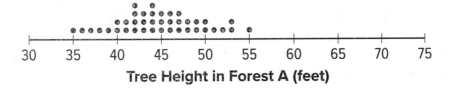

    mean: 56.03 feet, standard deviation: 7.87 feet

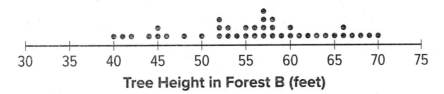

2. The number of minutes it takes Lin and Noah to finish their tests in German class is collected for the year.

mean: 29.48 minutes, standard deviation: 5.44 minutes

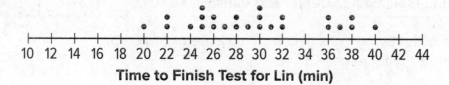

Time to Finish Test for Lin (min)

mean: 28.44 minutes, standard deviation: 7.40 minutes

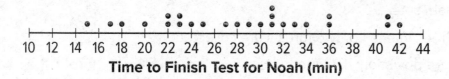

Time to Finish Test for Noah (min)

NAME _____ DATE _____ PERIOD _____

3. The number of raisins in a cereal with a name brand and the generic version of the same cereal are collected for several boxes.

mean: 289.1 raisins, standard deviation: 19.8 raisins

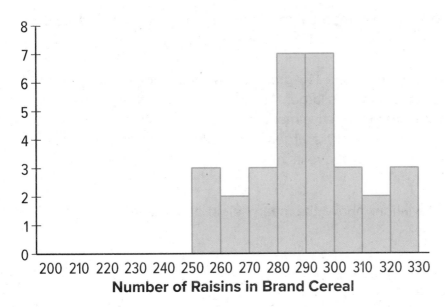

Number of Raisins in Brand Cereal

mean: 249.17 raisins, standard deviation: 26.35 raisins

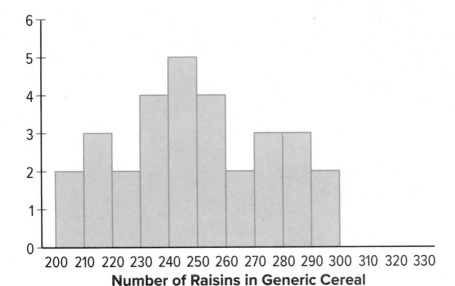

Number of Raisins in Generic Cereal

One use of standard deviation is it gives a natural scale as to how far above or below the mean a data point is. This is incredibly useful for comparing points from two different distributions.

For example, they say you cannot compare apples and oranges, but here is a way. The average weight of a granny smith apple is 128 grams with a standard deviation of about 10 grams. The average weight of a navel orange is 140 grams with a standard deviation of about 14 grams. If we have a 148-gram granny smith apple and a 161-gram navel orange, we might wonder which is larger for its species even though they are both about 20 grams above their respective mean. We could say that the apple, which is 2 standard deviations above its mean, is larger for its species than the orange, which is only 1.5 standard deviations above its mean.

1. How many standard deviations above the mean height of a tree in forest A is its tallest tree?

2. How many standard deviations above the mean height of a tree in forest B is its tallest tree?

3. Which tree is taller in its forest?

NAME _____ DATE _____ PERIOD _____

Summary
More Standard Deviation

The more variation a distribution has, the greater the standard deviation. A more compact distribution will have a lesser standard deviation.

The first dot plot shows the number of points that a player on a basketball team made during each of 15 games. The second dot plot shows the number of points scored by another player during the same 15 games.

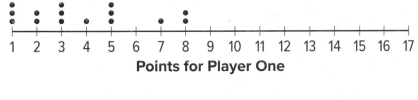

Points for Player One

Points for Player Two

The data in the first plot has a mean of approximately 3.87 points and standard deviation of about 2.33 points. The data in the second plot has a mean of approximately 7.73 points and a standard deviation of approximately 4.67 points. The second distribution has greater variability than the first distribution because the data is more spread out. This is shown in the standard deviation for the second distribution being greater than the standard deviation for the first distribution.

Standard deviation is calculated using the mean, so it makes sense to use it as a measure of variability when the mean is appropriate to use for the measure of center. In cases where the median is a more appropriate measure of center, the interquartile range is still a better measure of variability than standard deviation.

Practice
More Standard Deviation

1. Three drivers competed in the same fifteen drag races. The mean and standard deviation for the race times of each of the drivers are given.

   Driver A had a mean race time of 4.01 seconds and a standard deviation of 0.05 seconds.

   Driver B had a mean race time of 3.96 seconds and a standard deviation of 0.12 seconds.

   Driver C had a mean race time of 3.99 seconds and a standard deviation of 0.19 seconds.

   a. Which driver had the fastest typical race time?

   b. Which driver's race times were the most variable?

   c. Which driver do you predict will win the next drag race? Support your prediction using the mean and standard deviation.

2. The widths, in millimeters, of fabric produced at a ribbon factory are collected. The mean is approximately 23 millimeters and the standard deviation is approximately 0.06 millimeters. Interpret the mean and standard deviation in the context of the problem.

NAME _____ DATE _____ PERIOD _____

3. Select **all** the statements that are true about standard deviation.

(A.) It is a measure of center.

(B.) It is a measure of variability.

(C.) It is the same as the MAD.

(D.) It is calculated using the mean.

(E.) It is calculated using the median.

4. The number of different species of plants in some gardens is recorded. **(Lesson 1-12)**

1   2   3   4   4   5   5   6   7   8

   a. What is the mean?

   b. What is the standard deviation?

5. A set of data has ten numbers. The mean of the data is 12 and the standard deviation is 0. What values could make up a data set with these statistics? **(Lesson 1-12)**

6. Which box plot has the largest interquartile range? **(Lesson 1-11)**

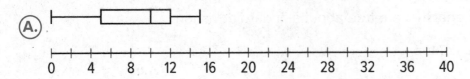

A.

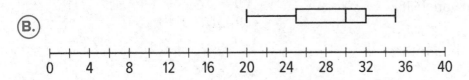

B.

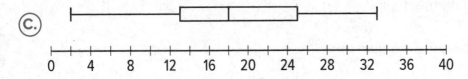

C.

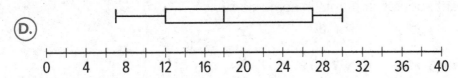

D.

7. Refer to the data. **(Lesson 1-9)**

1, 3, 3, 3, 4, 8, 9, 10, 10, 17

a. What is the five-number summary?

b. When the maximum, 17, is removed from the data set, what is the five-number summary?

Lesson 1-14

Outliers

NAME _____ DATE _____ PERIOD _____

Learning Goal Let's investigate outliers and how to deal with them.

 ## Warm Up
14.1 Health Care Spending

The histogram and box plot show the average amount of money, in thousands of dollars, spent on each person in the country (per capita spending) for health care in 34 countries.

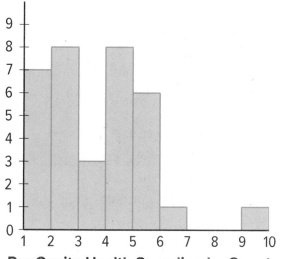

Per Capita Health Spending by Country
(thousands of dollars)

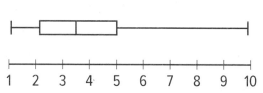

Per Capita Health Spending by Country
(thousands of dollars)

1. One value in the set is an **outlier**. Which one is it? What is its approximate value?

2. By one rule for deciding, a value is an outlier if it is more than 1.5 times the IQR greater than Q3. Show on the box plot whether or not your value meets this definition of outlier.

Activity

14.2 Investigating Outliers

Here is the data set used to create the histogram and box plot from the warm-up.

1.0803	1.0875	1.4663	1.7978	1.9702	1.9770	1.9890	2.1011
2.1495	2.2230	2.5443	2.7288	2.7344	2.8223	2.8348	3.2484
3.3912	3.5896	4.0334	4.1925	4.3763	4.5193	4.6004	4.7081
4.7528	4.8398	5.2050	5.2273	5.3854	5.4875	5.5284	5.5506
6.6475	9.8923						

1. Use technology to find the mean, standard deviation, and five-number summary.

2. The maximum value in this data set represents the spending for the United States. Should the per capita health spending for the United States be considered an outlier? Explain your reasoning.

3. Although outliers should not be removed without considering their cause, it is important to see how influential outliers can be for various statistics. Remove the value for the United States from the data set.

    a. Use technology to calculate the new mean, standard deviation, and five-number summary.

    b. How do the mean, standard deviation, median, and interquartile range of the data set with the outlier removed compare to the same summary statistics of the original data set?

NAME _____ DATE _____ PERIOD _____

Activity
14.3 Origins of Outliers

1. The number of property crime (such as theft) reports is collected for 50 colleges in California. Some summary statistics are given:

15	17	27	31	33	39	39	45
46	48	49	51	52	59	72	72
75	77	77	83	86	88	91	99
103	112	136	139	145	145	175	193
198	213	230	256	258	260	288	289
337	344	418	424	442	464	555	593
699	768						

- mean: 191.1 reports

- minimum: 15 reports

- Q1: 52 reports

- median: 107.5 reports

- Q3: 260 reports

- maximum: 768 reports

   a. Are any of the values outliers? Explain or show your reasoning.

   b. If there are any outliers, why do you think they might exist? Should they be included in an analysis of the data?

2. The situations described here each have an outlier. For each situation, how would you determine if it is appropriate to keep or remove the outlier when analyzing the data? Discuss your reasoning with your partner.

   a. A number cube has sides labelled 1–6. After rolling 15 times, Tyler records his data: 1, 1, 1, 1, 2, 2, 3, 3, 4, 4, 5, 5, 5, 6, 20.

   b. The dot plot represents the distribution of the number of siblings reported by a group of 20 people.

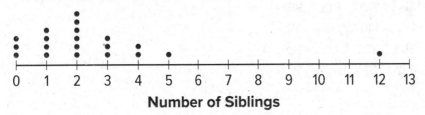

   Number of Siblings

   c. In a science class, 11 groups of students are synthesizing biodiesel. At the end of the experiment, each group recorded the mass in grams of the biodiesel they synthesized. The masses of biodiesel are 0, 1.245, 1.292, 1.375, 1.383, 1.412, 1.435, 1.471, 1.482, 1.501, 1.532.

Are you ready for more?

Look back at some of the numerical data you and your classmates collected in the first lesson of this unit.

   1. Are any of the values outliers? Explain or show your reasoning.

   2. If there are any outliers, why do you think they might exist? Should they be included in an analysis of the data?

NAME _____ DATE _____ PERIOD _____

Summary
Outliers

In statistics, an **outlier** is a data value that is unusual in that it differs quite a bit from the other values in the data set.

Outliers occur in data sets for a variety of reasons including, but not limited to:

- errors in the data that result from the data collection or data entry process

- results in the data that represent unusual values that occur in the population

Outliers can reveal cases worth studying in detail or errors in the data collection process. In general, they should be included in any analysis done with the data.

A value is an outlier if it is

- more than 1.5 times the interquartile range greater than

  Q3 (if $x > Q3 + 1.5 \cdot IQR$)

- more than 1.5 times the interquartile range less than

  Q1 (if $x < Q1 - 1.5 \cdot IQR$)

In this box plot, the minimum and maximum are at least two outliers.

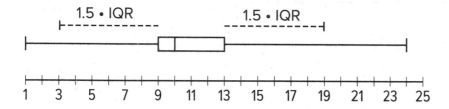

It is important to identify the source of outliers because outliers can impact measures of center and variability in significant ways.

The box plot displays the resting heart rate, in beats per minute (bpm), of

50 athletes taken five minutes after a workout.

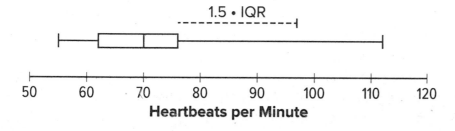

Some summary statistics include:

- mean: 69.78 bpm

- standard deviation: 10.71 bpm

- minimum: 55 bpm

- Q1: 62 bpm

- median: 70 bpm

- Q3: 76 bpm

- maximum: 112 bpm

It appears that the maximum value of 112 bpm may be an outlier. Since the interquartile range is 14 bpm ($76 - 62 = 14$) and $Q3 + 1.5 \cdot IQR = 97$, we should label the maximum value as an outlier. Searching through the actual data set, it could be confirmed that this is the only outlier.

After reviewing the data collection process, it is discovered that the athlete with the heart rate measurement of 112 bpm was taken one minute after a workout instead of five minutes after. The outlier should be deleted from the data set because it was not obtained under the right conditions.

Once the outlier is removed, the box plot and summary statistics are:

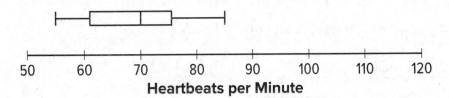

Heartbeats per Minute

- mean: 68.92 bpm

- standard deviation: 8.9 bpm

- minimum: 55 bpm

- Q1: 61 bpm

- median: 70 bpm

- Q3: 75.5 bpm

- maximum: 85 bpm

The mean decreased by 0.86 bpm and the median remained the same. The standard deviation decreased by 1.81 bpm which is about 17% of its previous value. Based on the standard deviation, the data set with the outlier removed shows much less variability than the original data set containing the outlier. Since the mean and standard deviation use all of the numerical values, removing one very large data point can affect these statistics in important ways.

NAME _____ DATE _____ PERIOD _____

The median remained the same after the removal of the outlier and the IQR increased slightly. These measures of center and variability are much more resistant to change than the mean and standard deviation. The median and IQR measure the middle of the data based on the number of values rather than the actual numerical values themselves, so the loss of a single value will not often have a great effect on these statistics.

The source of any possible errors should always be investigated. If the measurement of 112 beats per minute was found to be taken under the right conditions and merely included an athlete whose heart rate did not slow as much as the other athletes, it should not be deleted so that the data reflect the actual measurements. If the situation cannot be revisited to determine the source of the outlier, it should not be removed. To avoid tampering with the data and to report accurate results, data values should not be deleted unless they can be confirmed to be an error in the data collection or data entry process.

Glossary

outlier

1. The number of letters received in the mail over the past week is recorded.

   2 3 5 5 5 15

   Which value appears to be an outlier?

   (A.) 2   (C.) 5

   (B.) 3   (D.) 15

2. Elena collects 112 specimens of beetle and records their lengths for an ecology research project. When she returns to the laboratory, Elena finds that she incorrectly recorded one of lengths of the beetles as 122 centimeters (about 4 feet). What should she do with the outlier, 122 centimeters, when she analyzes her data?

3. Mai took a survey of students in her class to find out how many hours they spend reading each week. Here are some summary statistics for the data that Mai gathered:

   mean: 8.5 hours
   standard deviation: 5.3 hours
   median: 7 hours
   Q1: 5 hours
   Q3: 11 hours

   a. Give an example of a number of hours larger than the median which would be an outlier. Explain your reasoning.

   b. Are there any outliers below the median? Explain your reasoning.

NAME _____ DATE _____ PERIOD _____

4. The box plot shows the statistics for the weight, in pounds, of some dogs.

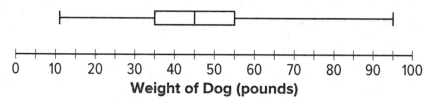

Weight of Dog (pounds)

Are there any outliers? Explain how you know.

5. The mean exam score for the first group of twenty examinees applying for a security job is 35.3 with a standard deviation of 3.6. The mean exam score for the second group of twenty examinees is 34.1 with a standard deviation of 0.5. Both distributions are close to symmetric in shape. (Lesson 1-13)

a. Use the mean and standard deviation to compare the scores of the two groups.

b. The minimum score required to get an in-person interview is 33. Which group do you think has more people get in-person interviews?

6. A group of pennies made in 2018 are weighed. The mean is approximately 2.5 grams with a standard deviation of 0.02 grams. Interpret the mean and standard deviation in terms of the context. (Lesson 1-13)

7. These values represent the expected number of paintings a person will produce over the next 10 days. **(Lesson 1-12)**

   0, 0, 0, 1, 1, 1, 2, 2, 3, 5

   a. What are the mean and standard deviation of the data?

   b. The artist is not pleased with these statistics. If the 5 is increased to a larger value, how does this impact the median, mean, and standard deviation?

8. List the four dot plots in order of variability from least to greatest. **(Lesson 1-11)**

   a.

   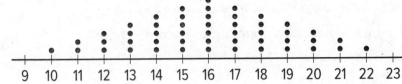

   b.

   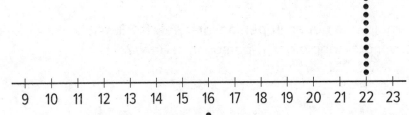

   c.

   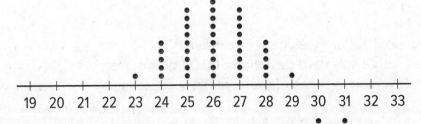

   d.

   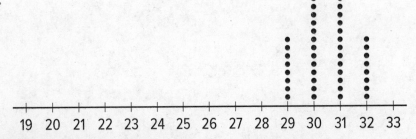

Lesson 1-15

Comparing Data Sets

NAME _____ DATE _____ PERIOD _____

Learning Goal Let's compare statistics for data sets.

Warm Up
15.1 Bowling Partners

Each histogram shows the bowling scores for the last 25 games played by each person. Choose 2 of these people to join your bowling team. Explain your reasoning.

Person A

mean: 118.96

median: 111

standard deviation: 32.96

interquartile range: 44

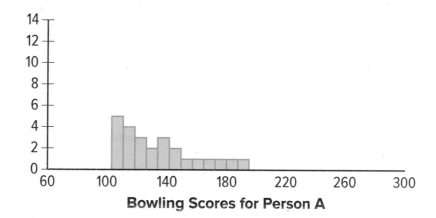

Person B

mean: 131.08

median: 129

standard deviation: 8.64

interquartile range: 8

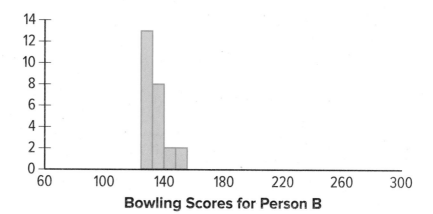

Person C

mean: 133.92

median: 145

standard deviation: 45.04

interquartile range: 74

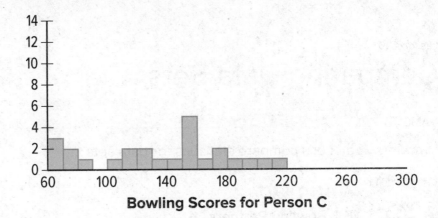

Bowling Scores for Person C

Person D

mean: 116.56

median: 103

standard deviation: 56.22

interquartile range: 31.5

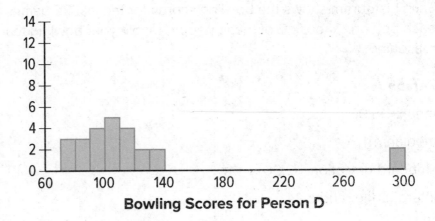

Bowling Scores for Person D

NAME _____ DATE _____ PERIOD _____

Activity
15.2 Comparing Marathon Times

All of the marathon runners from each of two different age groups have their finishing times represented in the dot plot.

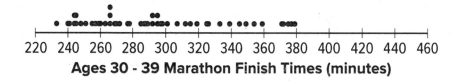

Ages 30 - 39 Marathon Finish Times (minutes)

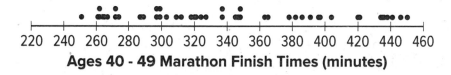

Ages 40 - 49 Marathon Finish Times (minutes)

1. Which age group tends to take longer to run the marathon? Explain your reasoning.

2. Which age group has more variable finish times? Explain your reasoning.

Are you ready for more?

1. How do you think finish times for a 20–29 age range will compare to these two distributions?

2. Find some actual marathon finish times for this group and make a box plot of your data to help compare.

For each group of data sets,

- Determine the best measure of center and measure of variability to use based on the shape of the distribution.

- Determine which set has the greatest measure of center.

- Determine which set has the greatest measure of variability.

- Be prepared to explain your reasoning.

| a | b |

1.

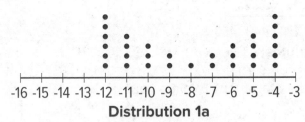

Distribution 1a

Distribution 1b

2.

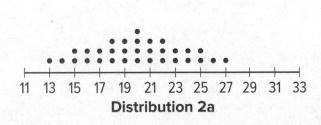

Distribution 2a

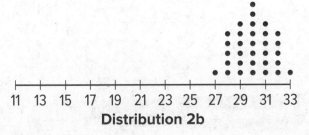

Distribution 2b

NAME _____ DATE _____ PERIOD _____

a	b

3.

Distribution 3a

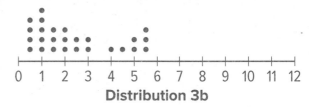

Distribution 3b

4.

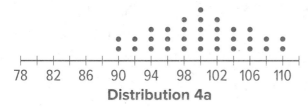

Distribution 4a

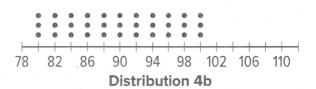

Distribution 4b

5.

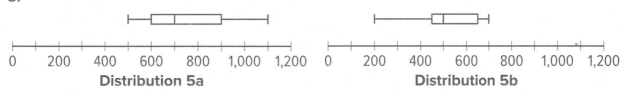

Distribution 5a   **Distribution 5b**

6.

a

A political podcast has mostly reviews that either love the podcast or hate it.

b

A cooking podcast has reviews that neither hate nor love the podcast.

7.

a

Stress testing concrete from site A has all 12 samples break at 450 pounds per square inch (psi).

b

Stress testing concrete from site B has samples break every 10 psi starting at 450 psi until the last core is broken at 560 psi.

c

Stress testing concrete from site C has 6 samples break at 430 psi and the other 6 break at 460 psi.

NAME _____ DATE _____ PERIOD _____

Summary
Comparing Data Sets

To compare data sets, it is helpful to look at the measures of center and measures of variability. The shape of the distribution can help choose the most useful measure of center and measure of variability.

When distributions are symmetric or approximately symmetric, the mean is the preferred measure of center and should be paired with the standard deviation as the preferred measure of variability. When distributions are skewed or when outliers are present, the median is usually a better measure of center and should be paired with the interquartile range (IQR) as the preferred measure of variability.

Once the appropriate measure of center and measure of variability are selected, these measures can be compared for data sets with similar shapes.

For example, let's compare the number of seconds it takes football players to complete a 40-yard dash at two different positions. First, we can look at a dot plot of the data to see that the tight end times do not seem symmetric, so we should probably find the median and IQR for both sets of data to compare information.

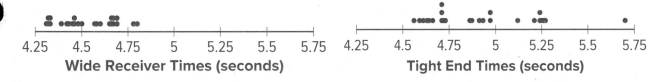

The median and IQR could be computed from the values, but can also be determined from a box plot.

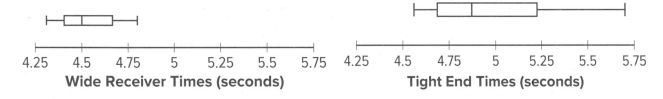

This shows that the tight end times have a greater median (about 4.9 seconds) compared to the median of wide receiver times (about 4.5 seconds). The IQR is also greater for the tight end times (about 0.5 seconds) compared to the IQR for the wide receiver times (about 0.25 seconds).

This means that the tight ends tend to be slower in the 40-yard dash when compared to the wide receivers. The tight ends also have greater variability in their times. Together, this can be taken to mean that, in general, a typical wide receiver is faster than a typical tight end, and the wide receivers tend to have more similar times to one another than the tight ends do to one another.

Practice
Comparing Data Sets

1. Twenty students participated in a psychology experiment which measured their heart rates in two different situations.

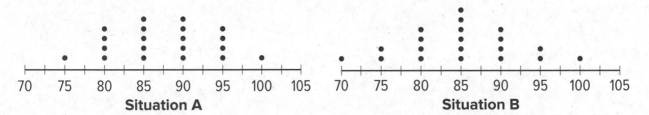

Situation A   Situation B

   a. What are the appropriate measures of center and variability to use with the data? Explain your reasoning.

   b. Which situation shows a greater typical heart rate?

   c. Which situation shows greater variability?

2 a. Invent two situations that you think would result in distributions with similar measures of variability. Explain your reasoning.

2 b. Invent two situations that you think would result in distributions with different measures of variability. Explain your reasoning.

3. The data set and some summary statistics are listed.

   11.5, 12.3, 13.5, 15.6, 16.7, 17.2, 18.4, 19, 19.5, 21.5

   mean: 16.52
   median: 16.95
   standard deviation: 3.11
   IQR: 5.5

NAME _____ DATE _____ PERIOD _____

a. How does adding 5 to each of the values in the data set impact the shape of the distribution?

b. How does adding 5 to each of the values in the data set impact the measures of center?

c. How does adding 5 to each of the values in the data set impact the measures of variability?

4. Here are two box plots:

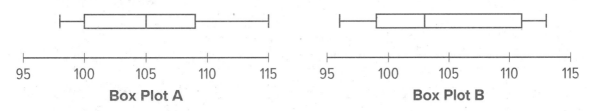

Box Plot A   **Box Plot B**

a. Which box plot has a greater median?

b. Which box plot has a greater measure of variability?

5. The depths of two lakes are measured at multiple spots. For the first lake, the mean depth is about 45 feet with a standard deviation of 8 feet. For the second lake, the mean depth is about 60 feet with a standard deviation of 27 feet. Noah says the second lake is generally deeper than the first lake. Do you agree with Noah? **(Lesson 1-13)**

6. The dot plots display the height, rounded to the nearest foot, of maple trees from two different tree farms. (Lesson 1-12)

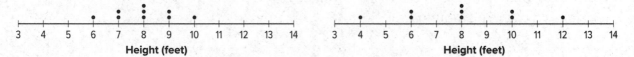

Height (feet)   Height (feet)

   a. Compare the mean and standard deviation of the two data sets.

   b. What does the standard deviation tell you about the trees at these farms?

7. Which box plot has an IQR of 10? (Lesson 1-11)

   (A.)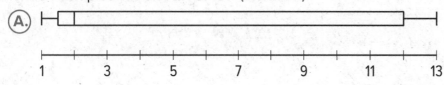

   (B.)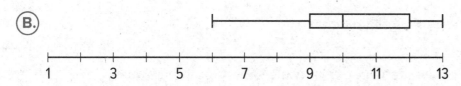

   (C.)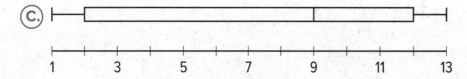

   (D.)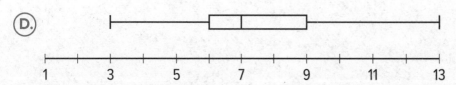

8. What effect does eliminating the lowest value, -6, from the data set -6, 3, 3, 3, 3, 5, 6, 6, 8, 10 have on the mean and median? (Lesson 1-9)

Lesson 1-16

Analyzing Data

NAME _____ DATE _____ PERIOD _____

Learning Goal Let's answer statistical questions by analyzing data, and comparing and contrasting their shape and measures of center and variability.

 ## Warm Up
16.1 Experimental Conditions

To test reaction time, the person running the test will hold a ruler at the 12-inch mark. The person whose reaction time is being tested will hold their thumb and forefinger open on either side of the flat side of the ruler at the 0-inch mark on the other side of the ruler. The person running the test will drop the ruler and the other person should close their fingers as soon as they notice the ruler moving to catch it. The distance that the ruler fell should be used as the data for this experiment.

With your partner, write a statistical question that can be answered by comparing data from two different conditions for the test.

Activity

16.2 Dropping the Ruler

Earlier, you and your partner agreed on a statistical question that can be answered using data collected in 2 different ruler-dropping conditions. With your partner, run the experiment to collect at least 20 results under each condition.

Analyze your 2 data sets to compare the statistical question. Next, create a visual display that includes:

- your statistical question

- the data you collected

- a data display

- the measure of center and variability you found that are appropriate for the data

- an answer to the statistical question with any supporting mathematical work

NAME _____ DATE _____ PERIOD _____

Activity
16.3 Heights and Handedness

Is there a connection between a student's dominant hand and their size? Use the table of information to compare the size of students with different dominant hands.

1. Here are the statistics for the high temperatures in a city during October:

   mean of 65.3 degrees Fahrenheit
   median of 63.5 degrees Fahrenheit
   standard deviation of 9.3 degrees Fahrenheit
   IQR of 7.1 degrees Fahrenheit

   Recall that the temperature C, measured in degrees Celsius, is related to the temperature F, measured in degrees Fahrenheit, by $C = \frac{5}{9}(F - 32)$.
   (Lesson 1-15)

   a. Describe how the value of each statistic changes when 32 is subtracted from the temperature in degrees Fahrenheit.

   b. Describe how the value of each statistic further changes when the new values are multiplied by $\frac{5}{9}$.

   c. Describe how to find the value of each statistic when the temperature is measured in degrees Celsius.

2. Here is a box plot. Give an example of a box plot that has a greater median and a greater measure of variability, but the same minimum and maximum values. (Lesson 1-15)

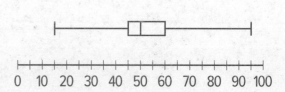

NAME _____ DATE _____ PERIOD _____

3. The mean vitamin C level for 20 dogs was 7.6 milligrams per liter, with a standard deviation of 2.1 milligrams per liter. One dog's vitamin C level was not in the normal range. It was 0.9 milligrams per liter, which is a very low level of vitamin C. **(Lesson 1-14)**

   a. If the value 0.9 is eliminated from the data set, does the mean increase or decrease?

   b. If the value 0.9 is eliminated from the data set, does the standard deviation increase or decrease?

4. The data set represents the number of hours that fifteen students walked during a two-week period.

   6   6   7   8   8   8   9   10   10   12   13   14   15   16   30

   The median is 10 hours, Q1 is 8, Q3 is 14, and the IQR is 6 hours. Are there any outliers in the data? Explain or show your reasoning. **(Lesson 1-14)**

5. Here are some summary statistics about the number of accounts that follow some bands on social media. **(Lesson 1-14)**

   mean: 15,976 followers   median: 16,432 followers
   standard deviation: 3,279 followers   Q1: 13,796 followers
   Q3: 19,070 followers   IQR: 5,274 followers

   a. Give an example of a number of followers that a very popular band might have that would be considered an outlier for this data. Explain or show your reasoning.

   b. Give an example of a number of followers that a relatively unknown band might have that would be considered an outlier for this data. Explain or show your reasoning.

6. The weights of one population of mountain gorillas have a mean of 203 pounds and standard deviation of 18 pounds. The weights of another population of mountain gorillas have a mean of 180 pounds and standard deviation of 25 pounds. Andre says the two populations are similar. Do you agree? Explain your reasoning. (Lesson 1-13)

7. The box plot represents the distribution of the amount of change, in cents, that 50 people were carrying when surveyed.

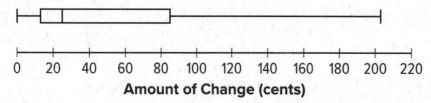

Amount of Change (cents)

The box plot represents the distribution of the same data set, but with the maximum, 203, removed.

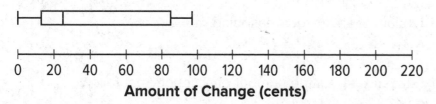

Amount of Change (cents)

The median is 25 cents for both plots. After examining the data, the value 203 is removed since it was an error in recording. (Lesson 1-10)

a. Explain why the median remains the same when 203 cents was removed from the data set.

b. When 203 cents is removed from the data set, does the mean remain the same? Explain your reasoning.

Learning Targets

Lesson	Learning Target(s)
1-1 Getting to Know You	• I can tell statistical questions from non-statistical questions and can explain the difference. • I can tell the difference between numerical and categorical data.
1-2 Data Representations	• I can find the five-number summary for data. • I can use a dot plot, histogram, or box plot to represent data.
1-3 A Gallery of Data	• I can graphically represent the data I collected and critique the representations of others.

(continued on the next page)

(continued from the previous page)

Lesson	Learning Target(s)
1-4 The Shape of Distributions	• I can describe the shape of a distribution using the terms "symmetric, skewed, uniform, bimodal, and bell-shaped." • I can use a graphical representation of data to suggest a situation that produced the data pictured.
1-5 Calculating Measures of Center and Variability	• I can calculate mean absolute deviation, interquartile range, mean, and median for a set of data.
1-6 Mystery Computations	• I can determine basic relationships between cell values in a spreadsheet by changing the values and noticing what happens in another cell.
1-7 Spreadsheet Computations	• I can use a spreadsheet as a calculator to find solutions to word problems.

(continued on the next page)

(continued from the previous page)

Lesson	Learning Target(s)
1-8 Spreadsheet Shortcuts	• I can use shortcuts to fill in cells on a spreadsheet.
1-9 Technological Graphing	• I can create graphic representations of data and calculate statistics using technology.
1-10 The Effect of Extremes	• I can describe how an extreme value will affect the mean and median. • I can use the shape of a distribution to compare the mean and median.
1-11 Comparing and Contrasting Data Distributions	• I can arrange data sets in order of variability given graphic representations.

(continued on the next page)

(continued from the previous page)

Lesson	Learning Target(s)
1-12 Standard Deviation	• I can describe standard deviation as a measure of variability.
	• I can use technology to compute standard deviation.
1-13 More Standard Deviation	• I can use standard deviation to say something about a situation.
1-14 Outliers	• I can find values that are outliers, investigate their source, and figure out what to do with them.
	• I can tell how an outlier will impact mean, median, IQR, or standard deviation.
1-15 Comparing Data Sets	• I can compare and contrast situations using measures of center and measures of variability.

(continued on the next page)

(continued from the previous page)

Lesson	Learning Target(s)
1-16 Analyzing Data	• I can collect data from an experiment and compare the results using measures of center and measures of variability.

Notes:

Linear Equations, Inequalities, and Systems

The volunteers who clean up trash after a parade could use linear equations to determine the most efficient route for trash collection. You will learn more about linear systems in this unit.

Syda Productions/Shutterstock

Topics
- Writing and Modeling with Equations
- Manipulating Equations and Understanding Their Structure
- Systems of Linear Equations in Two Variables
- Linear Inequalities in One Variable
- Linear Inequalities in Two Variables
- Systems of Linear Inequalities in Two Variables

Linear Equations, Inequalities, and Systems

Systems of Linear Inequalities in Two Variables

Lesson 2-1

Planning a Pizza Party

NAME _____ DATE _____ PERIOD _____

Learning Goal Let's write expressions to estimate the cost of a pizza party.

 ## Warm Up
1.1 A Main Dish and Some Side Dishes

Here are some letters and what they represent. All costs are in dollars.

- m represents the cost of a main dish.

- n represents the number of side dishes.

- s represents the cost of a side dish.

- t represents the total cost of a meal.

1. Discuss with a partner: What does each equation mean in this situation?

   a. $m = 7.50$

   b. $m = s + 4.50$

   c. $ns = 6$

   d. $m + ns = t$

2. Write a new equation that could be true in this situation.

Activity

1.2 How Much Will It Cost?

Imagine your class is having a pizza party.

Work with your group to plan what to order and to estimate what the party would cost.

1. Record your group's plan and cost estimate. What would it take to convince the class to go with your group's plan? Be prepared to explain your reasoning.

2. Write down one or more expressions that show how your group's cost estimate was calculated.

3. Respond to each question

   a. In your expression(s), are there quantities that might change on the day of the party? Which ones?

   b. Rewrite your expression(s), replacing the quantities that might change with letters. Be sure to specify what the letters represent.

NAME _____ DATE _____ PERIOD _____

Activity

1.3 What are the Constraints?

A **constraint** is something that limits what is possible or reasonable in a situation.

For example, one constraint in a pizza party might be the number of slices of pizza each person could have, s. We can write $s < 4$ to say that each person gets fewer than 4 slices.

1. Look at the expressions you wrote when planning the pizza party earlier.

   a. Choose an expression that uses one or more letters.

   b. For each letter, determine what values would be reasonable. (For instance, could the value be a non-whole number? A number greater than 50? A negative number? Exactly 2?)

2. Write equations or inequalities that represent some constraints in your pizza party plan. If a quantity must be an exact value, use the $=$ symbol. If it must be greater or less than a certain value to be reasonable, use the $<$ or $>$ symbol.

Expressions, equations, and inequalities are mathematical **models**. They are mathematical representations used to describe quantities and their relationships in a real-life situation. Often, what we want to describe are constraints. A **constraint** is something that limits what is possible or what is reasonable in a situation. For example, when planning a birthday party, we might be dealing with these quantities and constraints:

quantities	constraints
• the number of guests	• 20 people maximum
• the cost of food and drinks	• $5.50 per person
• the cost of birthday cake	• $40 for a large cake
• the cost of entertainment	• $15 for music and $27 for games
• the total cost	• no more than $180 total cost

We can use both numbers and letters to represent the quantities. For example, we can write 42 to represent the cost of entertainment, but we might use the letter n to represent the number of people at the party and the letter C for the total cost in dollars.

We can also write expressions using these numbers and letters. For instance, the expression $5.50n$ is a concise way to express the overall cost of food if it costs $5.50 per guest and there are n guests. Sometimes a constraint is an exact value. For instance, the cost of music is $15. Other times, a constraint is a boundary or a limit. For instance, the total cost must be no more than $180.

Symbols such as $<$, $>$, and $=$ can help us express these constraints.

quantities	constraints
• the number of guests	• $n \le 20$
• the cost of food and drinks	• $5.50n$
• the cost of birthday cake	• 40
• the cost of entertainment	• $15 + 27$
• the total cost	• $C \le 180$

NAME _____ DATE _____ PERIOD _____

Equations can show the relationship between different quantities and constraints. For example, the total cost of the party is the sum of the costs of food, cake, entertainment. We can represent this relationship with:

$$C = 5.50n + 40 + 15 + 27 \quad \text{or} \quad C = 5.50n + 82$$

Deciding how to use numbers and letters to represent quantities, relationships, and constraints is an important part of mathematical modeling. Making assumptions—about the cost of food per person, for example—is also important in modeling.

A model such as $C = 5.50n + 82$ can be an efficient way to make estimates or predictions. When a quantity or a constraint changes, or when we want to know something else, we can adjust the model and perform a simple calculation, instead of repeating a series of calculations.

Glossary

constraint

model

Practice
Planning a Pizza Party

1. The videography team entered a contest and won a monetary prize of $1,350. Which expression represents how much each person would get if there were x people on the team?

   A. $\dfrac{1350}{x}$

   B. $1350 + x$

   C. $\dfrac{1350}{5}$

   D. $1350 - x$

2. To support a local senior citizens center, a student club sent a flyer home to the n students in the school. The flyer said, "Please bring in money to support the senior citizens center. Paper money and coins accepted!" Their goal is to raise T dollars. Match each quantity to an expression, an equation, or an inequality that describes it.

   A. the dollar amount the club would have if they reached half of their goal

   B. the dollar amount the club would have if every student at the school donated 50 cents to the cause

   C. the dollar amount the club could donate if they made $50 more than their goal

   D. the dollar amount the club would still need to raise to reach its goal after every student at the school donated 50 cents

   E. the dollar amount the club would have if half of the students at the school each gave 50 cents

   1. $T + 50$

   2. $0.5T$

   3. $0.25n$

   4. $0.5n$

   5. $T - 0.5n$

3. Each of the 10 students in the baking club made 2 chocolate cakes for a fundraiser. They all used the same recipe, using C cups of flour in total. Write an expression that represents the amount of flour required for one cake.

NAME _____ DATE _____ PERIOD _____

4. A student club started a fundraising effort to support animal rescue organizations. The club sent an information flyer home to the *n* students in the school. It says, "We welcome donations of any amount, including any change you could spare!" Their goal is to raise *T* dollars, and to donate to a cat shelter and a dog shelter. Match each quantity to an expression, an equation, or an inequality that describes it.

A. The dollar amount the club would have if they reached one-fourth of their goal.

1. $\frac{3}{4}n \cdot \frac{1}{2}$

B. The dollar amount the club would have if every student at the school donated a quarter to the cause.

2. $\frac{1}{4}T$

C. The dollar amount the club could donate to the cat shelter if they reached their goal and gave a quarter of the total donation to a dog shelter.

3. $T - \frac{1}{4}n$

D. The dollar amount the club would still need to raise to reach its goal after every student at the school donated a quarter.

4. $\frac{3}{4}T$

E. The dollar amount the club would have if three-fourths of the students at the school each gave 50 cents.

5. $\frac{1}{4}n$

5. A softball team is ordering pizza to eat after their tournament. They plan to order cheese pizzas that cost $6 each and four-topping pizzas that cost $10 each. They order *c* cheese pizzas and *f* four-topping pizzas. Which expression represents the total cost of all of the pizzas they order?

(A.) $6 + 10$

(B.) $c + f$

(C.) $6c + 10f$

(D.) $6f + 10c$

6. The value of coins in the pockets of several students is recorded. What is the mean of the values: 10, 20, 35, 35, 35, 40, 45, 45, 50, 60 (Lesson 1-9)

Ⓐ 10 cents

Ⓑ 35 cents

Ⓒ 37.5 cents

Ⓓ 50 cents

7. The dot plot displays the number of hits a baseball team made in several games. The distribution is skewed to the left. (Lesson 1-10)

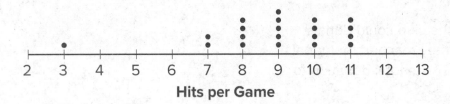

Hits per Game

If the game with 3 hits is considered to be recorded in error, it might be removed from the data set. If that happens:

a. What happens to the mean of the data set?

b. What happens to the median of the data set?

8. A set of data has MAD of 0 and one of the data values is 14. What can you say about the data values? (Lesson 1-11)

Lesson 2-2

Writing Equations to Model Relationships (Part 1)

NAME _____ DATE _____ PERIOD _____

Learning Goal Let's look at how equations can help us describe relationships and constraints.

Warm Up
2.1 Math Talk: Percent of 200

Evaluate mentally.

1. 25% of 200

2. 12% of 200

3. 8% of 200

4. p% of 200

Activity
2.2 A Platonic Relationship

These three figures are called Platonic solids.

Tetrahedron   **Cube**   **Dodecahedron**

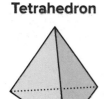 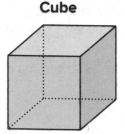 

The table shows the number of vertices, edges, and faces for the tetrahedron and dodecahedron.

	Faces	Vertices	Edges
Tetrahedron	4	4	6
Cube			
Dodecahedron	12	20	30

1. Complete the missing values for the cube. Then, make at least two observations about the number of faces, edges, and vertices in a Platonic solid.

2. There are some interesting relationships between the number of faces (F), edges (E), and vertices (V) in all Platonic solids. For example, the number of edges is always greater than the number of faces, or $E > F$. Another example: The number of edges is always less than the sum of the number of faces and the number of vertices, or $E < F + V$.

   There is a relationship that can be expressed with an equation. Can you find it? If so, write an equation to represent it.

There are two more Platonic solids: an octahedron which has 8 faces that are all triangles and an icosahedron which has 20 faces that are all triangles.

1. How many edges would each of these solids have? (Keep in mind that each edge is used in two faces.)

2. Use your discoveries from the activity to determine how many vertices each of these solids would have.

3. For all 5 Platonic solids, determine how many faces meet at each vertex.

NAME _____ DATE _____ PERIOD _____

Activity

2.3 Blueberries and Earnings

1. Write an equation to represent each situation.

   a. Blueberries are $4.99 a pound. Diego buys b pounds of blueberries and pays $14.95.

   b. Blueberries are $4.99 a pound. Jada buys p pounds of blueberries and pays c dollars.

   c. Blueberries are d dollars a pound. Lin buys q pounds of blueberries and pays t dollars.

   d. Noah earned n dollars over the summer. Mai earned $275, which is $45 more than Noah did.

   e. Noah earned v dollars over the summer. Mai earned m dollars, which is 45 dollars more than Noah did.

   f. Noah earned w dollars over the summer. Mai earned x dollars, which is y dollars more than Noah did.

2. How are the equations you wrote for the blueberry purchases like the equations you wrote for Mai and Noah's summer earnings? How are they different?

Activity

2.4 Car Prices

The tax on the sale of a car in Michigan is 6%. At a dealership in Ann Arbor, a car purchase also involves $120 in miscellaneous charges.

1. There are several quantities in this situation: the original car price, sales tax, miscellaneous charges, and total price. Write an equation to describe the relationship between all the quantities when:

   a. The original car price is $9,500.

   b. The original car price is $14,699.

NAME _____ DATE _____ PERIOD _____

c. The total price is $22,480.

d. The original price is p.

2. How would each equation you wrote change if the tax on car sales is $r\%$ and the miscellaneous charges are m dollars?

Summary

Writing Equations to Model Relationships (Part 1)

Suppose your class is planning a trip to a museum. The cost of admission is $7 per person and the cost of renting a bus for the day is $180.

- If 24 students and 3 teachers are going, we know the cost will be: $7(24) + 7(3) + 180$ or $7(24 + 3) + 180$.

- If 30 students and 4 teachers are going, the cost will be: $7(30 + 4) + 180$.

Notice that the numbers of students and teachers can vary. This means the cost of admission and the total cost of the trip can also vary, because they depend on how many people are going.

Letters are helpful for representing quantities that vary. If s represents the number of students who are going, t represents the number of teachers, and C represents the total cost, we can model the quantities and constraints by writing:

$$C = 7(s + t) + 180$$

Some quantities may be fixed. In this example, the bus rental costs $180 regardless of how many students and teachers are going (assuming only one bus is needed).

Letters can also be used to represent quantities that are constant. We might do this when we don't know what the value is, or when we want to understand the relationship between quantities (rather than the specific values).

For instance, if the bus rental is B dollars, we can express the total cost of the trip as $C = 7(s + t) + B$. No matter how many teachers or students are going on the trip, B dollars need to be added to the cost of admission.

NAME _____ DATE _____ PERIOD _____

Practice
Writing Equations to Model Relationships (Part 1)

1. Large cheese pizzas cost $5 each and large one-topping pizzas cost $6 each. Write an equation that represents the total cost, T, of c large cheese pizzas and d large one-topping pizzas.

2. Jada plans to serve milk and cookies for a book club meeting. She is preparing 12 ounces of milk and 4 cookies per person. Including herself, there are 15 people in the club. A package of cookies contains 24 cookies and costs $4.50.

   A 1-gallon jug of milk contains 128 ounces and costs $3. Let n represent number of people in the club, m represent the ounces of milk, c represent the number of cookies, and b represent Jada's budget in dollars.

   Select **all** of the equations that could represent the quantities and constraints in this situation.

   A. $m = 12(15)$

   B. $3m + 4.5c = b$

   C. $4n = c$

   D. $4(4.50) = c$

   E. $b = 2(3) + 3(4.50)$

3. A student on the track team runs 45 minutes each day as a part of her training. She begins her workout by running at a constant rate of 8 miles per hour for a minutes, then slows to a constant rate of 7.5 miles per hour for b minutes.

   Which equation describes the relationship between the distance she runs in miles, D, and her running speed, in miles per hour?

   A. $a + b = 45$

   B. $8a + 7.5b = D$

   C. $8\left(\dfrac{a}{60}\right) + 7.5\left(\dfrac{b}{60}\right) = D$

   D. $8(45 - b) + 7.5b = D$

4. Elena bikes 20 minutes each day for exercise. Write an equation to describe the relationship between her distance in miles, D, and her biking speed, in miles per hour, when she bikes:

   a. at a constant speed of 13 miles per hour for the entire 20 minutes

   b. at a constant speed of 15 miles per hour for the first 5 minutes, then at 12 miles per hour for the last 15 minutes

   c. at a constant speed of M miles per hour for the first 5 minutes, then at N miles per hour for the last 15 minutes

5. The dot plot displays the number of marshmallows added to hot cocoa by several kids. What is the MAD of the data represented in the dot plot?
   (Lesson 1-11)

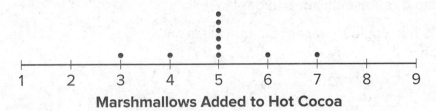

Marshmallows Added to Hot Cocoa

   A. 0.6 marshmallows

   B. 3 marshmallows

   C. 4 marshmallows

   D. 5 marshmallows

NAME _____ DATE _____ PERIOD _____

6. Here is a data set: **(Lesson 1-12)**

    5   10   10   10   15   100

a. After studying the data, the researcher realized that the value 100 was meant to be recorded as 15. What happens to the mean and standard deviation of the data set when the 100 is changed to a 15?

b. For the original data set, with the 100, would the median or the mean be a better choice of measure for the center? Explain your reasoning.

7. A coach for a little league baseball team is ordering trophies for the team. Players on the team are allowed to choose between 2 types of trophies. The gold baseball trophies cost $5.99 each and the uniform baseball trophies cost $6.49 each. The team orders g gold baseball trophies and u uniform baseball trophies. Write an expression that could represent the total cost of all of the trophies. **(Lesson 2-1)**

8. The robotics team needs to purchase $350 of new equipment. Each of the x students on the team plans to fundraise and contribute equally to the purchase. Which expression represents the amount that each student needs to fundraise? **(Lesson 2-1)**

  A. $350 - x$

  B. $350 + x$

  C. $\dfrac{350}{x}$

  D. $350 \cdot x$

9. In a trivia contest, players form teams and work together to earn as many points as possible for their team. Each team can have between 3 and 5 players. Each player can score up to 10 points in each round of the game. Elena and four of her friends decided to form a team and play a round. (Lesson 2-1)

Write an expression, an equation, or an inequality for each quantity described here. If you use a variable, specify what it represents.

a. the number of points that Elena's team earns in one round

b. the number of points Elena's team earns in one round if every player scores between 6 and 8 points

c. the number of points Elena's team earns if each player misses one point

d. the number of players in a game if there are 5 teams of 4 players each

e. the number of players in a game if there are at least 3 teams

Lesson 2-3

Writing Equations to Model Relationships (Part 2)

NAME _____ DATE _____ PERIOD _____

Learning Goal Let's use patterns to help us write equations.

Warm Up
3.1 Finding a Relationship

Here is a table of values. The two quantities, x and y, are related.

What are some strategies you could use to find a relationship between x and y? Brainstorm as many ways as possible.

x	y
1	0
3	8
5	24
7	48

Activity

3.2 Something about 400

1. Describe in words how the two quantities in each table are related.

- Table A

Number of Laps, x	0	1	2.5	6	9
Meters Run, y	0	400	1,000	2,400	3,600

- Table B

Meters from Home, x	0	75	128	319	396
Meters from School, y	400	325	272	81	4

- Table C

Electricity Bills in Dollars, x	85	124	309	816
Total Expenses in Dollars, y	485	524	709	1,216

- Table D

Monthly Salary in Dollars, x	872	998	1,015	2,110
Amount Deposited in Dollars, y	472	598	615	1,710

NAME _____ DATE _____ PERIOD _____

2. Match each table to an equation that represents the relationship.

Table A   Equation 1: $400 + x = y$

Table B   Equation 2: $x - 400 = y$

Table C   Equation 3: $x + y = 400$

Table D   Equation 4: $400 \cdot x = y$

Are you ready for more?

Express every number between 1 and 20 at least one way using exactly four 4's and any operation or mathematical symbol. For example, 1 could be written as $\frac{4}{4} + 4 - 4$.

Activity

3.3 What are the Relationships?

1. The table represents the relationship between the base length and the height of some parallelograms. All measurements are in inches.

Base Length (inches)	Height (inches)
1	48
2	24
3	16
4	12
6	8

What is the relationship between the base length and the height of these parallelograms?

2. Visitors to a carnival are invited to guess the number of beans in a jar. The person who guesses the correct number wins $300. If multiple people guess correctly, the prize will be divided evenly among them. What is the relationship between the number of people who guess correctly and the amount of money each person will receive?

NAME _____ DATE _____ PERIOD _____

3. A $\frac{1}{2}$-gallon jug of milk can fill 8 cups, while 32 fluid ounces of milk can fill 4 cups. What is the relationship between number of gallons and ounces? If you get stuck, try creating a table.

Summary
Writing Equations to Model Relationships (Part 2)

Sometimes, the relationship between two quantities is easy to see. For instance, we know that the perimeter of a square is always 4 times the side length of the square. If P represents the perimeter and s the side length, then the relationship between the two measurements (in the same unit) can be expressed as $P = 4s$, or $s = \frac{P}{4}$.

Other times, the relationship between quantities might take a bit of work to figure out—by doing calculations several times or by looking for a pattern. Here are two examples.

- A plane departed from New Orleans and is heading to San Diego. The table shows its distance from New Orleans, x, and its distance from San Diego, y, at some points along the way.

Miles from New Orleans, x	Miles from San Diego, y
100	1,500
300	1,300
500	1,100
	1,020
900	700
1,450	

What is the relationship between the two distances? Do you see any patterns in how each quantity is changing? Can you find out what the missing values are?

Notice that every time the distance from New Orleans increases by some number of miles, the distance from San Diego decreases by the same number of miles, and that the sum of the two values is always 1,600 miles.

The relationship can be expressed with any of these equations:

$x + y = 1,600$

$y = 1,600 - x$

$x = 1,600 - y$

- A company decides to donate $50,000 to charity. It will select up to 20 charitable organizations, as nominated by its employees. Each selected organization will receive an equal amount of donation.

What is the relationship between the number of selected organizations, n, and the dollar amount each of them will receive, d?

- If 5 organizations are selected, each one receives $10,000.

- If 10 organizations are selected, each one receives $5,000.

- If 20 organizations are selected, each one receives $2,500.

Do you notice a pattern here? 10,000 is $\frac{50,000}{5}$, 5,000 is $\frac{50,000}{10}$, and 2,500 is $\frac{50,000}{20}$.

We can generalize that the amount each organization receives is 50,000 divided by the number of selected organizations, or $d = \frac{50,000}{n}$.

NAME _____ DATE _____ PERIOD _____

Practice
Writing Equations to Model Relationships (Part 2)

1. A landscaping company is delivering crushed stone to a construction site. The table shows the total weight in pounds, W, of n loads of crushed stone. Which equation could represent the total weight, in pounds, for n loads of crushed stone?

Number of Loads of Crushed Stone	Total Weight in Pounds
0	0
1	2,000
2	4,000
3	6,000

Ⓐ $W = \dfrac{6,000}{n}$

Ⓑ $W = 6,000 - 2,000n$

Ⓒ $W = 2,000n$

Ⓓ $W = n + 2,000$

2. Members of the band sold soda and popcorn at a college football game to raise money for an upcoming trip. The band raised $2,000. The amount raised is divided equally among the m members of the band.

Which equation represents the amount, A, each member receives?

Ⓐ $A = \dfrac{m}{2,000}$

Ⓑ $A = \dfrac{2,000}{m}$

Ⓒ $A = 2,000m$

Ⓓ $A = 2,000 - m$

3. Tyler needs to complete this table for his consumer science class. He knows that 1 tablespoon contains 3 teaspoons and that 1 cup contains 16 tablespoons.

a. Complete the missing values in the table.

Number of Teaspoons	Number of Tablespoons	Number of Cups
		2
36	12	
	48	3

b. Write an equation that represents the number of teaspoons, t, contained in a cup, C.

4. The volume of dry goods, like apples or peaches, can be measured using bushels, pecks, and quarts. A bushel contains 4 pecks, and a peck contains 8 quarts.

What is the relationship between number of bushels, b, and the number of quarts, q? If you get stuck, try creating a table.

5. The data show the number of free throws attempted by a team in its first ten games.

2   11   11   11   12   12   13   14   14   15

The median is 12 attempts and the mean is 11.5 attempts. After reviewing the data, it is determined that 2 should not be included, since that was an exhibition game rather than a regular game during the season. **(Lesson 1-10)**

a. What happens to the median if 2 attempts is removed from the data set?

b. What happens to the mean if 2 attempts is removed from the data set?

NAME _____ DATE _____ PERIOD _____

6. The standard deviation for a data set is 0. What can you conclude about the data? (Lesson 1-12)

7. Elena has $225 in her bank account. She takes out $20 each week for w weeks. After w weeks she has d dollars left in her bank account. Write an equation that represents the amount of money left in her bank account after w weeks. (Lesson 2-2)

8. Priya is hosting a poetry club meeting this week and plans to have fruit punch and cheese for the meeting. She is preparing 8 ounces of fruit punch per person and 2 ounces of cheese per person. Including herself, there are 12 people in the club.

   A package of cheese contains 16 ounces and costs $3.99. A one-gallon jug of fruit punch contains 128 ounces and costs $2.50. Let p represent the number of people in the club, f represent the ounces of fruit punch, c represent the ounces of cheese, and b represent Priya's budget in dollars. (Lesson 2-2)

   Select **all** of the equations that could represent the quantities and constraints in this situation.

   Ⓐ $f = 8 \cdot 12$

   Ⓑ $c = 2 \cdot 3.99$

   Ⓒ $2 \cdot 3.99 + 2.50 = b$

   Ⓓ $2p = c$

   Ⓔ $8f + 2c = b$

9. The density of an object can be found by taking its mass and dividing by its volume. Write an equation to represent the relationship between the three quantities (density, mass, and volume) in each situation. Let the density, D, be measured in grams/cubic centimeters (or g/cm^3). (Lesson 2-2)

   a. The mass is 500 grams and the volume is 40 cubic centimeters.

   b. The mass is 125 grams and the volume is v cubic centimeters.

   c. The volume is 1.4 cubic centimeters and the density is 80 grams per cubic centimeter.

   d. The mass is m grams and the volume is v cubic centimeters.

Lesson 2-4

Equations and Their Solutions

NAME _____ DATE _____ PERIOD _____

Learning Goal Let's recall what we know about solutions to equations.

 ## Warm Up
4.1 What is a Solution?

A cookie contains 27 calories. Most of the calories come from c grams of carbohydrates. The rest come from other ingredients. One gram of carbohydrate contains 4 calories.

The equation $4c + 5 = 27$ represents the relationship between these quantities.

1. What could the 5 represent in this situation?

2. Priya said that neither 8 nor 3 could be the solution to the equation. Explain why she is correct.

3. Find the solution to the equation.

 ## Activity
4.2 Weekend Earnings

Jada has time on the weekends to earn some money. A local bookstore is looking for someone to help sort books and will pay $12.20 an hour. To get to and from the bookstore on a workday, however, Jada would have to spend $7.15 on bus fare.

Pixtal/AGE Fotostock

1. Write an equation that represents Jada's take-home earnings in dollars, E, if she works at the bookstore for h hours in one day.

2. One day, Jada takes home $90.45 after working h hours and after paying the bus fare. Write an equation to represent this situation.

3. Is 4 a solution to the last equation you wrote? What about 7? If so, be prepared to explain how you know one or both of them are solutions. If not, be prepared to explain why they are not solutions. Then, find the solution.

4. In this situation, what does the solution to the equation tell us?

Are you ready for more?

Jada has a second option to earn money—she could help some neighbors with errands and computer work for $11 an hour. After reconsidering her schedule, Jada realizes that she has about 9 hours available to work one day of the weekend.

Which option should she choose—sorting books at the bookstore or helping her neighbors? Explain your reasoning.

NAME _____ DATE _____ PERIOD _____

Activity

4.3 Calories from Protein and Fat

One gram of protein contains 4 calories. One gram of fat contains 9 calories. A snack has 60 calories from p grams of protein and f grams of fat.

The equation $4p + 9f = 60$ represents the relationship between these quantities.

1. Determine if each pair of values could be the number of grams of protein and fat in the snack. Be prepared to explain your reasoning.

   a. 5 grams of protein and 2 grams of fat

   b. 10.5 grams of protein and 2 grams of fat

   c. 8 grams of protein and 4 grams of fat

2. If there are 6 grams of fat in the snack, how many grams of protein are there? Show your reasoning.

3. In this situation, what does a solution to the equation $4p + 9f = 60$ tell us? Give an example of a solution.

Summary

Equations and Their Solutions

An equation that contains only one unknown quantity or one quantity that can vary is called an *equation in one variable*.

For example, the equation $2\ell + 2w = 72$ represents the relationship between the length, ℓ, and the width, w, of a rectangle that has a perimeter of 72 units. If we know that the length is 15 units, we can rewrite the equation as $2(15) + 2w = 72$. This is an equation in one variable, because w is the only quantity that we don't know. To solve this equation means to find a value of w that makes the equation true. In this case, 21 is the solution because substituting 21 for w in the equation results in a true statement.

$$2(15) + 2w = 72$$
$$2(15) + 2(21) = 72$$
$$30 + 42 = 72$$
$$72 = 72$$

An equation that contains two unknown quantities or two quantities that vary is called an *equation in two variables*. A solution to such an equation is a pair of numbers that makes the equation true.

Suppose Tyler spends $45 on T-shirts and socks. A T-shirt costs $10 and a pair of socks costs $2.50. If t represents the number of T-shirts and p represents the number of pairs of socks that Tyler buys, we can represent this situation with the equation:

$$10t + 2.50p = 45$$

This is an equation in two variables. More than one pair of values for t and p make the equation true.

$t = 3$ and $p = 6$

$$10(3) + 2.50(6) = 45$$
$$30 + 15 = 45$$
$$45 = 45$$

$t = 4$ and $p = 2$

$$10(4) + 2.50(2) = 45$$
$$40 + 5 = 45$$
$$45 = 45$$

$t = 2$ and $p = 10$

$$10(2) + 2.50(10) = 45$$
$$20 + 25 = 45$$
$$45 = 45$$

In this situation, one constraint is that the combined cost of shirts and socks must equal $45. Solutions to the equation are pairs of t and p values that satisfy this constraint.

Combinations such as $t = 1$ and $p = 10$ or $t = 2$ and $p = 7$ are not solutions because they don't meet the constraint. When these pairs of values are substituted into the equation, they result in statements that are false.

NAME _____ DATE _____ PERIOD _____

Practice
Equations and Their Solutions

1. An artist is selling children's crafts. Necklaces cost $2.25 each, and bracelets cost $1.50 each.

   Select **all** the combinations of necklaces and bracelets that the artist could sell for exactly $12.00.

   (A.) 5 necklaces and 1 bracelet

   (B.) 2 necklaces and 5 bracelets

   (C.) 3 necklaces and 3 bracelets

   (D.) 4 necklaces and 2 bracelets

   (E.) 3 necklaces and 5 bracelets

   (F.) 6 necklaces and no bracelets

   (G.) No necklaces and 8 bracelets

2. Diego is collecting dimes and nickels in a jar. He has collected $22.25 so far. The relationship between the numbers of dimes and nickels, and the amount of money in dollars is represented by the equation $0.10d + 0.05n = 22.25$.

   Select **all** the values (d, n) that could be solutions to the equation.

   (A.) (0, 445)

   (B.) (0.50, 435)

   (C.) (233, 21)

   (D.) (118, 209)

   (E.) (172, 101)

3. Volunteer drivers are needed to bring 80 students to the championship baseball game. Drivers either have cars, which can seat 4 students, or vans, which can seat 6 students. The equation $4c + 6v = 80$ describes the relationship between the number of cars, c, and number of vans, v, that can transport exactly 80 students.

Select **all** statements that are true about the situation.

(A.) If 12 cars go, then 2 vans are needed.

(B.) $c = 14$ and $v = 4$ are a pair of solutions to the equation.

(C.) The pair $c = 14$ and $v = 4$ is a solution to the equation.

(D.) 10 cars and 8 vans isn't enough to transport all the students.

(E.) If 20 cars go, no vans are needed.

(F.) 8 vans and 8 cars are numbers that meet the constraints in this situation.

4. The drama club is printing t-shirts for its members. The printing company charges a certain amount for each shirt plus a setup fee of $40. There are 21 students in the drama club.

   a. If there are 21 students in the club and the t-shirt order costs a total of $187, how much does each t-shirt cost? Show your reasoning.

   b. The equation $201.50 = f + 6.50(21)$ represents the cost of printing the shirts at a second printing company. Find the solution to the equation and state what it represents in this situation.

NAME _____ DATE _____ PERIOD _____

5. The box plot represents the distribution of the number of children in 30 different families.

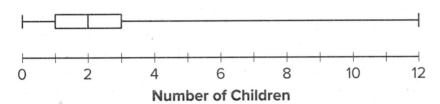

Number of Children

After further examination, the value of 12 is removed for having been recorded in error. The box plot represents the distribution of the same data set, but with the maximum, 12, removed.

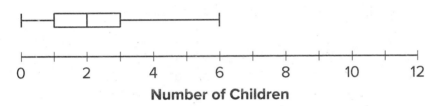

Number of Children

The median is 2 children for both plots. **(Lesson 1-10)**

a. Explain why the median remains the same when 12 was removed from the data set.

b. When 12 is removed from the data set, does the mean remain the same? Explain your reasoning.

6. The number of points Jada's basketball team scored in their games have a mean of about 44 and a standard deviation of about 15.7 points. Interpret the mean and standard deviation in the context of Jada's basketball team. (Lesson 1-13)

7. Kiran's family is having people over to watch a football game. They plan to serve soda and pretzels. They are preparing 12 ounces of soda and 3 ounces of pretzels per person. Including Kiran's family, there will be 10 people at the gathering.

   A bottle of soda contains 22 ounces and costs $1.50. A package of pretzels contains 16 ounces and costs $2.99. Let n represent number of people watching the football game, s represent the ounces of soda, p represent the ounces of pretzels, and b represent Kiran's budget in dollars. Which equation best represents Kiran's budget? (Lesson 2-2)

   (A.) $12s + 3p = b$

   (B.) $12 \cdot 10 + 3 \cdot 10 = b$

   (C.) $1.50s + 2.99p = b$

   (D.) $1.50 \cdot 6 + 2.99 \cdot 2 = b$

8. The speed of an object can be found by taking the distance it travels and dividing it by the time it takes to travel that distance. An object travels 100 feet in 2.5 seconds. Let the speed, S, be measured in feet per second. Write an equation to represent the relationship between the three quantities (speed, distance, and time). (Lesson 2-2)

9. A donut shop made 12 dozen donuts to give to a school's math club.

   Which expression represents how many donuts each student would get if the donuts were equally distributed and there were x students in math club? (Lesson 2-1)

   (A.) $\frac{x}{12}$ (C.) $(12 \cdot 12) \cdot x$

   (B.) $\frac{12 \cdot 12}{x}$ (D.) $12 \cdot 12$

Lesson 2-5

Equations and Their Graphs

NAME _____ DATE _____ PERIOD _____

Learning Goal Let's graph equations in two variables.

 Warm Up

5.1 Which One Doesn't Belong: Hours and Dollars

Which one doesn't belong?

A

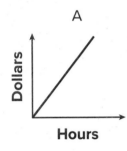

Hours

B

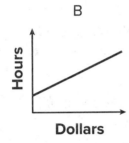

Dollars

C

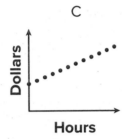

Hours

D

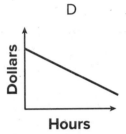

Hours

To get snacks for a class trip, Clare went to the "bulk" section of the grocery store, where she could buy any quantity of a product and the prices are usually good.

Clare purchased some salted almonds at $6 per pound and some dried figs at $9 per pound. She spent $75 before tax.

1. If she bought 2 pounds of almonds, how many pounds of figs did she buy?

2. If she bought 1 pound of figs, how many pounds of almonds did she buy?

3. Write an equation that describes the relationship between pounds of figs and pounds of almonds that Clare bought, and the dollar amount that she paid. Be sure to specify what the variables represent.

4. Here is a graph that represents the quantities in this situation.

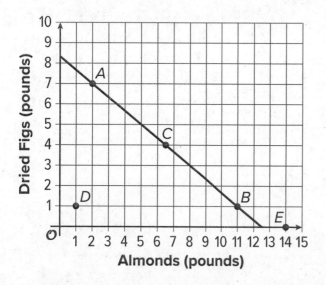

a. Choose any point on the line, state its coordinates, and explain what it tells us.

b. Choose any point that is *not* on the line, state its coordinates, and explain what it tells us.

NAME _____ DATE _____ PERIOD _____

Activity
5.3 Graph It!

1. A student has a savings account with $475 in it. She deposits $125 of her paycheck into the account every week. Her goal is to save $7,000 for college.

   a. How much will be in the account after 3 weeks?

   b. How long will it take before she has $1,350?

   c. Write an equation that represents the relationship between the dollar amount in her account and the number of weeks of saving.

   d. Graph your equation using graphing technology. Mark the points on the graph that represent the amount after 3 weeks and the week she has $1,350. Write down the coordinates.

   e. How long will it take her to reach her goal?

2. A 450-gallon tank full of water is draining at a rate of 20 gallons per minute.

    a. How many gallons will be in the tank after 7 minutes?

    b. How long will it take for the tank to have 200 gallons?

    c. Write an equation that represents the relationship between the gallons of water in the tank and minutes the tank has been draining.

    d. Graph your equation using graphing technology. Mark the points on the graph that represent the gallons after 7 minutes and the time when the tank has 200 gallons. Write down the coordinates.

    e. How long will it take until the tank is empty?

NAME _____ DATE _____ PERIOD _____

Are you ready for more?

1. Write an equation that represents the relationship between the gallons of water in the tank and hours the tank has been draining.

2. Write an equation that represents the relationship between the gallons of water in the tank and seconds the tank has been draining.

3. Graph each of your new equations. In what way are all of the graphs the same? In what way are they all different?

4. How would these graphs change if we used quarts of water instead of gallons? What would stay the same?

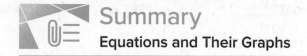
Like an equation, a graph can give us information about the relationship between quantities and the constraints on them.

Suppose we are buying beans and rice to feed a large gathering of people, and we plan to spend $120 on the two ingredients. Beans cost $2 a pound and rice costs $0.50 a pound.

If x represents pounds of beans and y represents pounds of rice, the equation $2x + 0.50y = 120$ can represent the constraints in this situation.

The graph of $2x + 0.50y = 120$ shows a straight line.

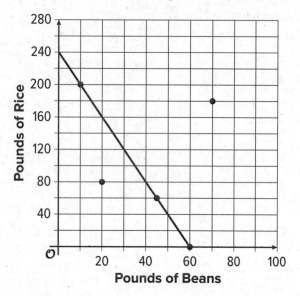

Each point on the line is a pair of x- and y-values that make the equation true and is thus a solution. It is also a pair of values that satisfy the constraints in the situation.

- The point (10, 200) is on the line. If we buy 10 pounds of beans and 200 pounds of rice, the cost will be $2(10) + 0.50(200)$, which equals $120.

- The points (60, 0) and (45, 60) are also on the line. If we buy only beans— 60 pounds of them—and no rice, we will spend $120. If we buy 45 pounds of beans and 60 pounds of rice, we will also spend $120.

What about points that are *not* on the line? They are not solutions because they don't satisfy the constraints, but they still have meaning in the situation.

- The point (20, 80) is not on the line. Buying 20 pounds of beans and 80 pounds of rice costs $2(20) + 0.50(80)$ or $80, which does not equal $120. This combination costs less than what we intend to spend.

- The point (70,180) means that we buy 70 pounds of beans and 180 pounds of rice. It will cost $2(70) + 0.50(180)$ or $230, which is over our budget of $120.

NAME _____ DATE _____ PERIOD _____

Practice
Equations and Their Graphs

1. Select **all** the points that are on the graph of the equation $4y - 6x = 12$.

   (A.) $(-4, -3)$

   (B.) $(-1, 1.5)$

   (C.) $(0, -2)$

   (D.) $(0, 3)$

   (E.) $(3, -4)$

   (F.) $(6, 4)$

2. Here is a graph of the equation $x + 3y = 6$. Select **all** coordinate pairs that represent a solution to the equation.

   (A.) $(0, 2)$

   (B.) $(0, 6)$

   (C.) $(2, 6)$

   (D.) $(3, 1)$

   (E.) $(4, 1)$

   (F.) $(6, 2)$

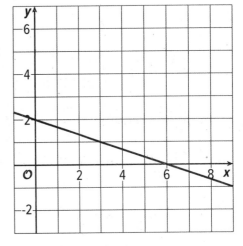

3. A theater is selling tickets to a play. Adult tickets cost $8 each and children's tickets cost $5 each. They collect $275 after selling x adult tickets and y children's tickets.

   What does the point $(30, 7)$ mean in this situation?

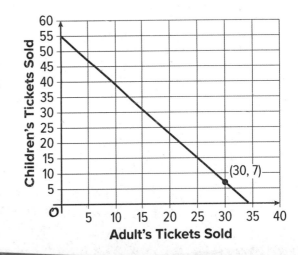

4. *Technology required.* Priya starts with $50 in her bank account. She then deposits $20 each week for 12 weeks.

   a. Write an equation that represents the relationship between the dollar amount in her bank account and the number of weeks of saving.

   b. Graph your equation using graphing technology. Mark the point on the graph that represents the amount after 3 weeks.

   c. How many weeks does it take her to have $250 in her bank account? Mark this point on the graph.

5. During the month of August, the mean of the daily rainfall in one city was 0.04 inches with a standard deviation of 0.15 inches. In another city, the mean of the daily rainfall was 0.01 inches with a standard deviation of 0.05 inches.

   Han says that both cities had a similar pattern of precipitation in the month of August. Do you agree with Han? Explain your reasoning. (Lesson 1-13)

NAME _____ DATE _____ PERIOD _____

6. In a video game, players form teams and work together to earn as many points as possible for their team. Each team can have between 2 and 4 players. Each player can score up to 20 points in each round of the game. Han and three of his friends decided to form a team and play a round.

Write an expression, an equation, or an inequality for each quantity described here. If you use a variable, specify what it represents. (Lesson 2-1)

   a. the allowable number of players on a team

   b. the number of points Han's team earns in one round if every player earns a perfect score

   c. the number of points Han's team earns in one round if no players earn a perfect score

   d. the number of players in a game with six teams of different sizes: two teams have 4 players each and the rest have 3 players each

   e. the possible number of players in a game with eight teams

7. A student on the cross-country team runs 30 minutes a day as a part of her training.

Write an equation to describe the relationship between the distance she runs in miles, D, and her running speed, in miles per hour, when she runs: (Lesson 2-2)

   a. at a constant speed of 4 miles per hour for the entire 30 minutes

   b. at a constant speed of 5 miles per hour the first 20 minutes, and then at 4 miles per hour the last 10 minutes

   c. at a constant speed of 6 miles per hour the first 15 minutes, and then at 5.5 miles per hour for the remaining 15 minutes

   d. at a constant speed of a miles per hour the first 6 minutes, and then at 6.5 miles per hour for the remaining 24 minutes

   e. at a constant speed of 5.4 miles per hour for m minutes, and then at b miles per hour for n minutes

8. In the 21st century, people measure length in feet and meters. At various points in history, people measured length in hands, cubits, and paces. There are 9 hands in 2 cubits. There are 5 cubits in 3 paces. (Lesson 2-3)

a. Write an equation to express the relationship between hands, h, and cubits, c.

b. Write an equation to express the relationship between hands, h, and paces, p.

9. The table shows the amount of money, A, in a savings account after m months.

Number of Months	Dollar Amount
5	1,200
6	1,300
7	1,400
8	1,500

Select **all** the equations that represent the relationship between the amount of money, A, and the number of months, m. (Lesson 2-3)

A. $A = 100m$

B. $A = 100(m - 5)$

C. $A - 700 = 100m$

D. $A - 1{,}200 = 100m$

E. $A = 700 + 100m$

F. $A = 1{,}200 + 100m$

G. $A = 1{,}200 + 100(m - 5)$

Lesson 2-6

Equivalent Equations

NAME _____ DATE _____ PERIOD _____

Learning Goal Let's investigate what makes two equations equivalent.

Warm Up
6.1 Two Expressions

Your teacher will assign you one of these expressions:

$$\frac{n^2 - 9}{2(4 - 3)} \quad \text{or} \quad (n + 3) \cdot \frac{n - 3}{8 - 3 \cdot 2}$$

Evaluate your expression when n is:

    a. 5   **b.** 7   **c.** 13   **d.** -1

Activity
6.2 Much Ado about Ages

1. Write as many equations as possible that could represent the relationship between the ages of the two children in each family described. Be prepared to explain what each part of your equation represents.

    a. In Family A, the youngest child is 7 years younger than the oldest, who is 18.

    b. In Family B, the middle child is 5 years older than the youngest child.

2. Tyler thinks that the relationship between the ages of the children in Family B can be described with $2m - 2y = 10$, where m is the age of the middle child and y is the age of the youngest. Explain why Tyler is right.

3. Are any of these equations **equivalent** to one another? If so, which ones? Explain your reasoning.

$$3a + 6 = 15 \qquad 3a = 9 \qquad a + 2 = 5 \qquad \frac{1}{3}a = 1$$

Are you ready for more?

Here is a puzzle:

$$m + m = N$$
$$N + N = p$$
$$m + p = Q$$
$$p + Q = ?$$

Which expressions could be equal to $p + Q$?

$$2p + m \qquad\qquad 4m + N \qquad\qquad 3N \qquad\qquad 9m$$

 Activity

6.3 What's Acceptable?

Noah is buying a pair of jeans and using a coupon for 10% off. The total price is $56.70, which includes $2.70 in sales tax. Noah's purchase can be modeled by the equation:

$$x - 0.1x + 2.70 = 56.70.$$

1. Discuss with a partner:

   a. What does the solution to the equation mean in this situation?

   b. How can you verify that 70 is not a solution but 60 is the solution?

NAME _____ DATE _____ PERIOD _____

2. Here are some equations that are related to $x - 0.1x + 2.70 = 56.70$. Each equation is a result of performing one or more moves on that original equation. Each can also be interpreted in terms of Noah's purchase.

For each equation, determine either what move was made or how the equation could be interpreted. (Some examples are given here.) Then, check if 60 is the solution of the equation.

	What Was Done?	Interpretation?	Same Solution?
Equation A $100x - 10x + 270$ $= 5,670$		The price is expressed in cents instead of dollars	
Equation B $x - 0.1x = 54$	Subtract 2.70 from both sides of the equation.		
Equation C $0.9x + 2.70$ $= 56.70$		10% off means paying 90% of the original price. 90% of the original price plus sales tax is $56.70	
Equation D $x - 0.1x = 56.70$		The price after using the coupon for 10% off and before sales tax is $56.70.	
Equation E $x - 0.1x = 59.40$	Subtract 2.70 from the left and add 2.70 to the right.		
Equation F $2(x - 0.1x + 2.70)$ $= 56.70$		The price of 2 pairs of jeans, after using the coupon for 10% off and paying sales tax, is $56.70.	

Which of the six equations are equivalent to the original equation? Why?

Suppose we bought two packs of markers and a $0.50 gluestick for $6.10. If p is the dollar cost of one pack of markers, the equation $2p + 0.50 = 6.10$ represents this purchase. The solution to this equation is 2.80.

Now suppose a friend bought six of the same packs of markers and three $0.50 gluesticks, and paid $18.30. The equation $6p + 1.50 = 18.30$ represents this purchase. The solution to this equation is also 2.80.

We can say that $2p + 0.50 = 6.10$ and $6p + 1.50 = 18.30$ are **equivalent equations** because they have exactly the same solution. Besides 2.80, no other values of p make either equation true. Only the price of $2.80 per pack of markers satisfies the constraint in each purchase.

$$2p + 0.50 = 6.10$$

$$6p + 1.50 = 18.30$$

How do we write equivalent equations like these?

There are certain moves we can perform!

In this example, the second equation, $6p + 1.50 = 18.30$, is a result of multiplying each side of the first equation by 3. Buying 3 times as many markers and glue sticks means paying 3 times as much money. The unit price of the markers hasn't changed.

Here are some other equations that are equivalent to $2p + 0.50 = 6.10$, along with the moves that led to these equations.

- $2p + 4 = 9.60$ Add 3.50 to each side of the original equation.
- $2p = 5.60$ Subtract 0.50 from each side of the original equation.
- $\frac{1}{2}(2p + 0.50) = 3.05$ Multiply each side of the original equation by $\frac{1}{2}$.
- $2(p + 0.25) = 6.10$ Apply the distributive property to rewrite the left side.

In each case:

The move is acceptable because it doesn't change the equality of the two sides of the equation. If $2p + 0.50$ has the same value as 6.10, then multiplying $2p + 0.50$ by $\frac{1}{2}$ and multiplying 6.10 by $\frac{1}{2}$ keep the two sides equal.

NAME _____ DATE _____ PERIOD _____

Only $p = 2.80$ makes the equation true. Any value of p that makes an equation false also makes the other equivalent equations false. (Try it!)

These moves—applying the distributive property, adding the same amount to both sides, dividing each side by the same number, and so on—might be familiar because we have performed them when solving equations. Solving an equation essentially involves writing a series of equivalent equations that eventually isolates the variable on one side.

Not all moves that we make on an equation would create equivalent equations, however!

For example, if we subtract 0.50 from the left side but add 0.50 to the right side, the result is $2p = 6.60$. The solution to this equation is 3.30, not 2.80. This means that $2p = 6.60$ is not equivalent to $2p + 0.50 = 6.10$.

Glossary

equivalent equations

1. Which equation is equivalent to the equation $6x + 9 = 12$?

   (A.) $x + 9 = 6$

   (B.) $2x + 3 = 4$

   (C.) $3x + 9 = 6$

   (D.) $6x + 12 = 9$

2. Select **all** the equations that have the same solution as the equation $3x - 12 = 24$.

   (A.) $15x - 60 = 120$

   (B.) $3x = 12$

   (C.) $3x = 36$

   (D.) $x - 4 = 8$

   (E.) $12x - 12 = 24$

3. Jada has a coin jar containing n nickels and d dimes worth a total of $3.65. The equation $0.05n + 0.1d = 3.65$ is one way to represent this situation.

   Which equation is equivalent to the equation $0.05n + 0.1d = 3.65$?

   (A.) $5n + d = 365$

   (B.) $0.5n + d = 365$

   (C.) $5n + 10d = 365$

   (D.) $0.05d + 0.1n = 365$

4. Select **all** the equations that have the same solution as $2x - 5 = 15$.

   (A.) $2x = 10$

   (B.) $2x = 20$

   (C.) $2(x - 5) = 15$

   (D.) $2x - 20 = 0$

   (E.) $4x - 10 = 30$

   (F.) $15 = 5 - 2x$

NAME _____ DATE _____ PERIOD _____

5. The number of hours spent in an airplane on a single flight is recorded on a dot plot. The mean is 5 hours and the standard deviation is approximately 5.82 hours. The median is 4 hours and the IQR is 3 hours. The value 26 hours is an outlier that should not have been included in the data.

Number of Hours Spent in an Airplane

When the outlier is removed from the data set: (Lesson 1-14)

a. What is the mean?

b. What is the standard deviation?

c. What is the median?

d. What is the IQR?

6. A basketball coach purchases bananas for the players on his team. The table shows total price in dollars, P, of n bananas.

Which equation could represent the total price in dollars for n bananas? (Lesson 2-3)

Number of Bananas	Total Price in Dollars
7	4.13
8	4.72
9	5.31
10	5.90

(A.) $P = 0.59n$

(B.) $P = 5.90 - 0.59n$

(C.) $P = \dfrac{5.90}{n}$

(D.) $P = n + 0.59$

7. Kiran is collecting dimes and quarters in a jar. He has collected $10.00 so far and has d dimes and q quarters. The relationship between the numbers of dimes and quarters, and the amount of money in dollars is represented by the equation $0.1d + 0.25q = 10$.

Select **all** the values (d, q) that could be solutions to the equation.
(Lesson 2-4)

(A.) (100, 0)

(B.) (20, 50)

(C.) (50, 20)

(D.) (0, 100)

(E.) (10, 36)

8. Here is a graph of the equation $3x - 2y = 12$.

Select **all** coordinate pairs that represent a solution to the equation.
(Lesson 2-5)

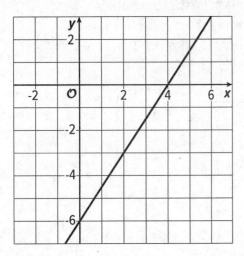

A. $(2, -3)$

B. $(4, 0)$

C. $(5, -1)$

D. $(0, -6)$

E. $(2, 3)$

9. Jada bought some sugar and strawberries to make strawberry jam. Sugar costs $1.80 per pound, and strawberries cost $2.50 per pound. Jada spent a total of $19.40.

Which point on the coordinate plane could represent the pounds of sugar and strawberries that Jada used to make jam? (Lesson 2-5)

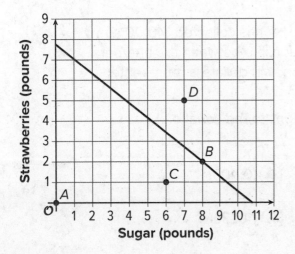

A. Point A

B. Point B

C. Point C

D. Point D

Lesson 2-7

Explaining Steps for Rewriting Equations

NAME _____ DATE _____ PERIOD _____

Learning Goal Let's think about why some steps for rewriting equations are valid but other steps are not.

 ## Warm Up
7.1 Math Talk: Could It be Zero?

Is 0 a solution to each equation?

$4(x + 2) = 10$

$12 - 8x = 3(x + 4)$

$5x = \frac{1}{2}x$

$\frac{6}{x} + 1 = 8$

 ## Activity
7.2 Explaining Acceptable Moves

Here are some pairs of equations. While one partner listens, the other partner should:

- Choose a pair of equations from column A. Explain why, if x is a number that makes the first equation true, then it also makes the second equation true.

- Choose a pair of equations from column B. Explain why the second equation is no longer true for a value of x that makes the first equation true.

Then, switch roles until you run out of time or you run out of pairs of equations.

	A		B	
1.	$16 = 4(9 - x)$ $16 = 36 - 4x$		$9x = 5x + 4$ $14x = 4$	
2.	$5x = 24 + 2x$ $3x = 24$		$\frac{1}{2}x - 8 = 9$ $x - 8 = 18$	
3.	$-3(2x + 9) = 12$ $2x + 9 = -4$		$6x - 6 = 3x$ $x - 1 = 3x$	

NAME _____ DATE _____ PERIOD _____

	A		B	
4.	$5x = 3 - x$ $5x = -x + 3$		$-11(x - 2) = 8$ $x - 2 = 8 + 11$	
5.	$18 = 3x - 6 + x$ $18 = 4x - 6$		$4 - 5x = 24$ $5x = 20$	

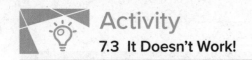

Activity

7.3 It Doesn't Work!

Noah is having trouble solving two equations. In each case, he took steps that he thought were acceptable but ended up with statements that are clearly not true.

Analyze Noah's work on each equation and the moves he made. Were they acceptable moves? Why do you think he ended up with a false equation?

Discuss your observations with your group and be prepared to share your conclusions. If you get stuck, consider solving each equation.

1. $x + 6 = 4x + 1 - 3x$ original equation

   $x + 6 = 4x - 3x + 1$ apply the commutative property

   $x + 6 = x + 1$ combine like terms

   $6 = 1$ subtract x from each side

2. $2(5 + x) - 1 = 3x + 9$ original equation

   $10 + 2x - 1 = 3x + 9$ apply the distributive property

   $2x - 1 = 3x - 1$ subtract 10 from each side

   $2x = 3x$ add 1 to each side

   $2 = 3$ divide each side by x

NAME _____ DATE _____ PERIOD _____

1. We can't divide the number 100 by zero because dividing by zero is undefined.

   a. Instead, try dividing 100 by 10, then 1, then 0.1, then 0.01. What happens as you divide by smaller numbers?

   b. Now try dividing the number -100 by 10, by 1, by 0.1, 0.01. What is the same and what is different?

2. In middle school, you used tape diagrams to represent division. This tape diagram shows that $6 \div 2 = 3$

   a. Draw a tape diagram that shows why $6 \div \frac{1}{2} = 12$.

   b. Try to draw a tape diagram that represents $6 \div 0$. Explain why this is so difficult.

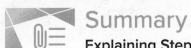

Summary
Explaining Steps for Rewriting Equations

When solving an equation, sometimes we end up with a false equation instead of a solution. Let's look at two examples.

Example 1: $4(x + 1) = 4x$

Here are two attempts to solve it.

$4(x + 1) = 4x$	original equation
$x + 1 = x$	divide each side by 4
$1 = 0$	subtract x from each side

$4(x + 1) = 4x$	original equation
$4x + 4 = 4x$	apply the distributive prop
$4 = 0$	subtract $4x$ from each side

Each attempt shows acceptable moves, but the final equation is a false statement. Why is that?

When solving an equation, we usually start by assuming that there is at least one value that makes the equation true. The equation $4(x + 1) = 4x$ can be interpreted as: 4 groups of $(x + 1)$ are equal to 4 groups of x. There are no values of x that can make this true.

For instance, if $x = 10$, then $x + 1 = 11$. It's not possible that 4 times 11 is equal to 4 times 10. Likewise, 1.5 is 1 more than 0.5, but 4 groups of 1.5 cannot be equal to 4 groups of 0.5.

Because of this, the moves made to solve the equation would not lead to a solution. The equation $4(x + 1) = 4x$ has no solutions.

NAME _____ DATE _____ PERIOD _____

Example 2: $2x - 5 = \dfrac{x - 20}{4}$

$$2x - 5 = \dfrac{x - 20}{4} \qquad \text{original equation}$$

$$8x - 20 = x - 20 \qquad \text{multiply each side by 4}$$

$$8x = x \qquad \text{add 20 to each side}$$

$$8 = 1 \qquad \text{divide each side by } x$$

Each step in the process seems acceptable, but the last equation is a false statement.

It is not easy to tell from the original equation whether it has a solution, but if we look at the equivalent equation $8x = x$, we can see that 0 could be a solution. When x is 0, the equation is $0 = 0$, which is a true statement. What is going on here?

The last move in the solving process was division by x. Because 0 could be the value of x and dividing by 0 gives an undefined number, we don't usually divide by the variable we're solving for. Doing this might make us miss a solution, namely $x = 0$.

Here are two ways to solve the equation once we get to $8x = x$:

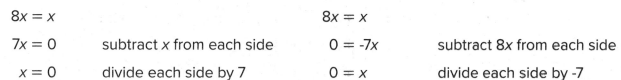

$8x = x$		$8x = x$	
$7x = 0$	subtract x from each side	$0 = -7x$	subtract $8x$ from each side
$x = 0$	divide each side by 7	$0 = x$	divide each side by -7

Practice

Explaining Steps for Rewriting Equations

1. Match each equation with an equivalent equation. Some of the answer choices are not used.

   A. $3x + 6 = 4x + 7$

   B. $3(x + 6) = 4x + 7$

   C. $4x + 3x = 7 - 6$

   1. $9x = 4x + 7$

   2. $3x + 18 = 4x + 7$

   3. $3x = 4x + 7$

   4. $3x - 1 = 4x$

   5. $7x = 1$

2. Mai says x is a number that makes Equation A true and also a number that makes Equation B true.

   Equation A: $-3(x + 7) = 24$

   Equation B: $x + 7 = -8$

   Which statement explains why this is true?

   A. Adding 3 to both sides of Equation A gives $x + 7 = -8$.

   B. Applying the distributive property to Equation A gives $x + 7 = -8$.

   C. Subtracting 3 from both sides of Equation A gives $x + 7 = -8$.

   D. Dividing both sides of Equation A by -3 gives $x + 7 = -8$.

3. Is 0 a solution to $2x + 10 = 4x + 10$? Explain or show your reasoning.

4. Kiran says that if x is a solution to the equation $x + 4 = 20$, then x is also a solution to the equation $5(x + 4) = 100$.

   Write a convincing explanation as to why this is true.

NAME _____ DATE _____ PERIOD _____

5. The entrepreneurship club is ordering potted plants for all 36 of its sponsors. One store charges $8.50 for each plant plus a delivery fee of $20. The equation $320 = x + 7.50(36)$ represents the cost of ordering potted plants at a second store.

   What does the x represent in this situation? (Lesson 2-4)

   (A.) The cost for each potted plant at the second store

   (B.) The delivery fee at the second store

   (C.) The total cost of ordering potted plants at the second store

   (D.) The number of sponsors of the entrepreneurship club

6. Which equation is equivalent to the equation $5x + 30 = 45$? (Lesson 2-6)

   (A.) $35x = 45$

   (B.) $5x = 75$

   (C.) $5(x + 30) = 45$

   (D.) $5(x + 6) = 45$

7. The environmental science club is printing T-shirts for its 15 members. The printing company charges a certain amount for each shirt plus a setup fee of $20.

   If the T-shirt order costs a total of $162.50, how much does the company charge for each shirt? (Lesson 2-4)

8. The graph shows the relationship between temperature in degrees Celsius and temperature in degrees Fahrenheit.

(Lesson 2-5)

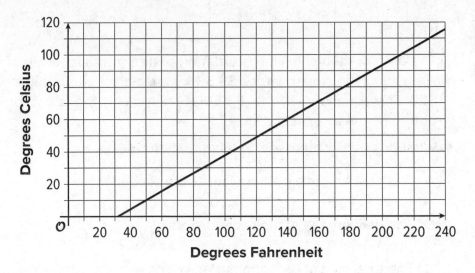

Degrees Fahrenheit

a. Mark the point on the graph that shows the temperature in Celsius when it is 60 degrees Fahrenheit.

b. Mark the point on the graph that shows the temperature in Fahrenheit when it is 60 degrees Celsius.

c. Water boils at 100 degrees Celsius. Use the graph to approximate the boiling temperature in Fahrenheit, or to confirm it, if you knew what it is.

d. The equation that converts Fahrenheit to Celsius is $C = \frac{5}{9}(F - 32)$.

Use it to calculate the temperature in Celsius when it is 60 degrees Fahrenheit. (This answer will be more exact than the point you found in the first part.)

9. Select **all** the equations that have the same solution as $2x - 5 = 15$.

(Lesson 2-6)

A. $2x = 10$

B. $2x = 20$

C. $2(x - 5) = 15$

D. $2x - 20 = 0$

E. $4x - 10 = 30$

F. $15 = 5 - 2x$

10. Diego's age d is 5 more than 2 times his sister's age s. This situation is represented by the equation $d = 2s + 5$. Which equation is equivalent to the equation $d = 2s + 5$? (Lesson 2-6)

A. $d = 2(s + 5)$

B. $d - 5 = 2s$

C. $d - 2 = s + 5$

D. $\frac{d}{2} = s + 5$

Lesson 2-8

Which Variable to Solve for? (Part 1)

NAME _____ DATE _____ PERIOD _____

Learning Goal Let's rearrange equations to pin down a certain quantity.

 ## Warm Up
8.1 Which Equations?

1. The table shows the relationship between the base length, b, and the area, A, of some parallelograms. All the parallelograms have the same height. Base length is measured in inches, and area is measured in square inches. Complete the table.

b (inches)	A (square inches)
1	3
2	6
3	9
4.5	
$\frac{11}{2}$	
	36
	46.5

2. Decide whether each equation could represent the relationship between b and A. Be prepared to explain your reasoning.

   a. $b = 3A$

   b. $b = \frac{A}{3}$

   c. $A = \frac{b}{3}$

   d. $A = 3b$

After a parade, a group of volunteers is helping to pick up the trash along a 2-mile stretch of a road.

The group decides to divide the length of the road so that each volunteer is responsible for cleaning up equal-length sections.

1. Find the length of a road section for each volunteer if there are the following numbers of volunteers. Be prepared to explain or show your reasoning.

    a. 8 volunteers

    b. 10 volunteers

    c. 25 volunteers

    d. 36 volunteers

2. Write an equation that would make it easy to find ℓ, the length of a road section in miles for each volunteer, if there are n volunteers.

3. Find the number of volunteers in the group if each volunteer cleans up a section of the following lengths. Be prepared to explain or show your reasoning.

    a. 0.4 mile

    b. $\frac{2}{7}$ mile

    c. 0.125 mile

    d. $\frac{6}{45}$ mile

4. Write an equation that would make it easy to find the number of volunteers, n, if each volunteer cleans up a section that is ℓ miles.

Illustrative Math

NAME _____ DATE _____ PERIOD _____

Are you ready for more?

Let's think about the graph of the equation $y = \frac{2}{x}$.

1. Make a table of (x, y) pairs that will help you graph the equation. Make sure to include some negative numbers for x and some numbers that are not integers.

2. Plot the graph on the coordinate axes. You may need to find a few more points to plot to make the graph look smooth.

x	y

3. The coordinate plane provided is **too** small to show the whole graph.

   What do you think the graph looks like when x is between 0 and $\frac{1}{2}$?

   Try some values of x to test your idea.

4. What is the largest value that y can ever be?

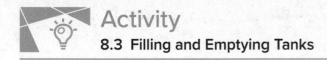

1. Tank A initially contained 124 liters of water. It is then filled with more water, at a constant rate of 9 liters per minute. How many liters of water are in Tank A after the following amounts of time have passed?

   a. 4 minutes

   b. 80 seconds

   c. m minutes

2. How many minutes have passed, m, when Tank A contains the following amounts of water?

   a. 151 liters

   b. 191.5 liters

   c. 270.25 liters

   d. p liters

3. Tank B, which initially contained 80 liters of water, is being drained at a rate of 2.5 liters per minute. How many liters of water remain in the tank after the following amounts of time?

   a. 30 seconds

   b. 7 minutes

   c. t minutes

4. For how many minutes, t, has the water been draining when Tank B contains the following amounts of water?

   a. 75 liters

   b. 32.5 liters

   c. 18 liters

   d. v liters

NAME _____ DATE _____ PERIOD _____

Summary
Which Variable to Solve for? (Part 1)

A relationship between quantities can be described in more than one way. Some ways are more helpful than others, depending on what we want to find out. Let's look at the angles of an isosceles triangle, for example.

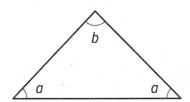

The two angles near the horizontal side have equal measurements in degrees, a.

The sum of angles in a triangle is 180°, so the relationship between the angles can be expressed as:

$a + a + b = 180$

Suppose we want to find a when b is 20°.

Let's substitute 20 for b and solve the equation.

$$
\begin{aligned}
a + a + b &= 180 \\
2a + 20 &= 180 \\
2a &= 180 - 20 \\
2a &= 160 \\
a &= 80
\end{aligned}
$$

What is the value of a if b is 45°?

$$
\begin{aligned}
a + a + b &= 180 \\
2a + 45 &= 180 \\
2a &= 180 - 45 \\
2a &= 135 \\
a &= 67.5
\end{aligned}
$$

Now suppose the bottom two angles are 34° each. How many degrees is the top angle?

Let's substitute 34 for a and solve the equation.

$$a + a + b = 180$$
$$34 + 34 + b = 180$$
$$68 + b = 180$$
$$b = 112$$

What is the value of b if a is 72.5°?

$$a + a + b = 180$$
$$72.5 + 72.5 + b = 180$$
$$145 + b = 180$$
$$b = 35$$

Notice that when b is given, we did the same calculation repeatedly to find a: we substituted b into the first equation, subtracted b from 180, and then divided the result by 2.

Instead of taking these steps over and over whenever we know b and want to find a, we can rearrange the equation to isolate a:

$$a + a + b = 180$$
$$2a + b = 180$$
$$2a = 180 - b$$
$$a = \frac{180 - b}{2}$$

This equation is equivalent to the first one. To find a, we can now simply substitute any value of b into this equation and evaluate the expression on right side.

Likewise, we can write an equivalent equation to make it easier to find b when we know a:

$$a + a + b = 180$$
$$2a + b = 180$$
$$b = 180 - 2a$$

Rearranging an equation to isolate one variable is called *solving for a variable*. In this example, we have solved for a and for b. All three equations are equivalent. Depending on what information we have and what we are interested in, we can choose a particular equation to use.

NAME _____ DATE _____ PERIOD _____

Practice
Which Variable to Solve for? (Part 1)

1. Priya is buying raisins and almonds to make trail mix. Almonds cost
   $5.20 per pound and raisins cost $2.75 per pound. Priya spent $11.70
   buying almonds and raisins. The relationship between pounds of almonds
   a, pounds of raisins r, and the total cost is represented by the equation
   $5.20a + 2.75r = 11.70$. How many pounds of raisins did Priya buy if she
   bought the following amounts of almonds?

   a. 2 pounds of almonds

   b. 1.06 pounds of almonds

   c. 0.64 pounds of almonds

   d. a pounds of almonds

2. Here is a linear equation in two variables: $2x + 4y - 31 = 123$. Solve the
   equation, first for x and then for y.

3. A chef bought $17.01 worth of ribs and chicken. Ribs cost $1.89 per pound
   and chicken costs $0.90 per pound. The equation $0.9c + 1.89r = 17.01$
   represents the relationship between the quantities in this situation. Show
   that each of the following equations is equivalent to $0.9c + 1.89r = 17.01$.
   Then, explain when it might be helpful to write the equation in these forms.

   a. $c = 18.9 - 2.1r$

b. $r = -\dfrac{10}{21}c + 9$

4. A car traveled 180 miles at a constant rate.

   a. Complete the table to show the rate at which the car was traveling if it completed the same distance in each number of hours.

   b. Write an equation that would make it easy to find the rate at which the car was traveling in miles per hour r, if it traveled for t hours.

Travel Time (hours)	Rate of Travel (miles per hour)
5	
4.5	
3	
2.25	

5. Bananas cost $0.50 each, and apples cost $1.00 each. Select **all** the combinations of bananas and apples that Elena could buy for exactly $3.50. (Lesson 2-4)

   A. 2 bananas and 2 apples

   B. 3 bananas and 2 apples

   C. 1 banana and 2 apples

   D. 1 banana and 3 apples

   E. 5 bananas and 2 apples

   F. 5 bananas and 1 apple

NAME _____ DATE _____ PERIOD _____

6. A group of 280 elementary school students and 40 adults are going on a field trip. They are planning to use two different types of buses to get to the destination. The first type of bus holds 50 people and the second type of bus holds 56 people.

   Andre says that 3 of the first type of bus and 3 of the second type of bus will hold all of the students and adults going on the field trip. Is Andre correct? Explain your reasoning. (Lesson 2-4)

7. Elena says *x* is a number that makes Equation A true, but that it does not make Equation B true.

   Equation A: $13 - 5x = 48$

   Equation B: $5x = 35$

   Write a convincing explanation as to why this is true. (Lesson 2-7)

8. To grow properly, each tomato plant needs 1.5 square feet of soil and each broccoli plant needs 2.25 square feet of soil. The graph shows the different combinations of broccoli and tomato plants in an 18 square foot plot of soil.

   Match each point to the statement that describes it. (Lesson 2-5)

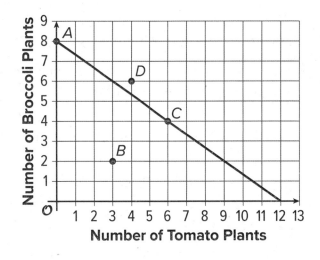

   A. Point A

   B. Point B

   C. Point C

   D. Point D

   1. The soil is fully used when 6 tomato plants and 4 broccoli plants are planted.

   2. Only broccoli was planted, but the plot is fully used and all plants can grow properly.

   3. After 3 tomato plants and 2 broccoli plants were planted, there is still extra space in the plot.

   4. With 4 tomato plants and 6 broccoli plants planted, the plot is overcrowded.

9. Select all the equations that are equivalent to the equation $3x - 4 = 5$.
(Lesson 2-6)

A. $3x = 9$

B. $3x - 4 + 4 = 5 + 4$

C. $x - 4 = 2$

D. $x = 9$

E. $-4 = 5 - 3x$

10. Han is solving an equation. He took steps that are acceptable but ended up with equations that are clearly not true.

$5x + 6 = 7x + 5 - 2x$	original equation
$5x + 6 = 7x - 2x + 5$	apply the commutative property
$5x + 6 = 5x + 5$	combine like terms
$6 = 5$	subtract $5x$ from each side

What can Han conclude as a result of these acceptable steps? (Lesson 2-7)

A. There's no value of x that can make the equation $5x + 6 = 7x + 5 - 2x$ true.

B. Any value of x can make the equation $5x + 6 = 7x + 5 - 2x$ true.

C. $x = 6$ is a solution to the equation $5x + 6 = 7x + 5 - 2x$.

D. $x = 5$ is a solution to the equation $5x + 6 = 7x + 5 - 2x$.

Lesson 2-9

Which Variable to Solve for? (Part 2)

NAME _____ DATE _____ PERIOD _____

Learning Goal Let's solve an equation for one of the variables.

 ## Warm Up
9.1 Faces, Vertices, and Edges

In an earlier lesson, you saw the equation $V + F - 2 = E$, which relates the number of vertices, faces, and edges in a Platonic solid.

1. Write an equation that makes it easier to find the number of vertices in each of the Platonic solids described:

   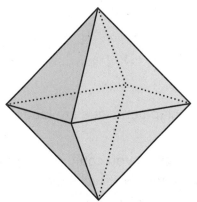

   a. An octahedron (shown here), which has 8 faces.

   b. An icosahedron, which has 30 edges.

2. A Buckminsterfullerene (also called a "Buckyball") is a polyhedron with 60 vertices. It is not a Platonic solid, but the numbers of faces, edges, and vertices are related the same way as those in a Platonic solid. Write an equation that makes it easier to find the number of faces a Buckyball has if we know how many edges it has.

Activity
9.2 Cargo Shipping

An automobile manufacturer is preparing a shipment of cars and trucks on a cargo ship that can carry 21,600 tons. The cars weigh 3.6 tons each and the trucks weigh 7.5 tons each.

1. Write an equation that represents the weight constraint of a shipment. Let c be the number of cars and t be the number of trucks.

2. For one shipment, trucks are loaded first and cars are loaded afterwards. (Even though trucks are bulkier than cars, a shipment can consist of all trucks as long as it is within the weight limit.)

   Find the number of cars that can be shipped if the cargo ship already has:

   a. 480 trucks

   b. 1,500 trucks

   c. 2,736 trucks

   d. t trucks

3. For a different shipment, cars are loaded first, and then trucks are loaded afterwards.

   a. Write an equation you could enter into a calculator or a spreadsheet tool to find the number of trucks that can be shipped if the number of cars is known.

   b. Use your equation and a calculator or a computer to find the number of trucks that can be shipped if the cargo ship already has 1,000 cars. What if the cargo ship already has 4,250 cars?

Are you ready for more?

For yet another shipment, the manufacturer is also shipping motorcycles, which weigh 0.3 ton each.

1. Write an equation that you could enter into a calculator or a spreadsheet tool to find the number of motorcycles that can be shipped, m, if the number of cars and trucks are known.

2. Use your equation to find the number of motorcycles that can be shipped if the cargo ship already contains 1,200 trucks and 3,000 cars.

NAME _____ DATE _____ PERIOD _____

Activity

9.3 Streets and Staffing

The Department of Streets of a city has a budget of $1,962,800 for resurfacing roads and hiring additional workers this year. The cost of resurfacing a mile of 2-lane road is estimated at $84,000. The average starting salary of a worker in the department is $36,000 a year.

1. Write an equation that represents the relationship between the miles of 2-lane roads the department could resurface, m, and the number of new workers it could hire, p, if it spends the entire budget.

2. Take the equation you wrote in the first question and:

   a. Solve for p. Explain what the solution represents in this situation.

   b. Solve for m. Explain what the solution represents in this situation.

3. The city is planning to hire 6 new workers and to use its entire budget.

   a. Which equation should be used to find out how many miles of 2-lane roads it could resurface? Explain your reasoning.

   b. Find the number of miles of 2-lane roads the city could resurface if it hires 6 new workers.

Solving for a variable is an efficient way to find out the values that meet the constraints in a situation. Here is an example.

An elevator has a capacity of 3,000 pounds and is being loaded with boxes of two sizes—small and large. A small box weighs 60 pounds and a large box weighs 150 pounds.

Let x be the number of small boxes and y the number of large boxes. To represent the combination of small and large boxes that fill the elevator to capacity, we can write:

$60x + 150y = 3{,}000$

If there are 10 large boxes already, how many small boxes can we load onto the elevator so that it fills it to capacity? What if there are 16 large boxes?

In each case, we can substitute 10 or 16 for y and perform acceptable moves to solve the equation. Or, we can first solve for x:

$$60x + 150y = 3{,}000 \qquad \text{original equation}$$
$$60x = 3{,}000 - 150y \qquad \text{subtract } 150y \text{ from each side}$$
$$x = \frac{3{,}000 - 150y}{60} \qquad \text{divide each side by 60}$$

This equation allows us to easily find the number of small boxes that can be loaded, x, by substituting any number of large boxes for y.

Now suppose we first load the elevator with small boxes, say, 30 or 42, and want to know how many large boxes can be added for the elevator to reach its capacity.

We can substitute 30 or 42 for x in the original equation and solve it. Or, we can first solve for y:

$$60x + 150y = 3{,}000 \qquad \text{original equation}$$
$$150y = 3{,}000 - 60x \qquad \text{subtract } 60x \text{ from each side}$$
$$y = \frac{3{,}000 - 60x}{150} \qquad \text{divide each side by 150}$$

Now, for any value of x, we can quickly find y by evaluating the expression on the right side of the equal sign.

Solving for a variable—before substituting any known values—can make it easier to test different values of one variable and see how they affect the other variable. It can save us the trouble of doing the same calculation over and over.

NAME _____ DATE _____ PERIOD _____

Practice
Which Variable to Solve for? (Part 2)

1. A car has a 16-gallon fuel tank. When driven on a highway, it has a gas mileage of 30 miles per gallon. The gas mileage (also called "fuel efficiency") tells us the number of miles the car can travel for a particular amount of fuel (one gallon of gasoline, in this case). After filling the gas tank, the driver got on a highway and drove for a while.

   a. How many miles has the car traveled if it has the following amounts of gas left in the tank?

      i. 15 gallons

      ii. 10 gallons

      iii. 2.5 gallons

   b. Write an equation that represents the relationship between the distance the car has traveled in miles, d, and the amount of gas left in the tank in gallons, x.

   c. How many gallons are left in the tank when the car has traveled the following distances on the highway?

      i. 90 miles

      ii. 246 miles

   d. Write an equation that makes it easier to find the amount of gas left in the tank, x, if we know the car has traveled d miles.

2. The area A of a rectangle is represented by the formula $A = lw$ where l is the length and w is the width. The length of the rectangle is 5.

   Write an equation that makes it easy to find the width of the rectangle if we know the area and the length.

3. Noah is helping to collect the entry fees at his school's sports game. Student entry costs $2.75 each and adult entry costs $5.25 each. At the end of the game, Diego collected $281.25.

Select **all** equations that could represent the relationship between the number of students, s, the number of adults, a, and the dollar amount received at the game.

A. $281.25 - 5.25a = 2.75s$

B. $a = 53.57 - \frac{2.75}{5.25}s$

C. $281.25 - 5.25s = a$

D. $281.25 + 2.75a = s$

E. $281.25 + 5.25s = a$

4. $V = \pi r^2 h$ is an equation to calculate the volume of a cylinder, V, where r represents the radius of the cylinder and h represents its height.

Which equation allows us to easily find the height of the cylinder because it is solved for h?

A. $r^2 h = \frac{V}{\pi}$

B. $h = V - \pi r^2$

C. $h = \frac{V}{\pi r^2}$

D. $\pi h = \frac{V}{r^2}$

5. The data represents the number of hours 10 students slept on Sunday night.

6   6   7   7   7   8   8   8   8   9

Are there any outliers? Explain your reasoning. (Lesson 1-14)

NAME _____ DATE _____ PERIOD _____

6. The table shows the volume of water in cubic meters, V, in a tank after water has been pumped out for a certain number of minutes.

   Which equation could represent the volume of water in cubic meters after t minutes of water being pumped out? (Lesson 2-4)

Time After Pumping Begins	Volume of Water (cubic meters)
0	30
5	27.5
10	20
15	7.5

   A. $V = 30 - 2.5t$

   B. $V = 30 - 0.5t$

   C. $V = 30 - 0.5t^2$

   D. $V = 30 - 0.1t^2$

7. A catering company is setting up for a wedding. They expect 150 people to attend. They can provide small tables that seat 6 people and large tables that seat 10 people. (Lesson 2-5)

   a. Find a combination of small and large tables that seats exactly 150 people.

   b. Let x represent the number of small tables and y represent the number of large tables. Write an equation to represent the relationship between x and y.

   c. Explain what the point (20,5) means in this situation.

   d. Is the point (20,5) a solution to the equation you wrote? Explain your reasoning.

8. Which equation has the same solution as $10x - x + 5 = 41$? (Lesson 2-6)

(A.) $10x + 5 = 41$

(B.) $10x - 5 + x = 41$

(C.) $9x = 46$

(D.) $9x + 5 = 41$

9. Noah is solving an equation and one of his moves is unacceptable. Here are the moves he made.

$2(x + 6) - 4 = 8 + 6x$ original equation

$2x + 12 - 4 = 8 + 6x$ apply the distributive property

$2x + 8 = 8 + 6x$ combine like terms

$2x = 6x$ subtract 8 from both sides

$2 = 6$ divide each side by x

Which answer best explains why the "divide each side by x step" is unacceptable? (Lesson 2-7)

(A.) When you divide both sides of $2x = 6x$ by x, you get $2x^2 = 6x^2$.

(B.) When you divide both sides of $2x = 6x$ by x, it could lead us to think that there is no solution while in fact the solution is $x = 0$.

(C.) When you divide both sides of $2x = 6x$ by x, you get $2 = 6x$.

(D.) When you divide both sides of $2x = 6x$ by x, it could lead us to think that there is no solution while in fact the solution is $x = 3$.

10. Lin says that if x is a solution to the equation $2x - 6 = 7x$, then x is also a solution to the equation $5x - 6 = 10x$. Write a convincing explanation about why this is true. (Lesson 2-7)

Lesson 2-10

Connecting Equations to Graphs (Part 1)

NAME _____ DATE _____ PERIOD _____

Learning Goal Let's investigate what graphs can tell us about the equations and relationships they represent.

 Warm Up

10.1 Games and Rides

Jada has $20 to spend on games and rides at the carnival. Games cost $1 each and rides are $2 each.

1. Which equation represents the relationship between the number of games, x, and the number of rides, y, that Jada could do if she spends all her money?

    (A.) $x + y = 20$

    (B.) $2x + y = 20$

    (C.) $x + 2y = 20$

2. Explain what each of the other two equations could mean in this situation.

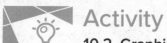
Here are three equations. Each represents the relationship between the number of games, x, the number of rides, y, and the dollar amount a student is spending on games and rides at a different amusement park.

Equation 1: $x + y = 20$

Equation 2: $2.50x + y = 15$

Equation 3: $x + 4y = 28$

Your teacher will assign to you (or ask you to choose) 1–2 equations. For each assigned (or chosen) equation, answer the questions.

First equation: _____

1. What's the number of rides the student could get on if they don't play any games? On the coordinate plane, mark the point that represents this situation and label the point with its coordinates.

2. What's the number of games the student could play if they don't get on any rides? On the coordinate plane, mark the point that represents this situation and label the point with its coordinates.

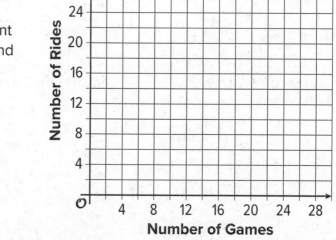

3. Draw a line to connect the two points you've drawn.

4. Complete the sentences: "If the student played no games,

   they can get on _____ rides. For every additional game that the student plays, x, the possible

   number of rides, y, _____ (increases or decreases) by _____."

5. What is the slope of your graph? Where does the graph intersect the vertical axis?

6. Rearrange the equation to solve for y.

NAME _____ DATE _____ PERIOD _____

7. What connections, if any, do you notice between your new equation and the graph?

Second equation: _____

1. What's the number of rides the student could get on if they don't play any games? On the coordinate plane, mark the point that represents this situation and label the point with its coordinates.

2. What's the number of games the student could play if they don't get on any rides? On the coordinate plane, mark the point that represents this situation and label the point with its coordinates.

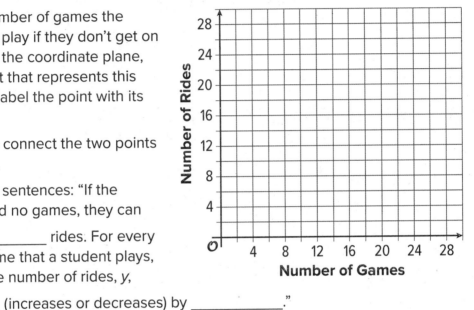

3. Draw a line to connect the two points you've drawn.

4. Complete the sentences: "If the student played no games, they can get on _____ rides. For every additional game that a student plays, x, the possible number of rides, y, _____ (increases or decreases) by _____."

5. What is the slope of your graph? Where does the graph intersect the vertical axis?

6. Rearrange the equation to solve for y.

7. What connections, if any, do you notice between your new equation and the graph?

Andre's coin jar contains 85 cents. There are no quarters or pennies in the jar, so the jar has all nickels, all dimes, or some of each.

1. Write an equation that relates the number of nickels, n, the number of dimes, d, and the amount of money, in cents, in the coin jar.

2. Graph your equation on the coordinate plane. Be sure to label the axes.

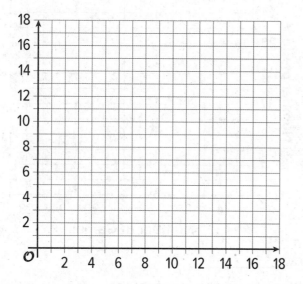

3. How many nickels are in the jar if there are no dimes?

4. How many dimes are in the jar if there are no nickels?

Are you ready for more?

What are all the different ways the coin jar could have 85 cents if it could also contain quarters?

NAME _____ DATE _____ PERIOD _____

Summary
Connecting Equations to Graphs (Part 1)

Linear equations can be written in different forms. Some forms allow us to better see the relationship between quantities or to predict the graph of the equation.

Suppose an athlete wishes to burn 700 calories a day by running and swimming. He burns 17.5 calories per minute of running and 12.5 calories per minute of freestyle swimming.

Let x represent the number of minutes of running and y the number of minutes of swimming. To represent the combination of running and swimming that would allow him to burn 700 calories, we can write:

$$17.5x + 12.5y = 700$$

We can reason that the more minutes he runs, the fewer minutes he has to swim to meet his goal. In other words, as x increases, y decreases. If we graph the equation, the line will slant down from left to right.

If the athlete only runs and doesn't swim, how many minutes would he need to run?

Let's substitute 0 for y to find x:

$$17.5x + 12.5(0) = 700$$
$$17.5x = 700$$
$$x = \frac{700}{17.5}$$
$$x = 40$$

On a graph, this combination of times is the point (40,0), which is the x-intercept.

If he only swims and doesn't run, how many minutes would he need to swim?

Let's substitute 0 for x to find y:

$$17.5(0) + 12.5y = 700$$
$$12.5y = 700$$
$$y = \frac{700}{12.5}$$
$$y = 56$$

On a graph, this combination of times is the point (0,56), which is the y-intercept.

If the athlete wants to know how many minutes he would need to swim if he runs for 15 minutes, 20 minutes, or 30 minutes, he can substitute each of these values for x in the equation and find y. Or, he can first solve the equation for y:

$$17.5x + 12.5y = 700$$

$$12.5y = 700 - 17.5x$$

$$y = \frac{700 - 17.5x}{12.5}$$

$$y = 56 - 1.4x$$

Notice that $y = 56 - 1.4x$, or $y = -1.4x + 56$, is written in **slope-intercept form**.

- The coefficient of x, -1.4, is the slope of the graph. It means that as x increases by 1, y falls by 1.4. For every additional minute of running, the athlete can swim 1.4 fewer minutes.

- The constant term, 56, tells us where the graph intersects the y-axis. It tells us the number minutes the athlete would need to swim if he does no running.

The first equation we wrote, $17.5x + 12.5y = 700$, is a linear equation in **standard form**. In general, it is expressed as $Ax + By = C$, where x and y are variables, and $A, B,$ and C are numbers.

The two equations, $17.5x + 12.5y = 700$ and $y = -1.4x + 56$, are equivalent, so they have the same solutions and the same graph.

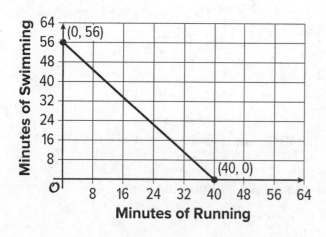

NAME _____ DATE _____ PERIOD _____

Practice
Connecting Equations to Graphs (Part 1)

1. Andre bought a new bag of cat food. The next day, he opened it to feed his cat. The graph shows how many ounces was left in the bag on the days after it was bought.

   a. How many ounces of food were in the bag 12 days after Andre bought it?

   b. How many days did it take for the bag to contain 16 ounces of food?

   c. How much did the bag weigh before it was opened?

   d. About how many days did it take for the bag to be empty?

2. A little league baseball team is ordering hats. The graph shows the relationship between the total cost, in dollars, and the number of hats ordered. What does the slope of the graph tell us in this situation?

   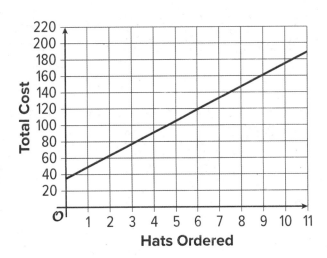

   Ⓐ It tells us that there is a fixed cost of approximately $35 for ordering hats.

   Ⓑ It tells us the amount that the total cost increases for each additional hat ordered.

   Ⓒ It tell us that when 9 hats are ordered, the total cost is approximately $160.

   Ⓓ It tells us that when the number of hats ordered increases by 10, the total cost increases by approximately $175.

3. A group of hikers is progressing steadily along an uphill trail. The graph shows their elevation (or height above sea level), in feet, at each distance from the start of the trail, in miles.

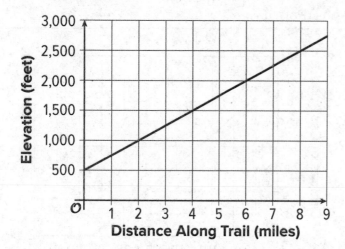

a. What is the slope of the graph? Show your reasoning.

b. What does the slope tell us about this situation?

c. Write an equation that represents the relationship between the hikers' distance from the start of the trail, x, and their elevation, y.

d. Does the equation $y - 250x = 500$ represent the same relationship between the distance from the start of trail and the elevation? Explain your reasoning.

NAME _____ DATE _____ PERIOD _____

4. A kindergarten teacher bought $21 worth of stickers and card stock for his class. The stickers cost $1.50 a sheet and the cardstock cost $3.50 per pack. The equation $1.5s + 3.5c = 21$ represents the relationship between sheets of stickers, s, packs of cardstocks, c, and the dollar amount a kindergarten teacher spent on these supplies.

a. Explain how we can tell that this graph represents the given equation.

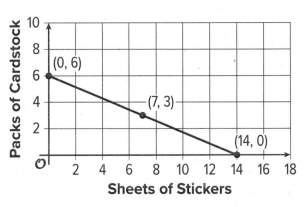

b. What do the vertical and horizontal intercepts, (0,6) and (14,0), mean in this situation?

5. In physics, the equation $PV = nRT$ is called the ideal gas law. It is used to approximate the behavior of many gases under different conditions. P, V, and T represent pressure, volume, and temperature, n represents the number of moles of gas, and R is a constant for the ideal gas. Which equation is solved for T? **(Lesson 2-9)**

A. $\frac{PV}{R} = nT$

B. $\frac{PV}{nR} = T$

C. $T = PV - nR$

D. $PVnR = T$

6. To raise funds for uniforms and travel expenses, the soccer team is holding a car wash in a part of town with a lot of car and truck traffic. The team spent $90 on supplies like sponges and soap. They plan to charge $10 per car and $20 per truck. Their goal is to raise $460. How many cars do they have to wash if they washed the following numbers of trucks? (Lesson 2-9)

   a. 4 trucks

   b. 15 trucks

   c. 21 trucks

   d. 27 trucks

   e. t trucks

7. During the Middle Ages, people often used grains, scruples, and drahms to measure the weights of different medicines. If 120 grains are equivalent to 6 scruples and 6 scruples are equivalent to 2 drahms, how many drahms are equivalent to 300 grains? Explain your reasoning. If you get stuck, try creating a table. (Lesson 2-3)

8. Explain why the equation $2(3x - 5) = 6x + 8$ has no solutions. (Lesson 2-7)

9. Consider the equation $3a + 0.1n = 123$. If we solve this equation for n, which equation would result? (Lesson 2-8)

   (A.) $0.1n = 123 - 3a$

   (B.) $n = 123 - 3a - 0.1$

   (C.) $n = 1,230 - 30a$

   (D.) $\dfrac{3a - 123}{0.1} = n$

10. Diego is buying shrimp and rice to make dinner. Shrimp costs $6.20 per pound and rice costs $1.25 per pound. Diego spent $10.55 buying shrimp and rice. The relationship between pounds of shrimp s, pounds of rice r, and the total cost is represented by the equation $6.20s + 1.25r = 10.55$. Write an equation that makes it easy to find the number of pounds of rice purchased if we know the number of pounds of shrimp purchased. (Lesson 2-8)

Lesson 2-11

Connecting Equations to Graphs (Part 2)

NAME _____ DATE _____ PERIOD _____

Learning Goal Let's analyze different forms of linear equations and how the forms relate to their graphs.

 Warm Up

11.1 Rewrite These!

Rewrite each quotient as a sum or a difference.

1. $\dfrac{4x - 10}{2}$

2. $\dfrac{1 - 50x}{-2}$

3. $\dfrac{5(x + 10)}{25}$

4. $\dfrac{-\frac{1}{5}x + 5}{2}$

Here are two graphs that represent situations you have seen in earlier activities.

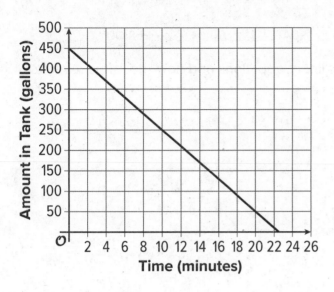

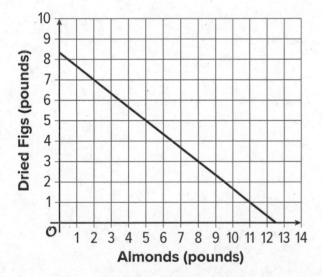

1. The first graph represents $a = 450 - 20t$, which describes the relationship between gallons of water in a tank and time in minutes.

   a. Where on the graph can we see the 450? Where can we see the -20?

   b. What do these numbers mean in this situation?

2. The second graph represents $6x + 9y = 75$. It describes the relationship between pounds of almonds and figs and the dollar amount Clare spent on them.

Suppose a classmate says, "I am not sure the graph represents $6x + 9y = 75$ because I don't see the 6, 9, or 75 on the graph." How would you show your classmate that the graph indeed represents this equation?

Activity

11.3 Slope Match

Match each of the equations with the slope m and y-intercept of its graph.

1. $-4x + 3y = 3$

2. $12x - 4y = 8$

3. $8x + 2y = 16$

4. $-x + \frac{1}{3}y = \frac{1}{3}$

5. $-4x + 3y = -6$

A. $m = 3$, y-int $= (0, 1)$

B. $m = \frac{4}{3}$, y-int $= (0, 1)$

C. $m = \frac{4}{3}$, y-int $= (0, -2)$

D. $m = -4$, y-int $= (0, 8)$

E. $m = 3$, y-int $= (0, -2)$

Are you ready for more?

Each equation in the statement is in the form $Ax + By = C$.

1. For each equation, graph the equation and on the same coordinate plane graph the line passing through $(0, 0)$ and (A, B). What is true about each pair of lines?

2. What are the coordinates of the x-intercept and y-intercept in terms of A, B, and C?

NAME _____ DATE _____ PERIOD _____

Summary
Connecting Equations to Graphs (Part 2)

Here are two situations and two equations that represent them.

Situation 1: Mai receives a $40 bus pass. Each school day, she spends $2.50 to travel to and from school.

Let d be the number of school days since Mai receives a pass and b the balance or dollar amount remaining on the pass.

$$b = 40 - 2.50d$$

Situation 2: A student club is raising money by selling popcorn and iced tea. The club is charging $3 per bag of popcorn and $1.50 per cup of iced tea, and plans to make $60.

Let p be the bags of popcorn sold and t the cups of iced tea sold.

$$3p + 1.50t = 60$$

Here are graphs of the equations. On each graph, the coordinates of some points are shown.

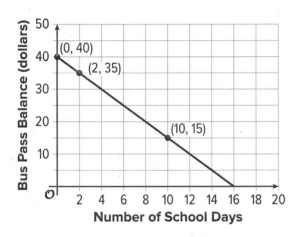

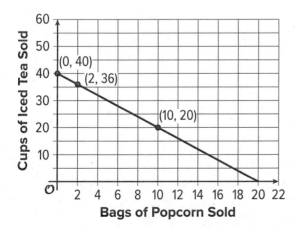

The 40 in the first equation can be observed on the graph and the -2.50 can be found with a quick calculation. The graph intersects the vertical axis at 40 and the -2.50 is the slope of the line. Every time d increases by 1, b decreases by 2.50. In other words, with each passing school day, the dollar amount in Mai's bus pass drops by $2.50.

The numbers in the second equation are not as apparent on the graph. The values where the line intersects the vertical and horizontal axes, 40 and 20, are not in the equation. We can, however, reason about where they come from.

- If p is 0 (no popcorn is sold), the club would need to sell 40 cups of iced tea to make $60 because $40(1.50) = 60$.

- If t is 0 (no iced tea is sold), the club would need to sell 20 bags of popcorn to make $60 because $20(3) = 60$.

What about the slope of the second graph? We can compute it from the graph, but it is not shown in the equation $3p + 1.50t = 60$.

Notice that in the first equation, the variable b was isolated. Let's rewrite the second equation and isolate t:

$$3p + 1.50t = 60$$
$$1.50t = 60 - 3p$$
$$t = \frac{60 - 3p}{1.50}$$
$$t = 40 - 2p$$

Now the numbers in the equation can be more easily related to the graph. The 40 is where the graph intersects the vertical axis and the -2 is the slope. The slope tells us that as p increases by 1, t falls by 2. In other words, for every additional bag of popcorn sold, the club can sell 2 fewer cups of iced tea.

NAME _____ DATE _____ PERIOD _____

Practice
Connecting Equations to Graphs (Part 2)

1. What is the slope of the graph of $5x - 2y = 20$?

   (A.) -10

   (B.) $-\frac{2}{5}$

   (C.) $\frac{5}{2}$

   (D.) 5

2. What is the *y*-intercept of each equation?

   a. $y = 6x + 2$

   b. $10x + 5y = 30$

   c. $y - 6 = 2(3x - 4)$

3. Han wanted to find the intercepts of the graph of the equation $10x + 4y = 20$. He decided to put the equation in slope-intercept form first. Here is his work:

$$10x + 4y = 20$$
$$4y = 20 - 10x$$
$$y = 5 - 10x$$

   He concluded that the *x*-intercept is $\left(\frac{1}{2}, 0\right)$ and the *y*-intercept is (0, 5).

   a. What error did Han make?

   b. What are the *x*- and *y*-intercepts of the line? Explain or show your reasoning.

4. Which graph represents the equation $12 = 3x + 4y$? Explain how you know.

A

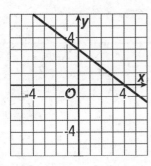

B

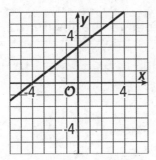

C

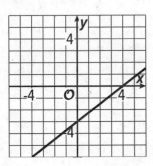

NAME _____ DATE _____ PERIOD _____

5. Clare knows that Priya has a bunch of nickels and dimes in her pocket and that the total amount is $1.25. (Lesson 2-5)

   a. Find one possibility for the number of nickels and number of dimes that could be in Priya's pocket.

   b. Write an equation that describes the relationship between the number of dimes and the number of nickels in Priya's pocket.

   c. Explain what the point (13, 6) means in this situation.

   d. Is the point (13, 6) a solution to the equation you wrote? Explain your reasoning.

6. A large company releases summary statistics about the annual salaries for its employees.

Mean	Standard Deviation	Minimum	Q1	Median	Q3	Maximum
$63,429	$38,439	$18,000	$50,000	$58,000	$68,000	$350,000

Based on this information, are there any outliers in the data? Explain your reasoning. (Lesson 1-14)

7. The graph shows how much money Priya has in her savings account weeks after she started saving on a regular basis. **(Lesson 2-10)**

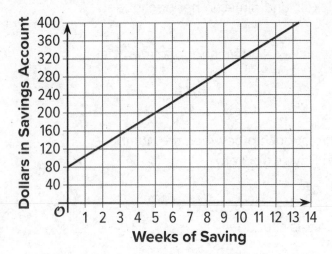

a. How much money does Priya have in the account after 10 weeks?

b. How long did it take her to save $200?

c. How much money did Priya have in her savings account when she started to save regularly?

d. Write an equation to represent the dollar amount in her savings account and the number of weeks of saving. Be sure to specify what each variable represents.

NAME _____ DATE _____ PERIOD _____

8. Noah has a coin jar containing d dimes and q quarters worth a total of $5.00.

Select **all** the equations that represent this situation. **(Lesson 2-6)**

(A.) $d + q = 5$

(B.) $d + q = 500$

(C.) $0.1d + 0.25q = 5$

(D.) $10d + 25q = 500$

(E.) $d = 50$

(F.) $q = 20$

9. Noah orders an extra-large pizza with t toppings that costs a total of d dollars. It costs $12.49 for the pizza plus $1.50 for each topping.

Select **all** of the equations that represent the relationship between the number of toppings t and total cost d of the pizza with t toppings. **(Lesson 2-9)**

(A.) $12.49 + t = d$

(B.) $12.49 + 1.50t = d$

(C.) $12.49 + 1.50d = t$

(D.) $12.49 = d + 1.50t$

(E.) $t = \dfrac{d - 12.49}{1.5}$

(F.) $t = d - \dfrac{12.49}{1.5}$

10. A school sells adult tickets and student tickets for a play. It collects $1,400 in total.

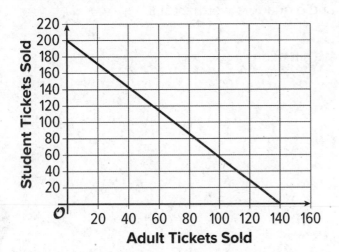

The graph shows the possible combinations of the number of adult tickets sold and the number of student tickets sold.

What does the vertical intercept (0, 200) tell us in this situation? (Lesson 2-10)

A. It tells us the decrease in the sale of adult tickets for each student ticket sold.

B. It tells us the decrease in the sale of student tickets for each adult ticket sold.

C. It tells us that if no adult tickets were sold, then 200 students tickets were sold.

D. It tells us that if no students tickets were sold, then 200 adult tickets were sold.

Lesson 2-12

Writing and Graphing Systems of Linear Equations

NAME _____ DATE _____ PERIOD _____

Learning Goal Let's recall what it means to solve a system of linear equations and how to do it by graphing.

 ## Warm Up
12.1 Math Talk: A Possible Mix?

Diego bought some raisins and walnuts to make trail mix. Raisins cost $4 a pound and walnuts cost $8 a pound. Diego spent $15 on both ingredients.

1. Decide if each pair of values could be a combination of raisins and walnuts that Diego bought.

   a. 4 pounds of raisins and 2 pounds of walnuts

   b. 1 pound of raisins and 1.5 pounds of walnuts

   c. 2.25 pounds of raisins and 0.75 pounds of walnuts

   d. 3.5 pounds of raisins and 1 pound of walnuts

1. Here is a situation you saw earlier: Diego bought some raisins and walnuts to make trail mix. Raisins cost $4 a pound and walnuts cost $8 a pound. Diego spent $15 on both ingredients.

   a. Write an equation to represent this constraint. Let x be the pounds of raisins and y be the pounds of walnuts.

   b. Use graphing technology to graph the equation.

   c. Complete the table with the amount of one ingredient Diego could have bought given the other. Be prepared to explain or show your reasoning.

Raisins (pounds)	Walnuts (pounds)
0	
0.25	
	1.375
	1.25
1.75	
3	

2. Here is a new piece of information: Diego bought a total of 2 pounds of raisins and walnuts combined.

a. Write an equation to represent this new constraint. Let x be the pounds of raisins and y be the pounds of walnuts.

b. Use graphing technology to graph the equation.

c. Complete the table with the amount of one ingredient Diego could have bought given the other. Be prepared to explain or show your reasoning.

Raisins (pounds)	Walnuts (pounds)
0	
0.25	
	1.375
	1.25
1.75	
3	

3. Diego spent $15 and bought exactly 2 pounds of raisins and walnuts. How many pounds of each did he buy? Explain or show how you know.

Activity

12.3 Meeting Constraints

Here are some situations that each relate two quantities and involve two constraints. For each situation, find the pair of values that meet both constraints and explain or show your reasoning.

1. A dining hall had a total of 25 tables—some long rectangular tables and some round ones. Long tables can seat 8 people. Round tables can seat 6 people. On a busy evening, all 190 seats at the tables are occupied. How many long tables, x, and how many round tables, y, are there?

2. A family bought a total of 16 adult and child tickets to a magic show. Adult tickets are $10.50 each and child tickets are $7.50 each. The family paid a total of $141. How many adult tickets, a, and child tickets, c, did they buy?

3. At a poster shop, Han paid $16.80 for 2 large posters and 3 small posters of his favorite band. Kiran paid $14.15 for 1 large poster and 4 small posters of his favorite TV shows. Posters of the same size have the same price. Find the price of a large poster, ℓ, and the price of a small poster, s.

Are you ready for more?

1. Make up equations for two lines that intersect at (4, 1).

2. Make up equations for three lines whose intersection points form a triangle with vertices at (-4, 0), (2, 9), and (6, 5).

NAME _____ DATE _____ PERIOD _____

Summary
Writing and Graphing Systems of Linear Equations

A costume designer needs some silver and gold thread for the costumes for a school play. She needs a total of 240 yards. At a store that sells thread by the yard, silver thread costs $0.04 a yard and gold thread costs $0.07 a yard. The designer has $15 to spend on the thread.

How many of each color should she get if she is buying exactly what is needed and spending all of her budget?

This situation involves two quantities and two constraints—length and cost. Answering the question means finding a pair of values that meets both constraints simultaneously. To do so, we can write two equations and graph them on the same coordinate plane.

Let x represents yards of silver thread and y yards of gold thread.

- The length constraint: $x + y = 240$

- The cost constraint: $0.04x + 0.07y = 15$

Every point on the graph of $x + y = 240$ is a pair of values that meets the length constraint.

Every point on the graph of $0.04x + 0.07y = 15$ is a pair of values that meets the cost constraint.

The point where the two graphs intersect gives the pair of values that meets *both* constraints.

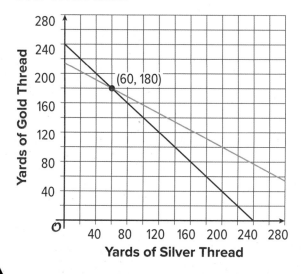

That point is (60, 180), which represents 60 yards of silver thread and 180 yards of gold thread.

If we substitute 60 for x and 180 for y in each equation, we find that these values make the equation true. (60, 180) is a solution to both equations simultaneously.

$$x + y = 240$$
$$60 + 180 = 240$$
$$240 = 240$$
$$0.04x + 0.07y = 15$$
$$0.04(60) + 0.07(180) = 15$$
$$2.40 + 12.60 = 15$$
$$15 = 15$$

Two or more equations that represent the constraints in the same situation form a **system of equations**. A curly bracket is often used to indicate a system.

$$\begin{cases} x + y = 240 \\ 0.04x + 0.07y = 15 \end{cases}$$

The **solution to a system of equations** is a pair of values that makes all of the equations in the system true. Graphing the equations is one way to find the solution to a system of equations.

Glossary

solution to a system of equations
system of equations

NAME _____ DATE _____ PERIOD _____

Practice

Writing and Graphing Systems of Linear Equations

1. The knitting club sold 40 scarves and hats at a winter festival and made $700 from the sales. They charged $18 for each scarf and $14 for each hat.

   If s represents the number of scarves sold and h represents the number of hats sold, which system of equations represents the constraints in this situation?

   (A.) $\begin{cases} 40s + h = 700 \\ 18s + 14h = 700 \end{cases}$

   (B.) $\begin{cases} 18s + 14h = 40 \\ s + h = 700 \end{cases}$

   (C.) $\begin{cases} s + h = 40 \\ 18s + 14h = 700 \end{cases}$

   (D.) $\begin{cases} 40(s + h) = 700 \\ 18s = 14h \end{cases}$

2. Here are two equations:

   Equation 1: $6x + 4y = 34$

   Equation 2: $5x - 2y = 15$

   a. Decide whether each (x, y) pair is a solution to one equation, both equations, or neither of the equations.

   i. $(3, 4)$

   ii. $(4, 2.5)$

   iii. $(5, 5)$

   iv. $(3, 2)$

   b. Is it possible to have more than one (x, y) pair that is a solution to both equations? Explain or show your reasoning.

3. Explain or show that the point $(5, -4)$ is a solution to this system of equations: $\begin{cases} 3x - 2y = 23 \\ 2x + y = 6 \end{cases}$

4. Diego is thinking of two positive numbers. He says, "If we triple the first number and double the second number, the sum is 34."

   a. Write an equation that represents this clue. Then, find two possible pairs of numbers Diego could be thinking of.

   b. Diego then says, "If we take half of the first number and double the second, the sum is 14." Write an equation that could represent this description.

   c. What are Diego's two numbers? Explain or show how you know. A coordinate plane is given here, in case helpful.

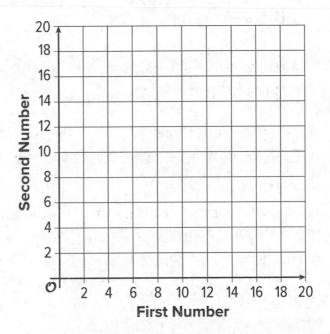

5. The table shows the volume of water in a tank after it has been filled to a certain height. Which equation could represent the volume of water in cubic inches, V, when the height is h inches? **(Lesson 2-4)**

Height of Water (inches)	Volume of Water (cubic inches)
0	0
1	1.05
2	8.40
3	28.35

   A. $h = V$

   B. $h = \dfrac{V}{4}$

   C. $V = h^2 + 0.05$

   D. $V = 1.05\,h^3$

NAME _____ DATE _____ PERIOD _____

6. Andre does not understand why a solution to the equation $3 - x = 4$ must also be a solution to the equation $12 = 9 - 3x$. Write a convincing explanation as to why this is true. **(Lesson 2-7)**

7. Volunteer drivers are needed to bring 80 students to the championship baseball game. Drivers either have cars, which can seat 4 students, or vans, which can seat 6 students. The equation $4c + 6v = 80$ describes the relationship between the number of cars c and number of vans v that can transport exactly 80 students. Explain how you know that this graph represents this equation. **(Lesson 2-10)**

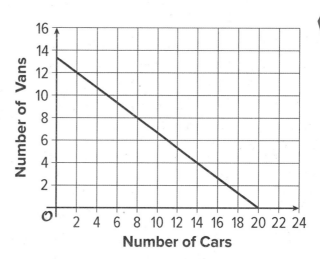

8. Three siblings are participating in a family-friendly running event.

- The oldest sibling begins at the start line of the race and runs 7 miles per hour the entire time.

- The middle sibling begins at the start line and walks at 3.5 miles per hour throughout the race.

- The youngest sibling joins the race 4 miles from the start line and runs 5 miles per hour the rest of the way.

Match each graph to the sibling whose running is represented by the graph. (Lesson 2-11)

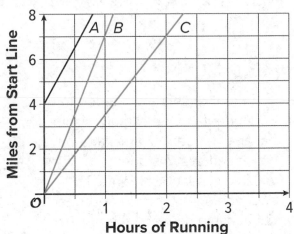

Oldest Sibling	Graph A
Middle Sibling	Graph B
Youngest Sibling	Graph C

9. What is the x-intercept of the graph of $y = 3 - 5x$? (Lesson 2-11)

Ⓐ. $(\frac{3}{5}, 0)$

Ⓑ. $(-5, 0)$

Ⓒ. $(0, 3)$

Ⓓ. $(0, \frac{5}{3})$

Lesson 2-13

Solving Systems by Substitution

NAME _____ DATE _____ PERIOD _____

Learning Goal Let's use substitution to solve systems of linear equations.

 ## Warm Up

13.1 Math Talk: Is It a Match?

Here are graphs of two equations in a system.

Determine if each of these systems could be represented by the graphs. Be prepared to explain how you know.

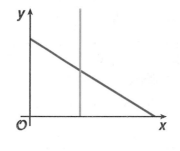

1. $\begin{cases} x + 2y = 8 \\ x = -5 \end{cases}$

2. $\begin{cases} y = -7x + 13 \\ y = -1 \end{cases}$

3. $\begin{cases} 3x = 8 \\ 3x + y = 15 \end{cases}$

4. $\begin{cases} y = 2x - 7 \\ 4 + y = 12 \end{cases}$

Activity

13.2 Four Systems

Here are four systems of equations you saw earlier. Solve each system. Then, check your solutions by substituting them into the original equations to see if the equations are true.

1. $\begin{cases} x + 2y = 8 \\ x = -5 \end{cases}$

2. $\begin{cases} y = -7x + 13 \\ y = -1 \end{cases}$

3. $\begin{cases} 3x = 8 \\ 3x + y = 15 \end{cases}$

4. $\begin{cases} y = 2x - 7 \\ 4 + y = 12 \end{cases}$

Activity

13.3 What about Now?

Solve each system without graphing.

1. $\begin{cases} 5x - 2y = 26 \\ y + 4 = x \end{cases}$

2. $\begin{cases} 2m - 2p = -6 \\ p = 2m + 10 \end{cases}$

3. $\begin{cases} 2d = 8f \\ 18 - 4f = 2d \end{cases}$

4. $\begin{cases} x + \frac{1}{7}y = 4 \\ y = 3x - 2 \end{cases}$

Are you ready for more?

Solve this system with four equations.
$\begin{cases} 3x + 2y - z + 5w = 20 \\ y = 2z - 3w \\ z = w + 1 \\ 2w = 8 \end{cases}$

NAME _____ DATE _____ PERIOD _____

Summary
Solving Systems by Substitution

The solution to a system can usually be found by graphing, but graphing may not always be the most precise or the most efficient way to solve a system.

Here is a system of equations:

$$\begin{cases} 3p + q = 71 \\ 2p - q = 30 \end{cases}$$

The graphs of the equations show an intersection at approximately 20 for p and approximately 10 for q.

Without technology, however, it is not easy to tell what the exact values are.

Instead of solving by graphing, we can solve the system algebraically. Here is one way.

If we subtract $3p$ from each side of the first equation, $3p + q = 71$, we get an equivalent equation: $q = 71 - 3p$. Rewriting the original equation this way allows us to isolate the variable q.

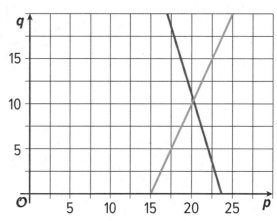

Because q is equal to $71 - 3p$, we can substitute the expression $71 - 3p$ in the place of q in the second equation. Doing this gives us an equation with only one variable, p, and makes it possible to find p.

$2p - q = 30$	original equation
$2p - (71 - 3p) = 30$	substitute $71 - 3p$ for q
$2p - 71 + 3p = 30$	apply distributive property
$5p - 71 = 30$	combine like terms
$5p = 101$	add 71 to both sides
$p = \dfrac{101}{5}$	divide both sides by 5
$p = 20.2$	

Now that we know the value of p, we can find the value of q by substituting 20.2 for p in either of the original equations and solving the equation.

$$3(20.2) + q = 71 \qquad\qquad 2(20.2) - q = 30$$
$$60.6 + q = 71 \qquad\qquad 40.4 - q = 30$$
$$q = 71 - 60.6 \qquad\qquad -q = 30 - 40.4$$
$$q = 10.4 \qquad\qquad -q = \text{-}10.4$$
$$q = \frac{\text{-}10.4}{\text{-}1}$$
$$q = 10.4$$

The solution to the system is the pair $p = 20.2$ and $q = 10.4$, or the point (20.2, 10.4) on the graph.

This method of solving a system of equations is called solving by **substitution**, because we substituted an expression for q into the second equation.

Glossary

substitution

NAME _____ DATE _____ PERIOD _____

Practice
Solving Systems by Substitution

1. Identify a solution to this system of equations: $\begin{cases} -4x + 3y = 23 \\ x - y = -7 \end{cases}$

(A.) (-5, 2)   (C.) (-3, 4)

(B.) (-2, 5)   (D.) (4, -3)

2. Lin is solving this system of equations: $\begin{cases} 6x - 5y = 34 \\ 3x + 2y = 8 \end{cases}$

She starts by rearranging the second equation to isolate the y variable: $y = 4 - 1.5x$. She then substituted the expression $4 - 1.5x$ for y in the first equation, as shown:

$$6x - 5(4 - 1.5x) = 34$$
$$6x - 20 - 7.5x = 34$$
$$-1.5x = 54$$
$$x = -36$$

$$y = 4 - 1.5x$$
$$y = 4 - 1.5 \cdot (-36)$$
$$y = 58$$

   a. Check to see if Lin's solution of (-36, 58) makes both equations in the system true.

   b. If your answer to the previous question is "no," find and explain her mistake. If your answer is "yes," graph the equations to verify the solution of the system.

3. Solve each system of equations.

   a. $\begin{cases} 2x - 4y = 20 \\ x = 4 \end{cases}$ b. $\begin{cases} y = 6x + 11 \\ 2x - 3y = 7 \end{cases}$

4. Tyler and Han are trying to solve this system by substitution: $\begin{cases} x + 3y = -5 \\ 9x + 3y = 3. \end{cases}$

   Tyler's first step is to isolate x in the first equation to get $x = -5 - 3y$. Han's first step is to isolate $3y$ in the first equation to get $3y = -5 - x$. Show that both first steps can be used to solve the system and will yield the same solution.

5. The dot plots show the distribution of the length, in centimeters, of 25 shark teeth for an extinct species of shark and the length, in centimeters, of 25 shark teeth for a closely related shark species that is still living.

   Compare the two dot plots using the shape of the distribution, measures of center, and measures of variability. Use the situation described in the problem in your explanation.

   (Lesson 1-15)

   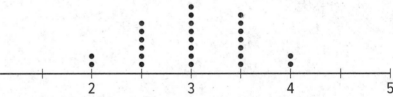

   Length of Teeth in Extinct Species in Centimeters

   mean: 3.02 cm

   standard deviation: 0.55 cm

   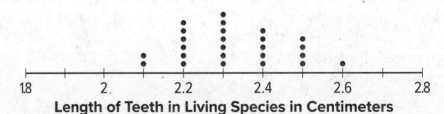

   Length of Teeth in Living Species in Centimeters

   mean: 2.32 cm

   standard deviation: 0.13 cm

NAME _____ DATE _____ PERIOD _____

6. Kiran buys supplies for the school's greenhouse. He buys f bags of fertilizer and p packages of soil. He pays \$5 for each bag of fertilizer and \$2 for each package of soil, and spends a total of \$90. The equation $5f + 2p = 90$ describes this relationship. If Kiran solves the equation for p, which equation would result? **(Lesson 2-8)**

(A.) $2p = 90 - 5f$

(B.) $p = \dfrac{5f - 90}{2}$

(C.) $p = 45 - 2.5f$

(D.) $p = \dfrac{85f}{2}$

7. Elena wanted to find the slope and y-intercept of the graph of $25x - 20y = 100$. She decided to put the equation in slope-intercept form first. Here is her work:

$$25x - 20y = 100$$
$$20y = 100 - 25x$$
$$y = 5 - \frac{5}{4}x$$

She concluded that the slope is $-\dfrac{5}{4}$ and the y-intercept is $(0, 5)$. **(Lesson 2-11)**

   a. What was Elena's mistake?

   b. What are the slope and y-intercept of the line? Explain or show your reasoning.

8. Find the x- and y-intercepts of the graph of each equation. **(Lesson 2-11)**

   a. $y = 10 - 2x$

   b. $4y + 9x = 18$

   c. $6x - 2y = 44$

   d. $2x = 4 + 12y$

9. Andre is buying snacks for the track and field team. He buys a pounds of apricots for \$6 per pound and b pounds of dried bananas for \$4 per pound. He buys a total of 5 pounds of apricots and dried bananas and spends a total of \$24.50. Which system of equations represents the constraints in this situation? **(Lesson 2-12)**

A. $\begin{cases} 6a + 4b = 5 \\ a + b = 24.50 \end{cases}$

B. $\begin{cases} 6a + 4b = 24.50 \\ a + b = 5 \end{cases}$

C. $\begin{cases} 6a = 4b \\ 5(a + b) = 24.50 \end{cases}$

D. $\begin{cases} 6a + b = 4 \\ 5a + b = 24.50 \end{cases}$

10. Here are two equations:

Equation 1: $y = 3x + 8$

Equation 2: $2x - y = -6$

Without using graphing technology:

(Lesson 2-12)

a. Find a point that is a solution to Equation 1 but not a solution to Equation 2.

b. Find a point that is a solution to Equation 2 but not a solution to Equation 1.

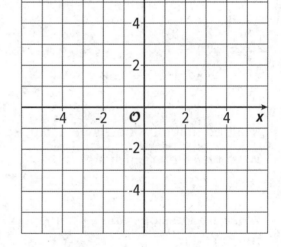

c. Graph the two equations.

d. Find a point that is a solution to both equations.

Lesson 2-14

Solving Systems by Elimination (Part 1)

NAME _____ DATE _____ PERIOD _____

Learning Goal Let's investigate how adding or subtracting equations can help us solve systems of linear equations.

 ## Warm Up
14.1 Notice and Wonder: Hanger Diagrams

What do you notice? What do you wonder?

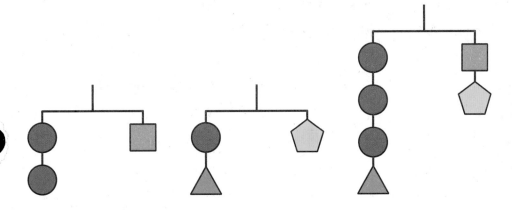

Diego is solving this system of equations:

$$\begin{cases} 4x + 3y = 10 \\ -4x + 5y = 6 \end{cases}$$

Here is his work:

$$4x + 3y = 10$$
$$\underline{-4x + 5y = 6} +$$
$$0 + 8y = 16$$
$$y = 2$$

$$4x + 3(2) = 10$$
$$4x + 6 = 10$$
$$4x = 4$$
$$x = 1$$

1. Make sense of Diego's work and discuss with a partner:

   a. What did Diego do to solve the system?

   b. Is the pair of x and y values that Diego found actually a solution to the system? How do you know?

2. Does Diego's method work for solving these systems? Be prepared to explain or show your reasoning.

   a. $\begin{cases} 2x + y = 4 \\ x - y = 11 \end{cases}$

   b. $\begin{cases} 8x + 11y = 37 \\ 8x + y = 7 \end{cases}$

NAME _____ DATE _____ PERIOD _____

Activity

14.3 Adding and Subtracting Equations to Solve Systems

Here are three systems of equations you saw earlier.

System A

$$\begin{cases} 4x + 3y = 10 \\ -4x + 5y = 6 \end{cases}$$

System B

$$\begin{cases} 2x + y = 4 \\ x - y = 11 \end{cases}$$

System C

$$\begin{cases} 8x + 11y = 37 \\ 8x + y = 7 \end{cases}$$

For each system:

1. Use graphing technology to graph the original two equations in the system. Then, identify the coordinates of the solution.

2. Find the sum or difference of the two original equations that would enable the system to be solved.

3. Graph the third equation on the same coordinate plane. Make an observation about the graph.

Mai wonders what would happen if we multiply equations. That is, we multiply the expressions on the left side of the two equations and set them equal to the expressions on the right side of the two equations.

1. In system B write out an equation that you would get if you multiply the two equations in this manner.

2. Does your original solution still work in this new equation?

3. Use graphing technology to graph this new equation on the same coordinate plane. Why is this approach not particularly helpful?

NAME _____ DATE _____ PERIOD _____

Summary
Solving Systems by Elimination (Part 1)

Another way to solve systems of equations algebraically is by **elimination**. Just like in substitution, the idea is to eliminate one variable so that we can solve for the other. This is done by adding or subtracting equations in the system.
Let's look at an example.

$$\begin{cases} 5x + 7y = 64 \\ 0.5x - 7y = \text{-}9 \end{cases}$$

Notice that one equation has $7y$ and the other has $-7y$.

If we add the second equation to the first, the $7y$ and $-7y$ add up to 0, which eliminates the y-variable, allowing us to solve for x.

$$\begin{aligned} 5x + 7y &= 64 \\ \underline{0.5x - 7y &= \text{-}9} + \\ 5.5x + 0 &= 55 \\ 5.5x &= 55 \\ x &= 10 \end{aligned}$$

Now that we know $x = 10$, we can substitute 10 for x in either of the equations and find y:

$$\begin{aligned} 5x + 7y &= 64 & 0.5x - 7y &= \text{-}9 \\ 5(10) + 7y &= 64 & 0.5(10) - 7y &= \text{-}9 \\ 50 + 7y &= 64 & 5 - 7y &= \text{-}9 \\ 7y &= 14 & -7y &= \text{-}14 \\ y &= 2 & y &= 2 \end{aligned}$$

In this system, the coefficient of y in the first equation happens to be the opposite of the coefficient of y in the second equation. The sum of the terms with y-variables is 0.

What if the equations don't have opposite coefficients for the same variable, like in the following system?

$$\begin{cases} 8r + 4s = 12 \\ 8r + s = \text{-}3 \end{cases}$$

Notice that both equations have $8r$ and if we subtract the second equation from the first, the variable r will be eliminated because $8r - 8r$ is 0.

$$8r + 4s = 12$$
$$\underline{8r + s = \text{-}3} -$$
$$0 + 3s = 15$$
$$3s = 15$$
$$s = 5$$

Substituting 5 for s in one of the equations gives us r:

$$8r + 4s = 12$$
$$8r + 4(5) = 12$$
$$8r + 20 = 12$$
$$8r = \text{-}8$$
$$r = \text{-}1$$

Adding or subtracting the equations in a system creates a new equation. How do we know the new equation shares a solution with the original system?

If we graph the original equations in the system and the new equation, we can see that all three lines intersect at the same point, but why do they?

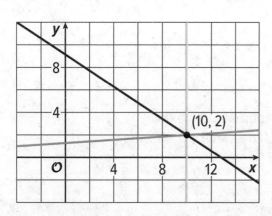
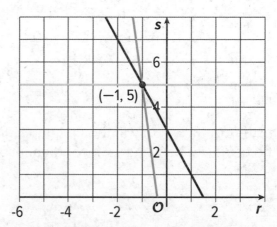

In future lessons, we will investigate why this strategy works.

Glossary

elimination

NAME _____ DATE _____ PERIOD _____

Practice
Solving Systems by Elimination (Part 1)

1. Which equation is the result of adding these two equations?

$$\begin{cases} -2x + 4y = 17 \\ 3x - 10y = -3 \end{cases}$$

(A.) $-5x - 6y = 14$

(B.) $-x - 6y = 14$

(C.) $x - 6y = 14$

(D.) $5x + 14y = 20$

2. Which equation is the result of subtracting the second equation from the first? $\begin{cases} 4x - 6y = 13 \\ -5x + 2y = 5 \end{cases}$

(A.) $-9x - 4y = 8$

(B.) $-x + 4y = 8$

(C.) $x - 4y = 8$

(D.) $9x - 8y = 8$

3. Solve this system of equations without graphing: $\begin{cases} 5x + 2y = 29 \\ 5x - 2y = 41 \end{cases}$

4. Here is a system of linear equations: $\begin{cases} 6x + 21y = 103 \\ -6x + 23y = 51 \end{cases}$

Would you rather use subtraction or addition to solve the system? Explain your reasoning.

5. Kiran sells *f* full boxes and *h* half-boxes of fruit to raise money for a band trip. He earns $5 for each full box and $2 for each half-box of fruit he sells and earns a total of $100 toward the cost of his band trip. The equation $5f + 2h = 100$ describes this relationship.

   Solve the equation for *f*. (Lesson 2-8)

6. Match each equation with the corresponding equation solved for *a*. (Lesson 2-8)

   A. $a + 2b = 5$

   B. $5a = 2b$

   C. $a + 5 = 2b$

   D. $5(a + 2b) = 0$

   E. $5a + 2b = 0$

   1. $a = \dfrac{2b}{5}$

   2. $a = \dfrac{-2b}{5}$

   3. $a = -2b$

   4. $a = 2b - 5$

   5. $a = 5 - 2b$

7. The volume of a cylinder is represented by the formula $V = \pi r^2 h$.

   Find each missing height and show your reasoning. (Lesson 2-9)

Volume (cubic inches)	Radius (inches)	Height (inches)
96π	4	
31.25π	2.5	
V	r	

NAME _____ DATE _____ PERIOD _____

8. Match each equation with the slope m and y-intercept of its graph. (Lesson 2-11)

A. $m = -6$, y-int $= (0, 12)$

B. $m = -6$, y-int $= (0, 5)$

C. $m = -\frac{5}{6}$, y-int $= (0, 1)$

D. $m = \frac{5}{6}$, y-int $= (0, 1)$

E. $m = \frac{5}{6}$, y-int $= (0, -1)$

F. $m = \frac{5}{6}$, y-int $= (0, -5)$

1. $5x - 6y = 30$

2. $y = 5 - 6x$

3. $y = \frac{5}{6}x + 1$

4. $5x - 6y = 6$

5. $5x + 6y = 6$

6. $6x + y = 12$

9. Solve each system of equations. (Lesson 2-13)

a. $\begin{cases} 2x + 3y = 4 \\ 2x = 7y + 24 \end{cases}$

b. $\begin{cases} 5x + 3y = 23 \\ 3y = 15x - 21 \end{cases}$

10. Elena and Kiran are playing a board game. After one round, Elena says, "You earned so many more points than I did. If you earned 5 more points, your score would be twice mine!"

Kiran says, "Oh, I don't think I did that much better. I only scored 9 points higher than you did." (Lesson 2-13)

a. Write a system of equations to represent each student's comment. Be sure to specify what your variables represent.

b. If both students were correct, how many points did each student score? Show your reasoning.

Lesson 2-15

Solving Systems by Elimination (Part 2)

NAME _____ DATE _____ PERIOD _____

Learning Goal Let's think about why adding and subtracting equations works for solving systems of linear equations.

 ## Warm Up
15.1 Is It Still True?

Here is an equation: $50 + 1 = 51$.

1. Perform each of the following operations and answer these questions: What does each resulting equation look like? Is it still a true equation?

   a. Add 12 to each side of the equation.

   b. Add $10 + 2$ to the left side of the equation and 12 to the right side.

   c. Add the equation $4 + 3 = 7$ to the equation $50 + 1 = 51$.

2. Write a new equation that, when added to $50 + 1 = 51$, gives a sum that is also a true equation.

3. Write a new equation that, when added to $50 + 1 = 51$, gives a sum that is a false equation.

Activity
15.2 Classroom Supplies

A teacher purchased 20 calculators and 10 measuring tapes for her class and paid $495. Later, she realized that she didn't order enough supplies. She placed another order of 8 of the same calculators and 1 more of the same measuring tape and paid $178.50.

This system represents the constraints in this situation:

$$\begin{cases} 20c + 10m = 495 \\ 8c + m = 178.50 \end{cases}$$

1. Discuss with a partner:

   a. In this situation, what do the solutions to the first equation mean?

   b. What do the solutions to the second equation mean?

   c. For each equation, how many possible solutions are there? Explain how you know.

   d. In this situation, what does the solution to the system mean?

NAME _____ DATE _____ PERIOD _____

2. Find the solution to the system. Explain or show your reasoning.

3. To be reimbursed for the cost of the supplies, the teacher recorded: "Items purchased: 28 calculators and 11 measuring tapes. Amount: $673.50."

   a. Write an equation to represent the relationship between the numbers of calculators and measuring tapes, the prices of those supplies, and the total amount spent.

   b. How is this equation related to the first two equations?

   c. In this situation, what do the solutions of this equation mean?

   d. How many possible solutions does this equation have? How many solutions make sense in this situation? Explain your reasoning.

Solve each system of equations without graphing and show your reasoning. Then, check your solutions.

1. $\begin{cases} 2x + 3y = 7 \\ -2x + 4y = 14 \end{cases}$

2. $\begin{cases} 2x + 3y = 7 \\ 3x - 3y = 3 \end{cases}$

3. $\begin{cases} 2x + 3y = 5 \\ 2x + 4y = 9 \end{cases}$

4. $\begin{cases} 2x + 3y = 16 \\ 6x - 5y = 20 \end{cases}$

Are you ready for more?

This system has three equations: $\begin{cases} 3x + 2y - z = 7 \\ -3x + y + 2z = -14 \\ 3x + y - z = 10 \end{cases}$

1. Add the first two equations to get a new equation.

2. Add the second two equations to get a new equation.

3. Solve the system of your two new equations.

4. What is the solution to the original system of equations?

NAME _____ DATE _____ PERIOD _____

Summary
Solving Systems by Elimination (Part 2)

When solving a system with two equations, why is it acceptable to add the two equations, or to subtract one equation from the other?

Remember that an equation is a statement that says two things are equal.

For example, the equation $a = b$ says a number a has the same value as another number b. The equation $10 + 2 = 12$ says that $10 + 2$ has the same value as 12.

If $a = b$ and $10 + 2 = 12$ are true statements, then adding $10 + 2$ to a and adding 12 to b means adding the same amount to each side of $a = b$. The result, $a + 10 + 2 = b + 12$, is also a true statement.

As long as we add an equal amount to each side of a true equation, the two sides of the resulting equation will remain equal.

We can reason the same way about adding variable equations in a system like this:

$$\begin{cases} e + f = 17 \\ -2e + f = -1 \end{cases}$$

In each equation, if (e, f) is a solution, the expression on the left of the equal sign and the number on the right are equal. Because $-2e + f$ is equal to -1:

- Adding $-2e + f$ to $e + f$ and adding -1 to 17 means adding an equal amount to each side of $e + f = 17$. The two sides of the new equation, $-e + 2f = 16$, stay equal.

  The e- and f-values that make the original equations true also make this equation true.

  $$\begin{array}{r} e + f = 17 \\ -2e + f = -1 \ + \\ \hline -e + 2f = 16 \end{array}$$

- Subtracting $-2e + f$ from $e + f$ and subtracting -1 from 17 means subtracting an equal amount from each side of $e + f = 17$. The two sides of the new equation, $3e = 18$, stay equal.

  $$\begin{array}{r} e + f = 17 \\ -2e + f = -1 \ - \\ \hline 3e \ \ \ \ = 18 \end{array}$$

The *f*-variable is eliminated, but the *e*-value that makes both the original equations true also makes this equation true.

From $3e = 18$, we know that $e = 6$. Because 6 is also the *e*-value that makes the original equations true, we can substitute it into one of the equations and find the *f*-value.

The solution to the system is $e = 6$, $f = 11$, or the point $(6, 11)$ on the graphs representing the system. If we substitute 6 and 11 for *e* and *f* in any of the equations, we will find true equations. (Try it!)

NAME _____ DATE _____ PERIOD _____

Practice
Solving Systems by Elimination (Part 2)

1. Solve this system of linear equations without graphing: $\begin{cases} 5x + 4y = 8 \\ 10x - 4y = 46 \end{cases}$

2. Select **all** the equations that share a solution with this system of equations.

$\begin{cases} 5x + 4y = 24 \\ 2x - 7y = 26 \end{cases}$

(A.) $7x + 3y = 50$

(B.) $7x - 3y = 50$

(C.) $5x + 4y = 2x - 7y$

(D.) $3x - 11y = \text{-}2$

(E.) $3x + 11y = \text{-}2$

3. Students performed in a play on a Friday and a Saturday. For both performances, adult tickets cost a dollars each and student tickets cost s dollars each.

   On Friday, they sold 125 adult tickets and 65 student tickets, and collected $1,200. On Saturday, they sold 140 adult tickets and 50 student tickets, and collected $1,230.

   This situation is represented by this system of equations:

   $\begin{cases} 125a + 65s = 1{,}200 \\ 140a + 50s = 1{,}230 \end{cases}$

   a. What could the equation $265a + 115s = 2{,}430$ mean in this situation?

   b. The solution to the original system is the pair $a = 7$ and $s = 5$. Explain why it makes sense that this pair of values is also the solution to the equation $265a + 115s = 2{,}430$.

4. Which statement explains why $13x - 13y = -26$ shares a solution with this system of equations: $\begin{cases} 10x - 3y = 29 \\ -3x + 10y = 55 \end{cases}$

A. Because $13x - 13y = -26$ is the product of the two equations in the system of equations, it then must share a solution with the system of equations.

B. The three equations all have the same slope but different y-intercepts. Equations with the same slope but different y-intercepts always share a solution.

C. Because $10x - 3y$ is equal to 29, I can add $10x - 3y$ to the left side of $-3x + 10y = 55$ and add 29 to the right side of the same equation. Adding equivalent expressions to each side of an equation does not change the solution to the equation.

D. Because $-3x + 10y$ is equal to 55, I can subtract $-3x + 10y$ from the left side of $10x - 3y = 29$ and subtract 55 from its right side. Subtracting equivalent expressions from each side of an equation does not change the solution to the equation.

5. Select **all** equations that can result from adding these two equations or subtracting one from the other. (Lesson 2-14)

$\begin{cases} x + y = 12 \\ 3x - 5y = 4 \end{cases}$

A. $-2x - 4y = 8$

B. $-2x + 6y = 8$

C. $4x - 4y = 16$

D. $4x + 4y = 16$

E. $2x - 6y = -8$

F. $5x - 4y = 28$

6. Solve each system of equations. (Lesson 2-13)

a. $\begin{cases} 7x - 12y = 180 \\ 7x = 84 \end{cases}$

b. $\begin{cases} -16y = 4x \\ 4x + 27y = 11 \end{cases}$

NAME _____ DATE _____ PERIOD _____

7. Here is a system of equations: $\begin{cases} 7x - 4y = -11 \\ 7x + 4y = -59 \end{cases}$

   Would you rather use subtraction or addition to solve the system? Explain your reasoning. **(Lesson 2-14)**

8. The box plot represents the distribution of the number of free throws that 20 students made out of 10 attempts.

   Free Throws Made

   After reviewing the data, the value recorded as 1 is determined to have been an error. The box plot represents the distribution of the same data set, but with the minimum, 1, removed.

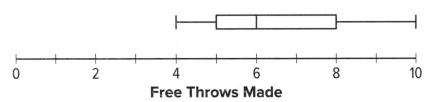

   Free Throws Made

   The median is 6 free throws for both plots. **(Lesson 1-10)**

   a. Explain why the median remains the same when 1 was removed from the data set.

   b. When 1 is removed from the data set, does the mean remain the same? Explain your reasoning.

9. In places where there are crickets, the outdoor temperature can be predicted by the rate at which crickets chirp. One equation that models the relationship between chirps and outdoor temperature is $f = \frac{1}{4}c + 40$, where c is the number of chirps per minute and f is the temperature in degrees Fahrenheit. **(Lesson 2-10)**

a. Suppose 110 chirps are heard in a minute. According to this model, what is the outdoor temperature?

b. If it is 75°*F* outside, about how many chirps can we expect to hear in one minute?

c. The equation is only a good model of the relationship when the outdoor temperature is at least 55°*F*. (Below that temperature, crickets aren't around or inclined to chirp.) How many chirps can we expect to hear in a minute at that temperature?

d. On the coordinate plane, draw a graph that represents the relationship between the number of chirps and the temperature.

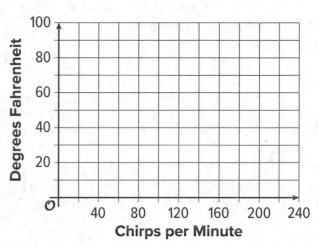

e. Explain what the coefficient $\frac{1}{4}$ in the equation tells us about the relationship.

f. Explain what the 40 in the equation tells us about the relationship.

Lesson 2-16

Solving Systems by Elimination (Part 3)

NAME _____ DATE _____ PERIOD _____

Learning Goal Let's find out how multiplying equations by a factor can help us solve systems of linear equations.

 Warm Up

16.1 Multiplying Equations By a Number

Consider two equations in a system:

$$\begin{cases} 4x + y = 1 & \text{Equation A} \\ x + 2y = 9 & \text{Equation B} \end{cases}$$

1. Use graphing technology to graph the equations. Then, identify the coordinates of the solution.

2. Write a few equations that are equivalent to equation A by multiplying both sides of it by the same number, for example, 2, -5, or $\frac{1}{2}$. Let's call the resulting equations A1, A2, and A3. Record your equations here:

   a. Equation A1:

   b. Equation A2:

   c. Equation A3:

3. Graph the equations you generated. Make a couple of observations about the graphs.

 ## Activity

16.2 Writing a New System to Solve a Given System

Here is a system you solved by graphing earlier.

$$\begin{cases} 4x + y = 1 & \text{Equation A} \\ x + 2y = 9 & \text{Equation B} \end{cases}$$

To start solving the system, Elena wrote:

$$\begin{cases} 4x + y = 1 \\ 4x + 8y = 36 \end{cases}$$

And then she wrote:

$$4x + y = 1$$
$$\underline{4x + 8y = 36 \; -}$$
$$-7y = -35$$

1. What were Elena's first two moves? What might be possible reasons for those moves?

2. Complete the solving process algebraically. Show that the solution is indeed $x = -1, y = 5$.

NAME _____ DATE _____ PERIOD _____

Activity

16.3 What Comes Next?

Your teacher will give you some slips of paper with systems of equations written on them. Each system represents a step in solving this system:

$$\begin{cases} \frac{4}{5}x + 6y = 15 \\ -x + 18y = 11 \end{cases}$$

Arrange the slips in the order that would lead to a solution. Be prepared to:

- Describe what move takes one system to the next system.

- Explain why each system is equivalent to the one before it.

This system of equations has solution (5, -2): $\begin{cases} Ax - By = 24 \\ Bx + Ay = 31 \end{cases}$

Find the missing values A and B.

Activity

16.4 Build Some Equivalent Systems

Here is a system of equations:

$\begin{cases} 12a + 5b = \text{-}15 \\ 8a + b = 11 \end{cases}$

1. To solve this system, Diego wrote these equivalent systems for his first two steps.

   Step 1:

   $\begin{cases} 12a + 5b = \text{-}15 \\ \text{-}40a + \text{-}5b = \text{-}55 \end{cases}$

   Step 2:

   $\begin{cases} 12a + 5b = \text{-}15 \\ \text{-}28a = \text{-}70 \end{cases}$

Describe the move that Diego made to get each equivalent system. Be prepared to explain how you know the systems in Step 1 and Step 2 have the same solution as the original system.

NAME _____ DATE _____ PERIOD _____

2. Write another set of equivalent systems (different than Diego's first two steps) that will allow one variable to be eliminated and enable you to solve the original system. Be prepared to describe the moves you make to create each new system and to explain why each one has the same solution as the original system.

3. Use your equivalent systems to solve the original system. Then, check your solution by substituting the pair of values into the original system.

Summary

Solving Systems by Elimination (Part 3)

We now have two algebraic strategies for solving systems of equations: by substitution and by elimination. In some systems, the equations may give us a clue as to which strategy to use. For example:

$$\begin{cases} y = 2x - 11 \\ 3x + 2y = 18 \end{cases}$$

In this system, y is already isolated in one equation. We can solve the system by substituting $2x - 11$ for y in the second equation and finding x.

$$\begin{cases} 3x - y = \text{-}17 \\ \text{-}3x + 4y = 23 \end{cases}$$

This system is set up nicely for elimination because of the opposite coefficients of the x-variable. Adding the two equations eliminates x so we can solve for y.

In other systems, which strategy to use is less straightforward, either because no variables are isolated, or because no variables have equal or opposite coefficients. For example:

$$\begin{cases} 2x + 3y = 15 & \text{Equation A} \\ 3x - 9y = 18 & \text{Equation B} \end{cases}$$

To solve this system by elimination, we first need to rewrite one or both equations so that one variable can be eliminated. To do that, we can multiply both sides of an equation by the same factor. Remember that doing this doesn't change the equality of the two sides of the equation, so the x- and y-values that make the first equation true also make the new equation true.

There are different ways to eliminate a variable with this approach. For instance, we could:

- Multiply Equation A by 3 to get $6x + 9y = 45$. Adding this equation to Equation B eliminates y.

$$\begin{cases} 6x + 9y = 45 & \text{Equation A1} \\ 3x - 9y = 18 & \text{Equation B} \end{cases}$$

- Multiply Equation B by $\frac{2}{3}$ to get $2x - 6y = 12$. Subtracting this equation from Equation A eliminates x.

$$\begin{cases} 2x + 3y = 15 & \text{Equation A} \\ 2x - 6y = 12 & \text{Equation B1} \end{cases}$$

- Multiply Equation A by $\frac{1}{2}$ to get $x + \frac{3}{2}y = 7\frac{1}{2}$ and multiply Equation B by $\frac{1}{3}$ to get $x - 3y = 6$. Subtracting one equation from the other eliminates x.

NAME _____ DATE _____ PERIOD _____

$$\begin{cases} x + \dfrac{3}{2}y = 7\dfrac{1}{2} & \text{Equation A2} \\ x - 3y = 6 & \text{Equation B2} \end{cases}$$

Each multiple of an original equation is equivalent to the original equation. So each new pair of equations is equivalent to the original system and has the same solution.

Let's solve the original system using the first equivalent system we found earlier.

$$\begin{cases} 6x + 9y = 45 & \text{Equation A1} \\ 3x - 9y = 18 & \text{Equation B} \end{cases}$$

- Adding the two equations eliminates y, leaving a new equation $9x = 63$, or $x = 7$.

$$6x + 9y = 45$$
$$\underline{3x - 9y = 18 \ +}$$
$$9x + 0 = 63$$
$$x = 7$$

- Putting together $x = 7$ and the original $3x - 9y = 18$ gives us another equivalent system.

$$\begin{cases} x = 7 \\ 3x - 9y = 18 \end{cases}$$

- Substituting 7 for x in the second equation allows us to solve for y.

$$3(7) - 9y = 18$$
$$21 - 9y = 18$$
$$-9y = -3$$
$$y = \dfrac{1}{3}$$

When we solve a system by elimination, we are essentially writing a series of **equivalent systems**, or systems with the same solution. Each equivalent system gets us closer and closer to the solution of the original system.

$$\begin{cases} 2x + 3y = 15 \\ 3x - 9y = 18 \end{cases} \quad \begin{cases} 6x + 9y = 45 \\ 3x - 9y = 18 \end{cases} \quad \begin{cases} x = 7 \\ 3x - 9y = 18 \end{cases} \quad \begin{cases} x = 7 \\ y = \dfrac{1}{3} \end{cases}$$

Glossary

equivalent systems

Practice
Solving Systems by Elimination (Part 3)

1. Solve each system of equations.

   a. $\begin{cases} 2x - 4y = 10 \\ x + 5y = 40 \end{cases}$

   b. $\begin{cases} 3x - 5y = 4 \\ -2x + 6y = 18 \end{cases}$

2. Tyler is solving this system of equations: $\begin{cases} 4p + 2q = 62 \\ 8p - q = 59 \end{cases}$

   He can think of two ways to eliminate a variable and solve the system:

   - Multiply $4p + 2q = 62$ by 2, then subtract $8p - q = 59$ from the result.

   - Multiply $8p - q = 59$ by 2, then add the result to $4p + 2q = 62$.

   Do both strategies work for solving the system? Explain or show your reasoning.

3. Andre and Elena are solving this system of equations: $\begin{cases} y = 3x \\ y = 9x - 30 \end{cases}$

   - Andre's first step is to write: $3x = 9x - 30$

   - Elena's first step is to create a new system: $\begin{cases} 3y = 9x \\ y = 9x - 30 \end{cases}$

   Do you agree with either first step? Explain your reasoning.

NAME _____ DATE _____ PERIOD _____

4. Select **all** systems that are equivalent to this system: $\begin{cases} 6d + 4.5e = 16.5 \\ 5d + 0.5e = 4 \end{cases}$

(A.) $\begin{cases} 6d + 4.5e = 16.5 \\ 45d + 4.5e = 4 \end{cases}$

(B.) $\begin{cases} 30d + 22.5e = 82.5 \\ 5d + 0.5e = 4 \end{cases}$

(C.) $\begin{cases} 30d + 22.5e = 82.5 \\ 30d + 3e = 24 \end{cases}$

(D.) $\begin{cases} 6d + 4.5e = 16.5 \\ 6d + 0.6e = 4.8 \end{cases}$

(E.) $\begin{cases} 12d + 9e = 33 \\ 10d + 0.5e = 8 \end{cases}$

(F.) $\begin{cases} 6d + 4.5e = 16.5 \\ 11d + 5e = 20.5 \end{cases}$

5. Here is a system of equations with a solution: $\begin{cases} p + 8q = -8 \\ \frac{1}{2}p + 5q = -5 \end{cases}$

   a. Write a system of equations that is equivalent to this system. Describe what you did to the original system to get the new system.

   b. Explain how you know the new system has the same solution as the original system.

6. The cost to mail a package is $5.00. Noah has postcard stamps that are worth $0.34 each and first-class stamps that are worth $0.49 each.

   (Lesson 2-8)

   a. Write an equation that relates the number of postcard stamps p, the number of first-class stamps f, and the cost of mailing the package.

   b. Solve the equation for f.

   c. Solve the equation for p.

   d. If Noah puts 7 first-class stamps on the package, how many postcard stamps will he need?

NAME _____ DATE _____ PERIOD _____

7. Here is a system of linear equations: $\begin{cases} 2x + 7y = 8 \\ y + 2x = 14 \end{cases}$

Find at least one way to solve the system by substitution and show your reasoning. How many ways can you find? (Regardless of the substitution that you do, the solution should be the same.) (Lesson 2-13)

8. Here is a system of equations: $\begin{cases} -7x + 3y = -65 \\ -7x + 10y = -135 \end{cases}$

   Write an equation that results from subtracting the two equations.
   (Lesson 2-14)

9. A grocery store sells bananas for b dollars a pound and grapes for g dollars a pound. Priya buys 2.2 pounds of bananas and 3.6 pounds of grapes for $9.35. Andre buys 1.6 pounds of bananas and 1.2 pounds of grapes for $3.68.

   This situation is represented by the system of equations:

   $\begin{cases} 2.2b + 3.6g = 9.35 \\ 1.6b + 1.2g = 3.68 \end{cases}$

   Explain why it makes sense in this situation that the solution of this system is also a solution to $3.8b + 4.8g = 13.03$. (Lesson 2-15)

Lesson 2-17

Systems of Linear Equations and Their Solutions

NAME _____ DATE _____ PERIOD _____

Learning Goal Let's find out how many solutions a system of equations could have.

Warm Up
17.1 A Curious System

Andre is trying to solve this system of equations: $\begin{cases} x + y = 3 \\ 4x = 12 - 4y \end{cases}$

Looking at the first equation, he thought, "The solution to the system is a pair of numbers that add up to 3. I wonder which two numbers they are."

1. Choose any two numbers that add up to 3. Let the first one be the x-value and the second one be the y-value.

2. The pair of values you chose is a solution to the first equation. Check if it is also a solution to the second equation. Then, pause for a brief discussion with your group.

3. How many solutions does the system have? Use what you know about equations or about solving systems to show that you are right.

Activity

17.2 What's the Deal?

A recreation center is offering special prices on its pool passes and gym memberships for the summer. On the first day of the offering, a family paid $96 for 4 pool passes and 2 gym memberships. Later that day, an individual bought a pool pass for herself, a pool pass for a friend, and 1 gym membership. She paid $72.

1. Write a system of equations that represents the relationships between pool passes, gym memberships, and the costs. Be sure to state what each variable represents.

2. Find the price of a pool pass and the price of a gym membership by solving the system algebraically. Explain or show your reasoning.

3. Use graphing technology to graph the equations in the system. Make 1-2 observations about your graphs.

NAME _____ DATE _____ PERIOD _____

Activity

17.3 Card Sort: Sorting Systems

Your teacher will give you a set of cards. Each card contains a system of equations.

Sort the systems into three groups based on the number of solutions each system has. Be prepared to explain how you know where each system belongs.

Are you ready for more?

1. In the cards, for each system with no solution, change a single constant term so that there are infinitely many solutions to the system.

2. For each system with infinitely many solutions, change a single constant term so that there are no solutions to the system.

3. Explain why in these situations it is impossible to change a single constant term so that there is exactly one solution to the system.

Activity

17.4 One, Zero, Infinitely Many

Here is an equation: $5x - 2y = 10$.

Create a second equation that would make a system of equations with:

1. One solution

2. No solutions

3. Infinitely many solutions

We have seen many examples of a system where one pair of values satisfies both equations. Not all systems, however, have one solution. Some systems have many solutions, and others have no solutions.

Let's look at three systems of equations and their graphs.

System 1: $\begin{cases} 3x + 4y = 8 \\ 3x - 4y = 8 \end{cases}$

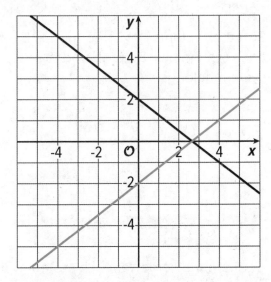

The graphs of the equations in System 1 intersect at one point. The coordinates of the point are the one pair of values that are simultaneously true for both equations. When we solve the equations, we get exactly one solution.

System 2: $\begin{cases} 3x + 4y = 8 \\ 6x + 8y = 16 \end{cases}$

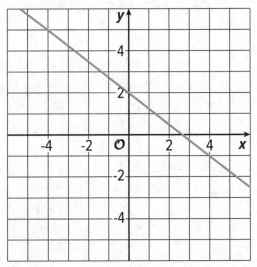

The graphs of the equations in System 2 appear to be the same line. This suggests that every point on the line is a solution to both equations, or that the system has infinitely many solutions.

NAME _____ DATE _____ PERIOD _____

System 3: $\begin{cases} 3x + 4y = 8 \\ 3x + 4y = \text{-}4 \end{cases}$

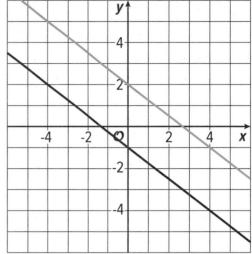

The graphs of the equations in System 3 appear to be parallel. If the lines never intersect, then there is no common point that is a solution to both equations and the system has no solutions.

How can we tell, without graphing, that System 2 indeed has many solutions?

- Notice that $3x + 4y = 8$ and $6x + 8y = 16$ are equivalent equations. Multiplying the first equation by 2 gives the second equation. Multiplying the second equation by $\frac{1}{2}$ gives the first equation. This means that any solution to the first equation is a solution to the second.

- Rearranging $3x + 4y = 8$ into slope-intercept form gives $y = \frac{8 - 3x}{4}$, or $y = 2 - \frac{3}{4}x$. Rearranging $6x + 8y = 16$ gives $y = \frac{16 - 6x}{8}$, which is also $y = 2 - \frac{3}{4}x$. Both lines have the same slope and the same y-value for the vertical intercept!

How can we tell, without graphing, that System 3 has no solutions?

- Notice that in one equation $3x + 4y$ equals 8, but in the other equation it equals -4. Because it is impossible for the same expression to equal 8 and -4, there must not be a pair of x- and y-values that are simultaneously true for both equations. This tells us that the system has no solutions.

- Rearranging each equation into slope-intercept form gives $y = 2 - \frac{3}{4}x$ and $y = \text{-}1 - \frac{3}{4}x$. The two graphs have the same slope but the y-values of their vertical intercepts are different. This tells us that the lines are parallel and will never cross.

Practice

Systems of Linear Equations and Their Solutions

1. Here is a system of equations: $\begin{cases} 3x - y = 17 \\ x + 4y = 10 \end{cases}$

   a. Solve the system by graphing the equations (by hand or using technology).

   b. Explain how you could tell, without graphing, that there is only one solution to the system.

2. Consider this system of linear equations: $\begin{cases} y = \frac{4}{5}x - 3 \\ y = \frac{4}{5}x + 1 \end{cases}$

   a. Without graphing, determine how many solutions you would expect this system of equations to have. Explain your reasoning.

   b. Try solving the system of equations algebraically and describe the result that you get. Does it match your prediction?

3. How many solutions does this system of equations have? Explain how you know.

   $\begin{cases} 9x - 3y = -6 \\ 5y = 15x + 10 \end{cases}$

NAME _____ DATE _____ PERIOD _____

4. Select **all** systems of equations that have no solutions.

(A.) $\begin{cases} y = 5 - 3x \\ y = -3x + 4 \end{cases}$

(B.) $\begin{cases} y = 4x - 1 \\ 4y = 16x - 4 \end{cases}$

(C.) $\begin{cases} 5x - 2y = 3 \\ 10x - 4y = 6 \end{cases}$

(D.) $\begin{cases} 3x + 7y = 42 \\ 6x + 14y = 50 \end{cases}$

(E.) $\begin{cases} y = 5 + 2x \\ y = 5x + 2 \end{cases}$

5. Solve each system of equations without graphing. **(Lesson 2-16)**

a. $\begin{cases} 2v + 6w = -36 \\ 5v + 2w = 1 \end{cases}$

b. $\begin{cases} 6t - 9u = 10 \\ 2t + 3u = 4 \end{cases}$

6. Select **all** the dot plots that appear to contain outliers. **(Lesson 1-14)**

(A.)

(B.)

(C.)
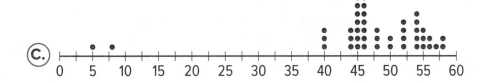

(D.)
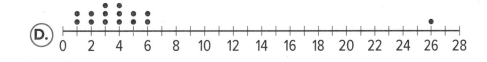

(E.)
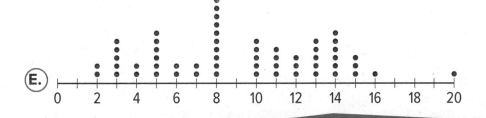

7. Here is a system of equations: $\begin{cases} -x + 6y = 9 \\ x + 6y = -3 \end{cases}$

Would you rather use subtraction or addition to solve the system? Explain your reasoning. (Lesson 2-14)

8. Here is a system of linear equations: $\begin{cases} 6x - y = 18 \\ 4x + 2y = 26 \end{cases}$

Select **all** the steps that would help to eliminate a variable and enable solving. (Lesson 2-16)

(A.) Multiply the first equation by 2, then subtract the second equation from the result.

(B.) Multiply the first equation by 4 and the second equation by 6, then subtract the resulting equations.

(C.) Multiply the first equation by 2, then add the result to the second equation.

(D.) Divide the second equation by 2, then add the result to the first equation.

(E.) Multiply the second equation by 6, then subtract the result from the first equation.

NAME _____ DATE _____ PERIOD _____

9. Consider this system of equations, which has one solution:

$$\begin{cases} 2x + 2y = 180 \\ 0.1x + 7y = 78 \end{cases}$$

Here are some equivalent systems. Each one is a step in solving the original system. **(Lesson 2-16)**

Step 1:

$$\begin{cases} 7x + 7y = 630 \\ 0.1x + 7y = 78 \end{cases}$$

Step 2:

$$\begin{cases} 6.9x = 552 \\ 0.1x + 7y = 78 \end{cases}$$

Step 3:

$$\begin{cases} x = 80 \\ 0.1x + 7y = 78 \end{cases}$$

a. Look at the original system and the system in Step 1.

    i. What was done to the original system to get the system in Step 1?

    ii. Explain why the system in Step 1 shares a solution with the original system.

b. Look at the system in Step 1 and the system in Step 2.

    i. What was done to the system in Step 1 to get the system in Step 2?

    ii. Explain why the system in Step 2 shares a solution with that in Step 1.

c. What is the solution to the original system?

Lesson 2-18

Representing Situations with Inequalities

NAME _____ DATE _____ PERIOD _____

Learning Goal Let's use inequalities to represent constraints in situations.

 ## Warm Up
18.1 What Do Those Symbols Mean?

1. Match each inequality to the meaning of a symbol within it.

   a. $h > 50$ • less than or equal to

   b. $h \leq 20$ • greater than

   c. $30 \geq h$ • greater than or equal to

2. Is 25 a solution to any of the inequalities? Which one(s)?

3. Is 40 a solution to any of the inequalities? Which one(s)?

4. Is 30 a solution to any of the inequalities? Which one(s)?

Seniors in a student council of a high school are trying to come up with a budget for the Senior Ball. Here is some information they have gathered:

- Last year, 120 people attended. It was a success and is expected to be even bigger this year. Anywhere up to 200 people might attend.

- There needs to be at least 1 chaperone for every 20 students.

- The ticket price cannot exceed $20 per person.

- The revenue from ticket sales needs to cover the cost of the meals and entertainment, and also make a profit of at least $200 to be contributed to the school.

Here are some inequalities the seniors wrote about the situation. Each letter stands for one quantity in the situation. Determine what is meant by each letter.

- $t \leq 20$

- $120 \leq p \leq 200$

- $pt - m \geq 200$

- $c \geq \dfrac{p}{20}$

Are you ready for more?

Kiran says we should add the constraint $t \geq 0$.

1. What is the reasoning behind this constraint?

2. What other "natural constraint" like this should be added?

NAME _____ DATE _____ PERIOD _____

Activity

18.3 Elevator Constraints

An elevator car in a skyscraper can hold at most 15 people. For safety reasons, each car can carry a maximum of 1,500 kg. On average, an adult weighs 70 kg and a child weighs 35 kg. Assume that each person carries 4 kg of gear with them.

1. Write as many equations and inequalities as you can think of to represent the constraints in this situation. Be sure to specify the meaning of any letters that you use. (Avoid using the letters *z*, *m*, or *g*.)

2. Trade your work with a partner and read each other's equations and inequalities.

    a. Explain to your partner what you think their statements mean, and listen to their explanation of yours.

    b. Make adjustments to your equations and inequalities so that they are communicated more clearly.

3. Rewrite your equations and inequalities so that they would work for a different building where:

    • an elevator car can hold at most *z* people

    • each car can carry a maximum of *m* kilograms

    • each person carries *g* kg of gear

Summary
Representing Situations with Inequalities

We have used equations and the equal sign to represent relationships and constraints in various situations. Not all relationships and constraints involve equality, however.

In some situations, one quantity is, or needs to be, greater than or less than another. To describe these situations, we can use inequalities and symbols such as $<$, $\leq$, $>$, or $\geq$.

When working with inequalities, it helps to remember what the symbol means, in words. For example:

- $100 < a$ means "100 is less than a."

- $y \leq 55$ means "y is less than or equal to 55," or "y is not more than 55."

- $20 > 18$ means "20 is greater than 18."

- $t \geq 40$ means "t is greater than or equal to 40," or "t is at least 40."

These inequalities are fairly straightforward. Each inequality states the relationship between two numbers ($20 > 18$), or they describe the limit or boundary of a quantity in terms of a number ($100 < a$).

Inequalities can also express relationships or constraints that are more complex. Here are some examples:

- The area of a rectangle, A, with a length of 4 meters and a width w meters is no more than 100 square meters.
  $A \leq 100$
  $4w \leq 100$

- To cover all the expenses of a musical production each week, the number of weekday tickets sold, d, and the number of weekend tickets sold, e, must be greater than 4,000.
  $d + e > 4,000$

- Elena would like the number of hours she works in a week, h, to be more than 5 but no more than 20.
  $h > 5$
  $h \leq 20$

- The total cost, T, of buying a adult shirts and c child shirts must be less than \$150. Adult shirts are \$12 each and children shirts are \$7 each.
  $T < 150$
  $12a + 7c < 150$

In upcoming lessons, we'll use inequalities to help us solve problems.

NAME _____ DATE _____ PERIOD _____

Practice
Representing Situations with Inequalities

1. Tyler goes to the store. His budget is $125. Which inequality represents x, the amount in dollars Tyler can spend at the store?

   (A.) $x \leq 125$

   (B.) $x \geq 125$

   (C.) $x > 125$

   (D.) $x < 125$

2. Jada is making lemonade for a get-together with her friends. She expects a total of 5 to 8 people to be there (including herself). She plans to prepare 2 cups of lemonade for each person.

   The lemonade recipe calls for 4 scoops of lemonade powder for each quart of water. Each quart is equivalent to 4 cups.

   Let n represent the number of people at the get-together, c the number of cups of water, and ℓ the number of scoops of lemonade powder.

   Select all the mathematical statements that represent the quantities and constraints in the situation.

   (A.) $5 < n < 8$

   (B.) $5 \leq n \leq 8$

   (C.) $c = 2n$

   (D.) $\ell = c$

   (E.) $10 < c < 16$

   (F.) $10 \leq \ell \leq 16$

3. A doctor sees between 7 and 12 patients each day. On Mondays and Tuesdays, the appointment times are 15 minutes. On Wednesdays and Thursdays, they are 30 minutes. On Fridays, they are one hour long. The doctor works for no more than 8 hours a day.

Here are some inequalities that represent this situation.

$$0.25 \leq y \leq 1 \qquad 7 \leq x \leq 12 \qquad xy \leq 8$$

a. What does each variable represent?

b. What does the expression xy in the last inequality mean in this situation?

4. Han wants to build a dog house. He makes a list of the materials needed:

- At least 60 square feet of plywood for the surfaces

- At least 36 feet of wood planks for the frame of the dog house

- Between 1 and 2 quarts of paint

Han's budget is $65. Plywood costs $0.70 per square foot, planks of wood cost $0.10 per foot, and paint costs $8 per quart.

Write inequalities to represent the material constraints and cost constraints in this situation. Be sure to specify what your variables represent.

NAME _____ DATE _____ PERIOD _____

5. The equation $V = \frac{1}{3}\pi r^2 h$ represents the volume of a cone, where r is the radius of the cone and h is the height of the cone.

   Which equation is solved for the height of the cone? (Lesson 2-9)

   (A.) $h = V - \pi r^2$

   (B.) $h = \frac{1}{3}\pi r^2 V$

   (C.) $3V - \pi r^2 = h$

   (D.) $h = \frac{3V}{\pi r^2}$

6. Solve each system of equations without graphing. (Lesson 2-14)

   a. $\begin{cases} 2x + 3y = 5 \\ 2x + 4y = 9 \end{cases}$

   b. $\begin{cases} \frac{2}{3}x + y = \frac{7}{3} \\ \frac{2}{3}x - y = 1 \end{cases}$

7. There is a pair of x and y values that make each equation true in this system of equations: $\begin{cases} 5x + 3y = 8 \\ 4x + 7y = 34 \end{cases}$

   Explain why the same pair of values also make $9x + 10y = 42$ true.

   (Lesson 2-15)

8. Which ordered pair is a solution to this system of equations?

   $\begin{cases} 7x + 5y = 59 \\ 3x - 9y = 159 \end{cases}$ (Lesson 2-16)

   (A.) $(-17, -12)$

   (B.) $(-17, 12)$

   (C.) $(17, -12)$

   (D.) $(17, 12)$

9. Which equation has exactly one solution in common with the equation
   $y = 6x - 2$? (Lesson 2-17)

   (A.) $18x - 3y = 6$

   (B.) $\frac{1}{2}y = 3x - 2$

   (C.) $2y = 4x - 12$

   (D.) $18x - 12 = 3y$

10. How many solutions does this system of equations have? Explain how you
    know. (Lesson 2-17)

    $$\begin{cases} y = -4x + 3 \\ 2x + 8y = 10 \end{cases}$$

Lesson 2-19

Solutions to Inequalities in One Variable

NAME _____ DATE _____ PERIOD _____

Learning Goal Let's find and interpret solutions to inequalities in one variable.

 Warm Up

19.1 Find a Value, Any Value

1. Write some solutions to the inequality $y \leq 9.2$. Be prepared to explain what makes a value a solution to this inequality.

2. Write one solution to the inequality $7(3 - x) > 14$. Be prepared to explain your reasoning.

A teacher is choosing between two options for a class field trip to an orchard.

- At Orchard A, admission costs $9 per person and 3 chaperones are required.

- At Orchard B, the cost is $10 per person, but only 1 chaperone is required.

- At each orchard, the same price applies to both chaperones and students.

1. Which orchard would be cheaper to visit if the class has:

   a. 8 students?

   b. 12 students?

   c. 30 students?

xalanx/iStock/Getty Images Plus

NAME _____ DATE _____ PERIOD _____

2. To help her compare the cost of her two options, the teacher first writes the equation $9(n + 3) = 10(n + 1)$, and then she writes the inequality $9(n + 3) < 10(n + 1)$.

   a. What does n represent in each statement?

   b. In this situation, what does the equation $9(n + 3) = 10(n + 1)$ mean?

   c. What does the solution to the inequality $9(n + 3) < 10(n + 1)$ tell us?

   d. Graph the solution to the inequality on the number line. Be prepared to show or explain your reasoning.

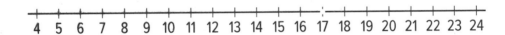

$$4 \quad 5 \quad 6 \quad 7 \quad 8 \quad 9 \quad 10 \quad 11 \quad 12 \quad 13 \quad 14 \quad 15 \quad 16 \quad 17 \quad 18 \quad 19 \quad 20 \quad 21 \quad 22 \quad 23 \quad 24$$

 Activity

19.3 Part-Time Work

To help pay for his tuition, a college student plans to work in the evenings and on weekends. He has been offered two part-time jobs: working in the guest-services department at a hotel and waiting tables at a popular restaurant.

- The job at the hotel pays $18 an hour and offers $33 in transportation allowance per month.

- The job at the restaurant pays $7.50 an hour plus tips. The entire waitstaff typically collects about $50 in tips each hour. Tips are divided equally among the 4 waitstaff members who share a shift.

1. The equation $7.50h + \frac{50}{4}h = 18h + 33$ represents a possible constraint about a situation.

   a. Solve the equation and check your solution.

   b. Here is a graph on a number line.

   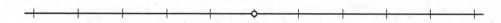

   Put a scale on the number line so that the point marked with a circle represents the solution to the equation.

   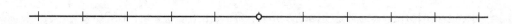

2. Does one job pay better if:

   a. The student works fewer hours than the solution you found earlier? If so, which job?

   b. The student works more hours than the solution you found earlier? If so, which job?

      Be prepared to explain or show how you know.

NAME _____ DATE _____ PERIOD _____

3. Here are two inequalities and two graphs that represent the solutions to the inequalities.

- Inequality 1: $7.50h + \frac{50}{4}h < 18h + 33$

- Inequality 2: $7.50h + \frac{50}{4}h > 18h + 33$

A

B

a. Put the same scale on each number line so that the circle represents the number of hours that you found earlier.

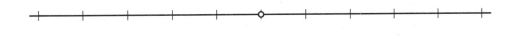

b. Match each inequality with a graph that shows its solution. Be prepared to explain or show how you know.

Activity
19.4 Equality and Inequality

1. Solve this equation and check your solution: $\dfrac{-4(x+3)}{5} = 4x - 12$.

2. Consider the inequality: $-\dfrac{4(x+3)}{5} \le 4x - 12$.

   a. Choose a couple of values less than 2 for x. Are they solutions to the inequality?

   b. Choose a couple of values greater than 2 for x. Are they solutions to the inequality?

   c. Choose 2 for x. Is it a solution?

   d. Graph the solution to the inequality on the number line.

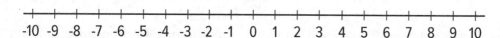

Are you ready for more?

Here is a different type of inequality: $x^2 \le 4$.

1. Is 1 a solution to the inequality? Is 3 a solution? How about -3?

2. Describe all solutions to this inequality. (If you like, you can graph the solutions on a number line.)

3. Describe all solutions to the inequality $x^2 \ge 9$. Test several numbers to make sure your answer is correct.

NAME _____ DATE _____ PERIOD _____

Activity

19.5 More or Less?

Consider the inequality $-\frac{1}{2}x + 6 < 4x - 3$. Let's look at another way to find its solutions.

1. Use graphing technology to graph $y = -\frac{1}{2}x + 6$ and $y = 4x - 3$ on the same coordinate plane.

2. Use your graphs to answer the following questions:

   a. Find the values of $-\frac{1}{2}x + 6$ and $4x - 3$ when x is 1.

   b. What value of x makes $-\frac{1}{2}x + 6$ and $4x - 3$ equal?

   c. For what values of x is $-\frac{1}{2}x + 6$ less than $4x - 3$?

   d. For what values of x is $-\frac{1}{2}x + 6$ greater than $4x - 3$?

3. What is the solution to the inequality $-\frac{1}{2}x + 6 < 4x - 3$? Be prepared to explain how you know.

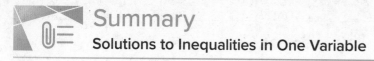
The equation $\frac{1}{2}t = 10$ is an equation in one variable. Its solution is any value of t that makes the equation true. Only $t = 20$ meets that requirement, so 20 is the only solution.

The inequality $\frac{1}{2}t > 10$ is an inequality in one variable. Any value of t that makes the inequality true is a solution. For instance, 30 and 48 are both solutions because substituting these values for t produces true inequalities. $\frac{1}{2}(30) > 10$ is true, as is $\frac{1}{2}(48) > 10$. Because the inequality has a range of values that make it true, we sometimes refer to *all* the solutions as the **solution set**.

One way to find the solutions to an inequality is by reasoning. For example, to find the solution to $2p < 8$, we can reason that if 2 times a value is less than 8, then that value must be less than 4. So a solution to $2p < 8$ is any value of p that is less than 4.

Another way to find the solutions to $2p < 8$ is to solve the related equation $2p = 8$. In this case, dividing each side of the equation by 2 gives $p = 4$. This point, where p is 4, is the *boundary* of the solution to the inequality.

To find out the range of values that make the inequality true, we can try values less than and greater than 4 in our inequality and see which ones make a true statement.

Let's try some values less than 4:

- If $p = 3$, the inequality is $2(3) < 8$ or $6 < 8$, which is true.

- If $p = -1$, the inequality is $2(-1) < 8$ or $-2 < 8$, which is also true.

Let's try values greater than 4:

- If $p = 5$, the inequality is $2(5) < 8$ or $10 < 8$, which is false.

- If $p = 12$, the inequality is $2(12) < 8$ or $24 < 8$, which is also false.

In general, the inequality is false when p is greater than or equal to 4 and true when p is less than 4.

We can represent the solution set to an inequality by writing an inequality, $p < 4$, or by graphing on a number line. The ray pointing to the left represents all values less than 4.

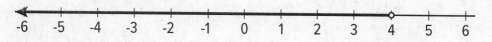

Glossary

solution set

NAME _____ DATE _____ PERIOD _____

Practice
Solutions to Inequalities in One Variable

1. Here is an inequality: $\dfrac{7x + 6}{2} \leq 3x + 2$

   Select **all** of the values that are a solution to the inequality.

   (A.) $x = -3$

   (B.) $x = -2$

   (C.) $x = -1$

   (D.) $x = 0$

   (E.) $x = 1$

   (F.) $x = 2$

   (G.) $x = 3$

2. Find the solution set to this inequality: $2x - 3 > \dfrac{2x - 5}{2}$

   (A.) $x < \dfrac{1}{2}$

   (B.) $x > \dfrac{1}{2}$

   (C.) $x \leq \dfrac{1}{2}$

   (D.) $x \geq \dfrac{1}{2}$

3. Here is an inequality: $\dfrac{-10 + x}{4} + 5 \geq \dfrac{7x - 5}{3}$

   What value of x will produce equality (or make the two sides equal)?

4. Noah is solving the inequality $7x + 5 > 2x + 35$. First, he solves the equation $7x + 5 = 2x + 35$ and gets $x = 6$.

   How does the solution to the equation $7x + 5 = 2x + 35$ help Noah solve the inequality $7x + 5 > 2x + 35$? Explain your reasoning.

5. Which graph represents the solution to $5 + 8x < 3(2x + 4)$?

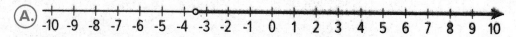

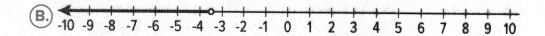

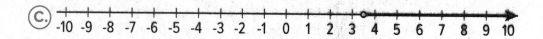

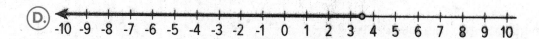

6. Solve this system of linear equations without graphing: $\begin{cases} 7x + 11y = -2 \\ 7x + 3y = 30 \end{cases}$

   (Lesson 2-15)

7. Kiran has 27 nickels and quarters in his pocket, worth a total of $2.75. (Lesson 2-12)

   a. Write a system of equations to represent the relationships between the number of nickels n, the number of quarters q, and the dollar amount in this situation.

   b. How many nickels and quarters are in Kiran's pocket? Show your reasoning.

NAME _____ DATE _____ PERIOD _____

8. How many solutions does this system of equations have? Explain how you know. **(Lesson 2-17)**

$$\begin{cases} y + \frac{2}{3}x = 4 \\ 2x = 12 - 3y \end{cases}$$

9. The principal of a school is hosting a small luncheon for her staff. She plans to prepare two sandwiches for each person. Some staff members offer to bring salads and beverages.

The principal has a budget of $225 and expects at least 16 people to attend. Sandwiches cost $3 each.

Select **all** of the equations and inequalities that could represent the constraints in the situation, where n is number of people attending and s is number of sandwiches. **(Lesson 2-18)**

(A.) $n \geq 16$

(B.) $n \geq 32$

(C.) $s < 2n$

(D.) $s = 2n$

(E.) $3n \leq 225$

(F.) $3s \leq 225$

10. Students at the college are allowed to work on campus no more than 20 hours per week. The jobs that are available pay different rates, starting from $8.75 an hour. Students can earn a maximum of $320 per week.

Write at least two inequalities that could represent the constraints in this situation. Be sure to specify what your variables represent. **(Lesson 2-18)**

Lesson 2-20

Writing and Solving Inequalities in One Variable

NAME _____ DATE _____ PERIOD _____

Learning Goal Let's solve problems by writing and solving inequalities in one variable.

Warm Up
20.1 Dinner for Drama Club

Kiran is getting dinner for his drama club on the evening of their final rehearsal. The budget for dinner is $60.

Kiran plans to buy some prepared dishes from a supermarket. The prepared dishes are sold by the pound, at $5.29 a pound. He also plans to buy two large bottles of sparkling water at $2.49 each.

1. Represent the constraints in the situation mathematically. If you use variables, specify what each one means.

2. How many pounds of prepared dishes can Kiran buy? Explain or show your reasoning.

Activity

20.2 Gasoline in the Tank

Han is about to mow some lawns in his neighborhood. His lawn mower has a 5-gallon fuel tank, but Han is not sure how much gasoline is in the tank.

He knows, however, that the lawn mower uses 0.4 gallon of gasoline per hour of mowing.

What are all the possible values for x, the number of hours Han can mow without refilling the lawn mower?

Write one or more inequalities to represent your response. Be prepared to explain or show your reasoning.

NAME _____ DATE _____ PERIOD _____

 Activity

20.3 Different Ways of Solving

Andre and Priya used different strategies to solve the following inequality but reached the same solution.

$2(2x + 1.5) < 18 - x$

1. Make sense of each strategy until you can explain what each student has done.

Andre

$$2(2x + 1.5) = 18 - x$$

$$4x + 3 = 18 - x$$

$$4x - 15 = \text{-}x$$

$$\text{-}15 = \text{-}5x$$

$$3 = x$$

Testing to see if $x = 4$ is a solution:

$$2(2 \cdot 4 + 1.5) < 18 - 4$$

$$2(9.5) < 14$$

$$19 < 14$$

The inequality is false, so 4 is not a solution. If a number greater than 3 is not a solution, the solution must be less than 3, or $3 > x$.

Priya

$$2(2x + 1.5) = 18 - x$$

$$4x + 3 = 18 - x$$

$$5x + 3 = 18$$

$$5x = 15$$

$$x = 3$$

In $4x + 3 = 18 - x$, there is $4x$ on the left and $\text{-}x$ on the right.

If x is a negative number, $4x + 3$ could be positive or negative, but $18 - x$ will always be positive.

For $4x + 3 < 18 - x$ to be true, x must include negative numbers or x must be less than 3.

2. Here are four inequalities.

   Work with a partner to decide on at least two inequalities to solve. Solve one inequality using Andre's strategy (by testing values on either side the given solution), while your partner uses Priya's strategy (by reasoning about the parts of the inequality). Switch strategies for the other inequality.

   a. $\frac{1}{5}p > -10$.

   b. $4(x + 7) \leq 4(2x + 8)$

   c. $-9n < 36$

NAME _____ DATE _____ PERIOD _____

d. $\dfrac{c}{3} < -2(c - 7)$

Are you ready for more?

Using positive integers between 1 and 9 and each positive integer at most once, fill in values to get two constraints so that $x = 7$ is the only integer that will satisfy both constraints at the same time.

$\boxed{}x + \boxed{} < \boxed{}x + \boxed{}$

$\boxed{}x + \boxed{} > \boxed{}x + \boxed{}$

 ## Activity

20.4 Matching Inequalities and Solutions

Match each inequality to a graph that represents its solutions. Be prepared to explain or show your reasoning.

1. $6x \le 3x$

A. ![number line from -10 to 10, open circle at 0, shaded right]

2. $\dfrac{1}{4}x > -\dfrac{1}{2}$

B. ![number line from -10 to 10, closed circle at 2, shaded left]

3. $5x + 4 \ge 7x$

C. ![number line from -10 to 10, closed circle at 2, shaded right]

4. $8x - 2 < -4(x - 1)$

D. ![number line from -10 to 10, open circle at 1, shaded left]

5. $\dfrac{4x - 1}{3} > -1$

E. ![number line from -10 to 10, open circle at -2, shaded right]

6. $\dfrac{12}{5} - \dfrac{x}{5} \le x$

F. ![number line from -10 to 10, closed circle at -1, shaded left]

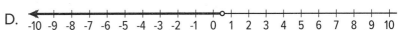

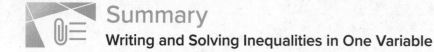
Writing and solving inequalities can help us make sense of the constraints in a situation and solve problems. Let's look at an example.

Clare would like to buy a video game that costs $130. She has saved $48 so far and plans on saving $5 of her allowance each week. How many weeks, w, will it be until she has enough money to buy the game?

To represent the constraints, we can write $48 + 5w \geq 130$. Let's reason about the solutions:

- Because Clare has $48 already and needs to have at least $130 to afford the game, she needs to save at least $82 more.

- If she saves $5 each week, it will take at least $\frac{82}{5}$ weeks to reach $82.

- $\frac{82}{5}$ is 16.4. Any time shorter than 16.4 weeks won't allow her to save enough.

- Assuming she saves $5 at the end of each week (instead of saving smaller amounts throughout a week), it will be at least 17 weeks before she can afford the game.

We can also solve by writing and solving a related equation to find the boundary value for w, and then determine whether the solutions are less than or greater than that value.

$$48 + 5w = 130$$
$$5w = 82$$
$$w = \frac{82}{5}$$
$$w = 16.4$$

Substituting 16.4 for w in the original inequality gives a true statement. (When $w = 16.4$, we get $130 \geq 130$.)

Substituting a value greater than 16.4 for w also gives a true statement. (When $w = 17$, we get $133 \geq 130$.)

NAME _____ DATE _____ PERIOD _____

Substituting a value less than 16.4 for w gives a false statement.
(When $w = 16$, we get $128 \geq 130$.)

The solution set is therefore $w \geq 16.4$.

Sometimes the structure of an inequality can help us see whether the solutions are less than or greater than a boundary value.

For example, to find the solutions to $3x > 8x$, we can solve the equation $3x = 8x$, which gives us $x = 0$. Then, instead of testing values on either side of 0, we could reason as follows about the inequality:

- If x is a positive value, then $3x$ would be less than $8x$.

- For $3x$ to be *greater* than $8x$, x must include negative values.

- For the solutions to include negative values, they must be less than 0, so the solution set would be $x < 0$.

Practice
Writing and Solving Inequalities in One Variable

1. Solve $2x < 10$. Explain how to find the solution set.

2. Lin is solving the inequality $15 - x < 14$. She knows the solution to the equation $15 - x = 14$ is $x = 1$.

   How can Lin determine whether $x > 1$ or $x < 1$ is the solution to the inequality?

3. A cell phone company offers two texting plans. People who use plan A pay 10 cents for each text sent or received. People who use plan B pay 12 dollars per month, and then pay an additional 2 cents for each text sent or received.

   a. Write an inequality to represent the fact that it is cheaper for someone to use plan A than plan B. Use x to represent the number of texts they send.

   b. Solve the inequality.

4. Clare made an error when solving $-4x + 3 < 23$.

   Describe the error that she made.

   $-4x + 3 < 23$

   $-4x < 20$

   $x < -5$

NAME _____ DATE _____ PERIOD _____

5. Diego's goal is to walk more than 70,000 steps this week. The mean number of steps that Diego walked during the first 4 days of this week is 8,019.

a. Write an inequality that expresses the mean number of steps that Diego needs to walk during the last 3 days of this week to walk more than 70,000 steps. Remember to define any variables that you use.

b. If the mean number of steps Diego walks during the last 3 days of the week is 12,642, will Diego reach his goal of walking more that 70,000 steps this week?

6. Here are statistics for the length of some frog jumps in inches:

How does each statistic change if the length of the jumps are measured in feet instead of inches? **(Lesson 1-15)**

- the mean is 41 inches

- the median is 39 inches

- the standard deviation is about 9.6 inches

- the IQR is 5.5 inches

7. Solve this system of linear equations without graphing: $\begin{cases} 3y + 7 = 5x \\ 7x - 3y = 1 \end{cases}$

(Lesson 2-15)

8. Solve each system of equations without graphing. (Lesson 2-16)

   a. $\begin{cases} 5x + 14y = -5 \\ -3x + 10y = 72 \end{cases}$

   b. $\begin{cases} 20x - 5y = 289 \\ 22x + 9y = 257 \end{cases}$

9. Noah and Lin are solving this system: $\begin{cases} 8x + 15y = 58 \\ 12x - 9y = 150 \end{cases}$

   Noah multiplies the first equation by 12 and the second equation by 8, which gives: $\begin{cases} 96x + 180y = 696 \\ 96x - 72y = 1{,}200 \end{cases}$

   Lin says, "I know you can eliminate x by doing that and then subtracting the second equation from the first, but I can use smaller numbers. Instead of what you did, try multiplying the first equation by 6 and the second equation by 4." (Lesson 2-16)

   a. Do you agree with Lin that her approach also works? Explain your reasoning.

   b. What are the smallest whole-number factors by which you can multiply the equations in order to eliminate x?

10. What is the solution set of the inequality $\frac{x + 2}{2} \geq -7 - \frac{x}{2}$? (Lesson 2-19)

    (A.) $x \leq -8$

    (C.) $x \geq -\frac{9}{2}$

    (B.) $x \geq -8$

    (D.) $x \geq 8$

Lesson 2-21

Graphing Linear Inequalities in Two Variables (Part 1)

NAME _____ DATE _____ PERIOD _____

Learning Goal Let's find out how to use graphs to represent solutions to inequalities in two variables.

Warm Up
21.1 Math Talk: Less Than, Equal to, or More Than 12?

Here is an expression: $2x + 3y$.

Decide if the values in each ordered pair, (x, y), make the value of the expression less than, greater than, or equal to 12.

(0, 5)

(6, 0)

(-1, -1)

(-5, 10)

Here are four inequalities. Study each inequality assigned to your group and work with your group to:

- Find some coordinate pairs that represent solutions to the inequality and some coordinate pairs that do not represent solutions.

- Plot both sets of points. Either use two different colors or two different symbols like X and O.

- Plot enough points until you start to see the region that contains solutions and the region that contains non-solutions. Look for a pattern describing the region where solutions are plotted.

$x \geq y$

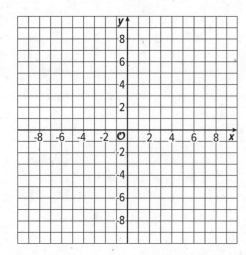

$-2y \geq -4$

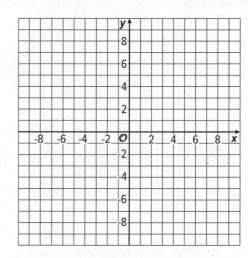

$3x < 0$

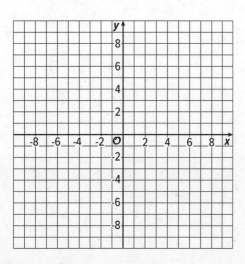

$x + y > 10$

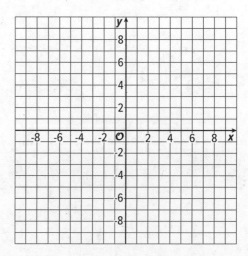

NAME _____ DATE _____ PERIOD _____

Activity

21.3 Sketching Solutions to Inequalities

1. Here is a graph that represents solutions to the equation $x - y = 5$.

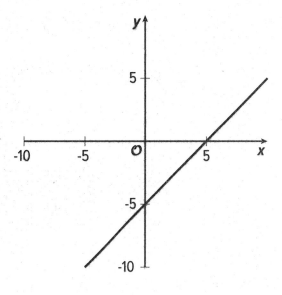

Sketch 4 quick graphs representing the solutions to each of these inequalities:

a. $x - y < 5$

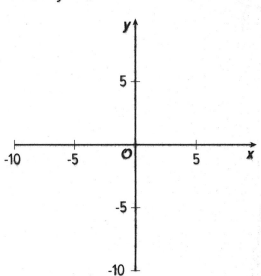

b. $x - y \leq 5$

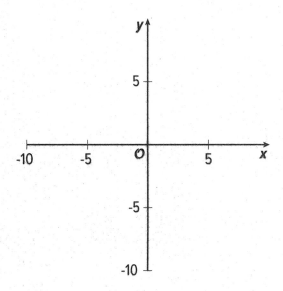

c. $x - y > 5$

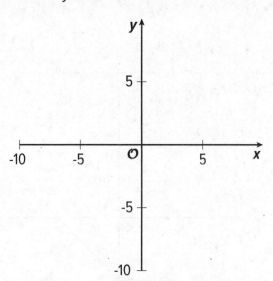

d. $x - y \geq 5$

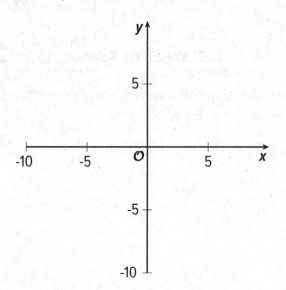

NAME _____ DATE _____ PERIOD _____

2. For each graph, write an inequality whose solutions are represented by the shaded part of the graph.

a.
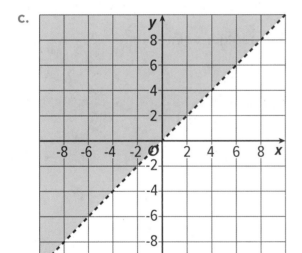

b.
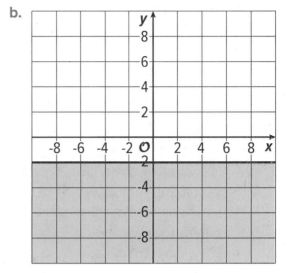

c.

d.
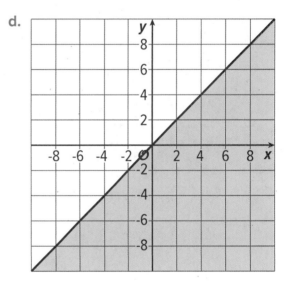

1. The points (7, 3) and (7, 5) are both in the solution region of the inequality $x - 2y < 3$.

   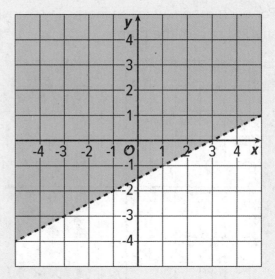

   a. Compute $x - 2y$ for both of these points

   b. Which point comes closest to satisfying the equation $x - 2y = 3$? That is, for which (x, y) pair is $x - 2y$ closest to 3?

2. The points (3, 2) and (5, 2) are also in the solution region. Which of these points comes closest to satisfying the equation $x - 2y = 3$?

3. Find a point in the solution region that comes even closer to satisfying the equation $x - 2y = 3$. What is the value of $x - 2y$?

4. For the points (5, 2) and (7, 3), $x - 2y = 1$. Find another point in the solution region for which $x - 2y = 1$.

5. Find $x - 2y$ for the point (5, 3). Then find two other points that give the same answer.

NAME _____ DATE _____ PERIOD _____

Summary

Graphing Linear Inequalities in Two Variables (Part 1)

The equation $x + y = 7$ is an equation in two variables. Its solution is any pair of x and y whose sum is 7. The pairs $x = 0, y = 7$ and $x = 5$, $y = 2$ are two examples.

We can represent all the solutions to $x + y = 7$ by graphing the equation on a coordinate plane.

The graph is a line. All the points on the line are solutions to $x + y = 7$.

The inequality $x + y \leq 7$ is an inequality in two variables. Its solution is any pair of x and y whose sum is 7 or less than 7.

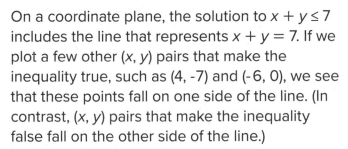

This means it includes all the pairs that are solutions to the equation $x + y = 7$, but also many other pairs of x and y that add up to a value less than 7. The pairs $x = 4, y = -7$ and $x = -6, y = 0$ are two examples.

On a coordinate plane, the solution to $x + y \leq 7$ includes the line that represents $x + y = 7$. If we plot a few other (x, y) pairs that make the inequality true, such as $(4, -7)$ and $(-6, 0)$, we see that these points fall on one side of the line. (In contrast, (x, y) pairs that make the inequality false fall on the other side of the line.)

We can shade that region on one side of the line to indicate that all points in it are solutions.

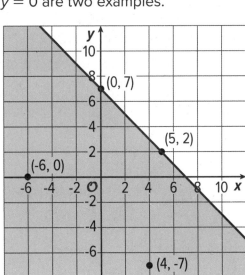

What about the inequality $x + y < 7$?

The solution is any pair of x and y whose sum is less than 7. This means pairs like $x = 0, y = 7$ and $x = 5, y = 2$ are *not* solutions.

On a coordinate plane, the solution does not include points on the line that represent $x + y = 7$ (because those points are x and y pairs whose sum is 7).

To exclude points on that boundary line, we can use a dashed line.

All points below that line are (x, y) pairs that make $x + y < 7$ true. The region on that side of the line can be shaded to show that it contains the solutions.

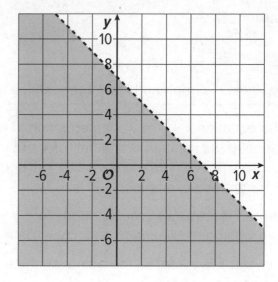

NAME _____ DATE _____ PERIOD _____

Practice
Graphing Linear Inequalities in Two Variables (Part 1)

1. Here is a graph of the equation
   $2y - x = 1$.

   a. Are the points $\left(0, \frac{1}{2}\right)$ and $(-7, -3)$ solutions to
   the equation? Explain or show how you know.

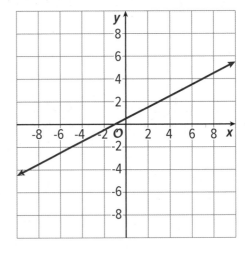

   b. Check if each of these points is a solution to
   the inequality $2y - x > 1$:

   $(0, 2)$ $\left(8, \frac{1}{2}\right)$ $(-6, 3)$ $(-7, -3)$

   c. Shade the region that represents the solution set to the inequality
   $2y - x > 1$.

   d. Are the points on the line included in the solution set? Explain how you
   know.

2. Select **all** coordinate pairs that are solutions to the inequality
   $5x + 9y < 45$.

   (A.) $(0, 0)$ (E.) $(0, 9)$

   (B.) $(5, 0)$ (F.) $(5, 9)$

   (C.) $(9, 0)$ (G.) $(-5, -9)$

   (D.) $(0, 5)$

3. Consider the linear equation $2y - 3x = 5$.

   a. The pair (-1, 1) is a solution to the equation. Find another (x, y) pair that is a solution to the equation.

   b. Are (-1, 1) and (4, 1) solutions to the inequality $2y - 3x < 5$? Explain how you know.

   c. Explain how to use the answers to the previous questions to graph the solution set to the inequality $2y - 3x < 5$.

4. The boundary line on the graph represents the equation $5x + 2y = 6$. Write an inequality that is represented by the graph.

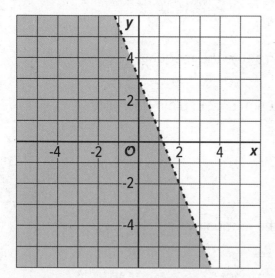

NAME _____ DATE _____ PERIOD _____

5. Choose the inequality whose solution set is represented by this graph.

(A.) $x - 3y < 5$

(B.) $x - 3y \leq 5$

(C.) $x - 3y > 5$

(D.) $x - 3y \geq 5$

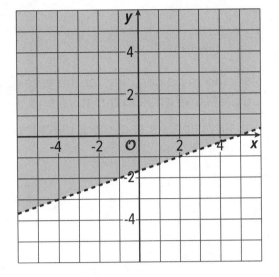

6. Solve each system of equations without graphing. **(Lesson 2-14)**

a.
$$\begin{cases} 4d + 7e = 68 \\ -4d - 6e = -72 \end{cases}$$

b.
$$\begin{cases} \frac{1}{4}x + y = 1 \\ \frac{3}{2}x - y = \frac{4}{3} \end{cases}$$

7. Mai and Tyler are selling items to earn money for their elementary school. The school earns w dollars for every wreath sold and p dollars for every potted plant sold. Mai sells 14 wreaths and 3 potted plants and the school earns $70.50. Tyler sells 10 wreaths and 7 potted plants and the school earns $62.50.

This situation is represented by this system of equations:

$$\begin{cases} 14w + 3p = 70.50 \\ 10w + 7p = 62.50 \end{cases}$$

Explain why it makes sense in this situation that the solution of this system is also a solution to $4w + (-4p) = 8.00$. **(Lesson 2-15)**

8. Elena is planning to go camping for the weekend and has already spent $40 on supplies. She goes to the store and buys more supplies.

   Which inequality represents d, the total amount in dollars that Elena spends on supplies? (Lesson 2-18)

   (A.) $d > 40$

   (B.) $d \geq 40$

   (C.) $d < 40$

   (D.) $d \leq 40$

9. Solve this inequality: $\dfrac{x-4}{3} \geq \dfrac{x+3}{2}$ (Lesson 2-19)

10. Which graph represents the solution to $\dfrac{4x-8}{3} \leq 2x-5$? (Lesson 2-19)

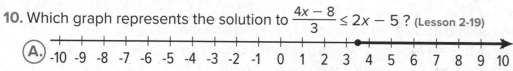

11. Solve $-x < 3$. Explain how to find the solution set. (Lesson 2-20)

Lesson 2-22

Graphing Linear Inequalities in Two Variables (Part 2)

NAME _____ DATE _____ PERIOD _____

Learning Goal Let's write inequalities in two variables and make sense of the solutions by reasoning and by graphing.

Warm Up
22.1 Landscaping Options

A homeowner is making plans to landscape her yard. She plans to hire professionals to install grass sod in some parts of the yard and flower beds in other parts.

Grass sod installation costs $2 per square foot and flower bed installation costs $12 square foot. Her budget for the project is $3,000.

1. Write an equation that represents the square feet of grass sod, *x*, and the square feet of flower beds, *y*, that she could afford if she used her entire budget.

2. On the coordinate plane, sketch a graph that represents your equation. Be prepared to explain your reasoning.

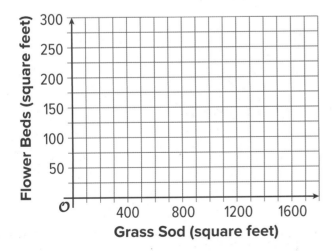

David Madison/Photodisc/Getty Images

Activity

22.2 Rethinking Landscaping

The homeowner is worried about the work needed to maintain a grass lawn and flower beds, so she is now looking at some low-maintenance materials.

She is considering artificial turf, which costs $15 per square foot to install, and gravel, which costs $3 per square foot. She may use a combination of the two materials in different parts of the yard. Her budget is still $3,000.

Here is a graph representing some constraints in this situation.

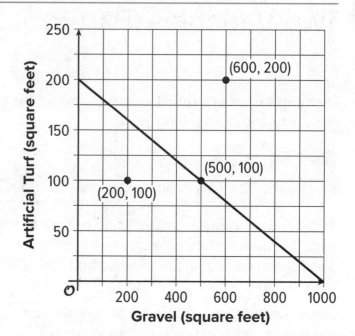

1. The graph shows a line going through (500, 100).

   a. In this situation, what does the point (500, 100) mean?

   b. Write an equation that the line represents.

   c. What do the solutions to the equation mean?

2. The point (600, 200) is located to the right and above the line.

   a. Does that combination of turf and gravel meet the homeowner's constraints? Explain or show your reasoning.

   b. Choose another point in the same region (to the right and above the line). Check if the combination meets the homeowner's constraints.

NAME _____ DATE _____ PERIOD _____

3. The point (200, 100) is located to the left and below the line.

   a. Does that combination of turf and gravel meet the homeowner's constraints? Explain or show your reasoning.

   b. Choose another point in the same region (to the left and below the line). Check if the combination meets the homeowner's constraints.

4. Write an inequality that represents the constraints in this situation. Explain what the solutions mean and show the solution region on the graph.

Activity

22.3 The Saturday Market

A vendor at the Saturday Market makes $9 profit on each necklace she sells and $5 profit on each bracelet.

1. Find a combination of necklaces and bracelets that she could sell and make:

   a. exactly $100 profit

   b. more than $100 profit

2. Write an equation whose solution is the combination of necklaces and bracelets she could sell and make exactly $100 profit.

3. Write an inequality whose solutions are the combinations of necklaces and bracelets she could sell and make more than $100 profit.

4. Graph the solutions to your inequality.

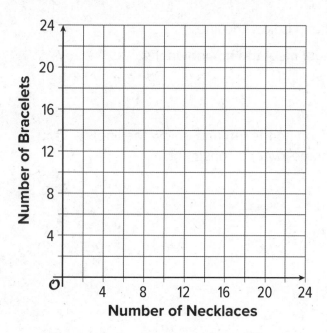

Number of Bracelets (vertical axis)

Number of Necklaces (horizontal axis)

5. Is (3, 18.6) a solution to the inequality? Explain your reasoning.

Are you ready for more?

1. Write an inequality using two variables x and y where the solution would be represented by shading the entire coordinate plane.

2. Write an inequality using two variables x and y where the solution would be represented by not shading any of the coordinate plane.

Activity

22.4 Charity Concerts

A popular band is trying to raise at least $20,000 for charity by holding multiple concerts at a park. It plans to sell tickets at $25 each. For each 2-hour concert, the band would need to pay the park $1,250 in fees for security, cleaning, and traffic services.

The band needs to find the combinations of number of tickets sold, t, and number of concerts held, c, that would allow it to reach its fundraising goal.

NAME _____ DATE _____ PERIOD _____

1. Write an inequality to represent the constraints in this situation.

2. Graph the solutions to the inequality on the coordinate plane.

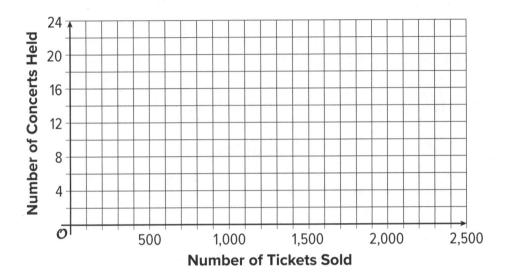

3. Name two possible combinations of number of tickets sold and number of concerts held that would allow the band to meet its goal.

4. Which combination of tickets and concerts would mean *more* money for charity:

   a. 1,300 tickets and 10 concerts, or 1,300 tickets and 5 concerts?

   b. 1,600 tickets and 16 concerts, or 1,200 tickets and 9 concerts?

   c. 2,000 tickets and 4 concerts, or 2,500 tickets and 10 concerts?

Inequalities in two variables can represent constraints in real-life situations. Graphing their solutions can enable us to solve problems.

Suppose a café is purchasing coffee and tea from a supplier and can spend up to $1,000. Coffee beans cost $12 per kilogram and tea leaves costs $8 per kilogram.

Buying c kilograms of coffee beans and t kilograms of tea leaves will therefore cost $12c + 8t$. To represent the budget constraints, we can write: $12c + 8t \leq 1,000$.

The solution to this inequality is any pair of c and t that makes the inequality true. In this situation, it is any combination of the kilograms of coffee and tea that the café can order without going over the $1,000 budget.

We can try different pairs of c and t to see what combinations satisfy the constraint, but it would be difficult to capture all the possible combinations this way. Instead, we can graph a related equation, $12c + 8t = 1,000$, and then find out which region represents all possible solutions.

Here is the graph of that equation.

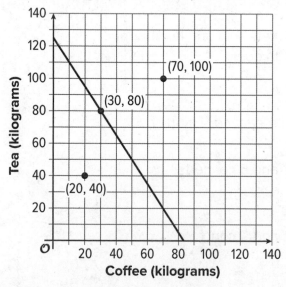

To determine the solution region, let's take one point on the line and one point on each side of the line, and see if the pairs of values produce true statements.

A point on the line: $(30, 80)$

$$12(30) + 8(80) \leq 1,000$$

$$360 + 640 \leq 1,000$$

$$1,000 \leq 1,000$$

This is true.

NAME _____ DATE _____ PERIOD _____

A point below the line: (20, 40)

12(20) + 8(40) ≤ 1,000

240 + 320 ≤ 1,000

560 ≤ 1,000

This is true.

A point above the line: (70, 100)

12(70) + 8(100) ≤ 1,000

840 + 800 ≤ 1,000

1,640 ≤ 1,000

This is false.

The points on the line and in the region below the line are solutions to the inequality. Let's shade the solution region.

It is easy to read solutions from the graph. For example, without any computation, we can tell that (50, 20) is a solution because it falls in the shaded region. If the café orders 50 kilograms of coffee and 20 kilograms of tea, the cost will be less than $1,000.

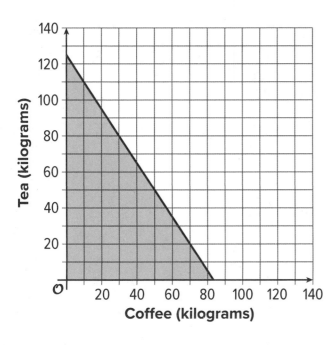

1. To qualify for a loan from a bank, the total in someone's checking and savings accounts together must be $500 or more.

   a. Which of these inequalities best represents this situation?

   - $x + y < 500$

   - $x + y \leq 500$

   - $x + y > 500$

   - $x + y \geq 500$

   b. Complete the graph so that it represents solutions to an inequality representing this situation. (Be clear about whether you want to use a solid or dashed line.)

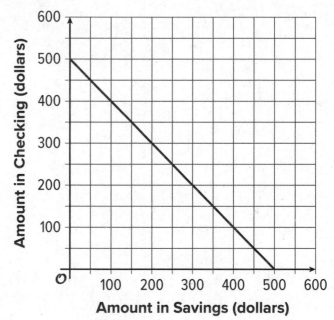

2. The soccer team is selling bags of popcorn for $3 each and cups of lemonade for $2 each. To make a profit, they must collect a total of more than $120.

   a. Write an inequality to represent the number of bags of popcorn sold, p, and the number of cups of lemonade sold, c, in order to make a profit.

   b. Graph the solution set to the inequality on the coordinate plane.

   c. Explain how we could check if the boundary is included or excluded from the solution region.

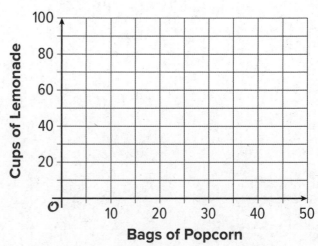

NAME _____ DATE _____ PERIOD _____

3. Tickets to the aquarium are $11 for adults and $6 for children. An after-school program has a budget of $200 for a trip to the aquarium.

If the boundary line in each graph represents the equation $11x + 6y = 200$, which graph represents the cost constraint in this situation?

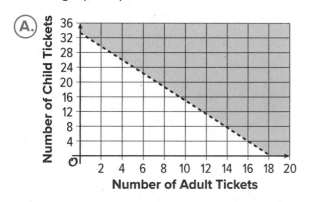

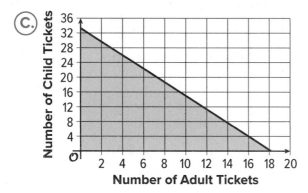

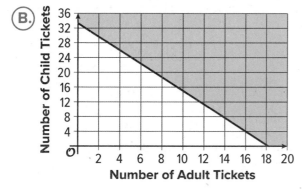

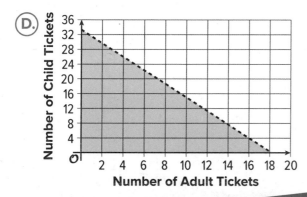

4. Tyler filled a small jar with quarters and dimes and donated it to his school's charity club. The club member receiving the jar asked, "Do you happen to know how much is in the jar?" Tyler said, "I know it's at least $8.50, but I don't know the exact amount."

a. Write an inequality to represent the relationship between the number of dimes, d, the number of quarters, q, and the dollar amount of the money in the jar.

b. Graph the solution set to the inequality and explain what a solution means in this situation.

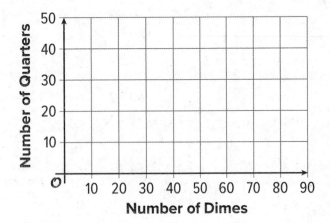

c. Suppose Tyler knew there are 25 dimes in the jar. Write an inequality that represents how many quarters could be in the jar.

5. Andre is solving the inequality $14x + 3 \leq 8x + 3$. He first solves a related equation.

$$14x + 3 = 8x + 3$$
$$14x = 8x$$
$$8 = 14$$

This seems strange to Andre. He thinks he probably made a mistake. What was his mistake? (Lesson 2-20)

NAME _____ DATE _____ PERIOD _____

6. Kiran says, "I bought 2.5 pounds of red and yellow lentils. Both were $1.80 per pound. I spent a total of $4.05." (Lesson 2-17)

a. Write a system of equations to describe the relationships between the quantities in Kiran's statement. Be sure to specify what each variable represents.

b. Elena says, "That can't be right." Explain how Elena can tell that something is wrong with Kiran's statement.

c. Kiran says, "Oops, I meant to say I bought 2.25 pounds of lentils." Revise your system of equations to reflect this correction.

d. Is it possible to tell for sure how many pounds of each kind of lentil Kiran might have bought? Explain your reasoning.

7. Here is an inequality: $-7 - (3x + 2) < -8(x + 1)$

Select **all** the values of x that are solutions to the inequality. (Lesson 2-19)

A. $x = -0.2$

B. $x = -0.1$

C. $x = 0$

D. $x = 0.1$

E. $x = 0.2$

F. $x = 0.3$

8. Here is a graph of the equation $6x + 2y = -8$. (Lesson 2-21)

   a. Are the points (1.5, -4) and (0, -4) solutions to the equation? Explain or show how you know.

   b. Check if each of these points is a solution to the inequality $6x + 2y \leq -8$:

      (-2, 2)   (4, -2)   (0,0)   (-4, -4)

   c. Shade the solutions to the inequality.

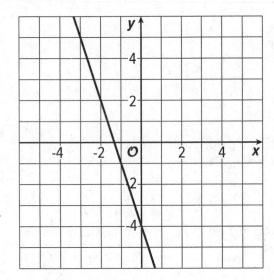

   d. Are the points on the line included in the solution region? Explain how you know.

Lesson 2-23

Solving Problems with Inequalities in Two Variables

NAME _____ DATE _____ PERIOD _____

Learning Goal Let's practice writing, interpreting, and graphing solutions to inequalities in two variables.

Warm Up

23.1 Graphing Inequalities with Technology

Use graphing technology to graph the solution region of each inequality and sketch each graph. Adjust the graphing window as needed to show meaningful information.

$y > x$

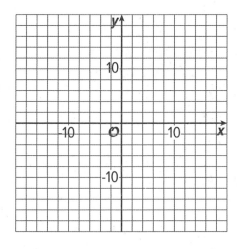

$y \geq x$

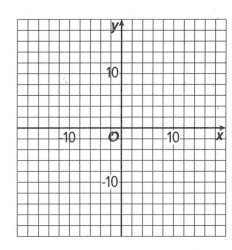

$y < -8$

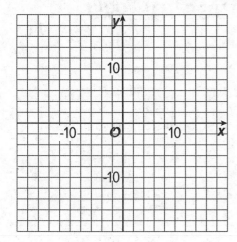

$-x + 8 \leq y$

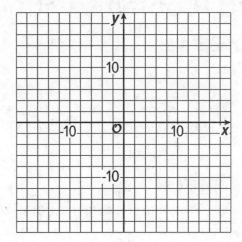

$y < 10x - 200$

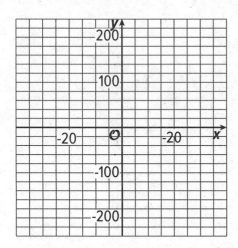

$2x + 3y > 60$

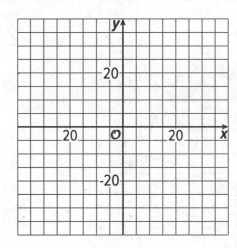

NAME _____ DATE _____ PERIOD _____

Activity

23.2 Solving Problems with Inequalities in Two Variables

Here are three situations. There are two questions about each situation.
For each question that you work on:

- a. Write an inequality to describe the constraints. Specify what each variable represents.

- b. Use graphing technology to graph the inequality. Sketch the solution region on the coordinate plane and label the axes.

- c. Name one solution to the inequality and explain what it represents in that situation.

- d. Answer the question about the situation.

Bank Accounts

1. A customer opens a checking account and a savings account at a bank. They will deposit a maximum of $600, some in the checking account and some in the savings account. (They might not deposit all of it and keep some of the money as cash.)

   If the customer deposits $200 in their checking account, what can you say about the amount they deposit in their savings account?

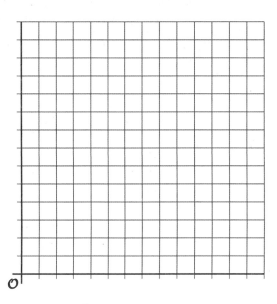

2. The bank requires a minimum balance of $50 in the savings account. It does not matter how much money is kept in the checking account.

If the customer deposits no money in the checking account but is able to maintain both accounts without penalty, what can you say about the amount deposited in the savings account?

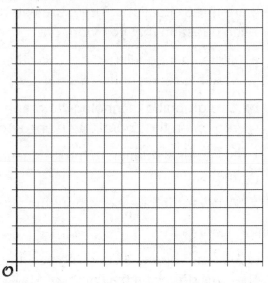

Concert Tickets

1. Two kinds of tickets to an outdoor concert were sold: lawn tickets and seat tickets. Fewer than 400 tickets in total were sold.

If you know that exactly 100 lawn tickets were sold, what can you say about the number of seat tickets?

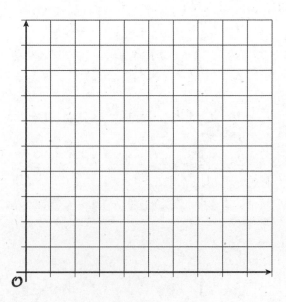

NAME _____ DATE _____ PERIOD _____

2. Lawn tickets cost $30 each and seat tickets cost $50 each. The organizers want to make at least $14,000 from ticket sales.

If you know that exactly 200 seat tickets were sold, what can you say about the number of lawn tickets?

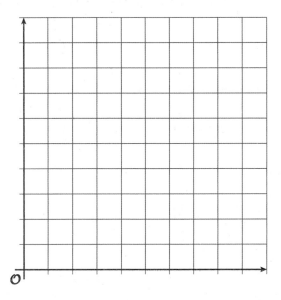

Advertising Packages

1. An advertising agency offers two packages for small businesses who need advertising services. A basic package includes only design services. A premium package includes design and promotion. The agency's goal is to sell at least 60 packages in total.

   If the agency sells exactly 45 basic packages, what can you say about the number of premium packages it needs to sell to meet its goal?

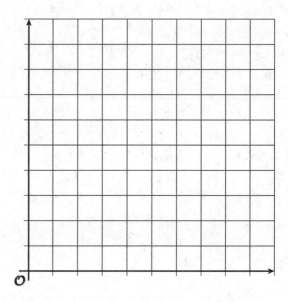

NAME _____ DATE _____ PERIOD _____

2. The basic advertising package has a value of $1,000 and the premium package has a value of $2,500. The goal of the agency is to sell more than $60,000 worth of small-business advertising packages.

If you know that exactly 10 premium packages were sold, what can you say about the number of basic packages the agency needs to sell to meet its goal?

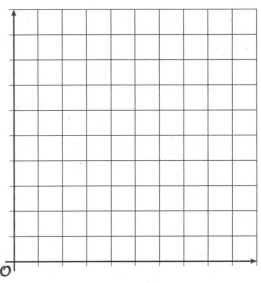

This activity will require a partner.

1. Without letting your partner see it, write an equation of a line so that both the *x*-intercept and the *y*-intercept are each between -3 and 3. Graph your equation on one of the coordinate systems.

Your Inequality   Your Partner's Inequality

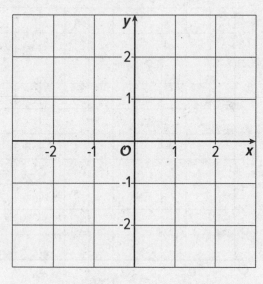 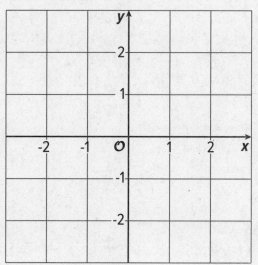

2. Still without letting your partner see it, write an inequality for which your equation is the related equation. In other words, your line should be the boundary between solutions and non-solutions. Shade the solutions on your graph.

3. Take turns stating coordinates of points. Your partner will tell you whether your guess is a solution to their inequality. After each partner has stated a point, each may guess what the other's inequality is. If neither guesses correctly, play continues. Use the other coordinate system to keep track of your guesses.

NAME _____ DATE _____ PERIOD _____

Activity

23.3 Card Sort: Representations of Inequalities

Your teacher will give you a set of cards. Take turns with your partner to match a group of 4 cards that contain: a situation, an inequality that represents it, a graph that represents the solution region, and a solution written as a coordinate pair.

For each match that you find, explain to your partner how you know it's a match.

For each match that your partner finds, listen carefully to their explanation. If you disagree, discuss your thinking and work to reach an agreement.

Record your matches.

Group 1

- situation: perimeter of a rectangle

- inequality:

- a solution:

- sketch of graph:

Group 3

- situation: honey and jam

- inequality:

- a solution:

- sketch of graph:

Group 2

- situation: jar of coins

- inequality:

- a solution:

- sketch of graph:

Group 4

- situation: a school trip

- inequality:

- a solution:

- sketch of graph:

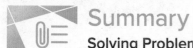

Suppose we want to find the solution to $x - y > 5$. We can start by graphing the related equation $x - y = 5$.

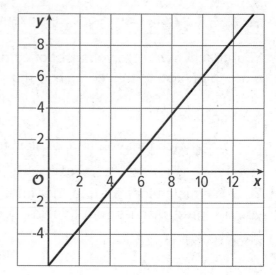

When identifying the solution region, it is important *not* to assume that the solution will be above the line because of a ">" symbol or below the line because of a "<" symbol.

Instead, test the points on the line and on either side of the line and see if they are solutions.

For $x - y > 5$, points on the line and above the line are *not* solutions to the inequality because the (x, y) pairs make the inequality false. Points that are below the lines are solutions, so we can shade that lower region.

Graphing technology can help us graph the solution to an inequality in two variables.

Many graphing tools allow us to enter inequalities such as $x - y > 5$ and will show the solution region, as shown here.

Some tools, however, may require the inequalities to be in slope-intercept form or another form before displaying the solution region. Be sure to learn how to use the graphing technology available in your classroom.

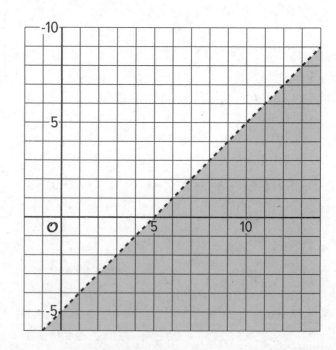

NAME _____ DATE _____ PERIOD _____

Although graphing using technology is efficient, we still need to analyze the graph with care. Here are some things to consider:

- The graphing window. If the graphing window is too small, we may not be able to really see the solution region or the boundary line, as shown here.

- The meaning of solution points in the situation. For example, if *x* and *y* represent the lengths of two sides of a rectangle, then only positive values of *x* and *y* (or points in the first quadrant) make sense in the situation.

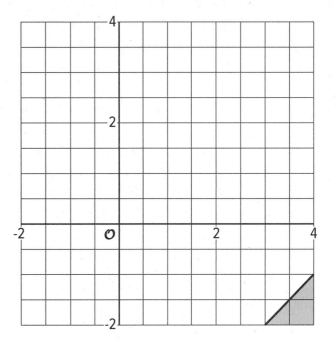

1. This year, students in the 9th grade are collecting dimes and quarters for a school fundraiser. They are trying to collect more money than the students who were in the 9th grade last year. The students in 9th grade last year collected $143.88.

   Using d to represent the number of dimes collected and q to represent the number of quarters, which statement best represents this situation?

   A. $0.25d + 0.1q \geq 143.88$

   B. $0.25q + 0.1d \geq 143.88$

   C. $0.25d + 0.1q > 143.88$

   D. $0.25q + 0.1d > 143.88$

2. A farmer is creating a budget for planting soybeans and wheat. Planting soybeans costs $200 per acre and planting wheat costs $500 per acre. He wants to spend no more than $100,000 planting soybeans and wheat.

   a. Write an inequality to describe the constraints. Specify what each variable represents.

   b. Name one solution to the inequality and explain what it represents in that situation.

NAME _____ DATE _____ PERIOD _____

3. Elena is ordering dried chili peppers and corn husks for her cooking class. Chili peppers cost $16.95 per pound and corn husks cost $6.49 per pound.

Elena spends less than $50 on d pounds of dried chili peppers and h pounds of corn husks.

Here is a graph that represents this situation.

a. Write an inequality that represents this situation.

b. Can Elena purchase 2 pounds of dried chili peppers and 4 pounds of corn husks and spend less than $50? Explain your reasoning.

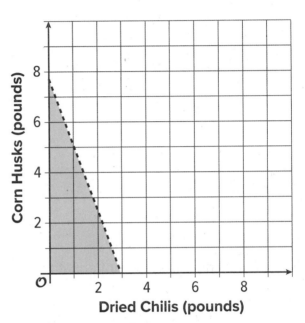

c. Can Elena purchase 1.5 pounds of dried chili peppers and 3 pounds of corn husks and spend less than $50? Explain your reasoning.

4. Which inequality is represented by the graph?

(A.) $4x - 2y > 12$

(B.) $4x - 2y < 12$

(C.) $4x + 2y > 12$

(D.) $4x + 2y < 12$

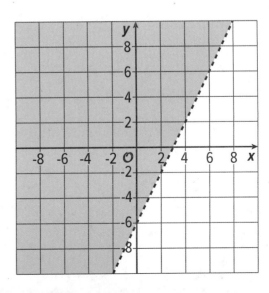

5. Here are some statistics for the number of minutes it took each of 9 members of the track team to run 1 mile.

When a tenth time is added to the list, the standard deviation increases to 1.5. Is the tenth time likely an outlier? Explain your reasoning. (Lesson 1-14)

- mean: 7.3

- median: 7.1

- standard deviation: 1.1

- Q1: 6.8

- Q3: 7.4

6. Elena is solving this system of equations: $\begin{cases} 10x - 6y = 16 \\ 5x - 3y = 8 \end{cases}$

She multiplies the second equation by 2, then subtracts the resulting equation from the first. To her surprise, she gets the equation $0 = 0$.

What is special about this system of equations? Why does she get this result and what does it mean about the solutions? (If you are not sure, try graphing them.) (Lesson 2-17)

NAME _____ DATE _____ PERIOD _____

7. Jada has a sleeping bag that is rated for 30°F. This means that if the temperature outside is at least 30°F, Jada will be able to stay warm in her sleeping bag. (Lesson 2-18)

   a. Write an inequality that represents the outdoor temperature at which Jada will be able to stay warm in her sleeping bag.

   b. Write an inequality that represents the outdoor temperature at which a thicker or warmer sleeping bag would be needed to keep Jada warm.

8. What is the solution set to this inequality:
   $6x - 20 > 3(2 - x) + 6x - 2$? (Lesson 2-19)

9. Here is a graph of the equation $2x - 3y = 15$. (Lesson 2-21)

   a. Are the points (1.5, -4) and (4, -4) solutions to the equation? Explain or show how you know.

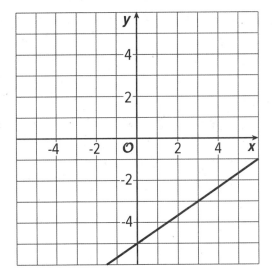

   b. Check if each of these points is a solution to the inequality $2x - 3y < 15$:

      (0, -5) (4, -2) (2, -4) (5, -1)

   c. Shade the solutions to the inequality.

   d. Are the points on the line included in the solution region? Explain how you know.

10. A store sells notepads in packages of 24 and packages of 6. The organizers of a conference need to prepare at least 200 notepads for the event. (Lesson 2-22)

    a. Would they have enough notepads if they bought these quantities?

        i. Seven packages of 24 and one package of 6

        ii. Five packages of 24 and fifteen packages of 6

    b. Write an inequality to represent the relationship between the number of large and small packages of notepads and the number of notepads needed for the event.

    c. Use graphing technology to graph the solution set to the inequality. Then, use the graph to name two other possible combinations of large and small packages of notepads that will meet the number of notepads needed for the event.

Lesson 2-24

Solutions to Systems of Linear Inequalities in Two Variables

NAME _____ DATE _____ PERIOD _____

Learning Goal Let's look at situations where two constraints (that can be expressed by inequalities) must be met simultaneously.

Warm Up
24.1 A Silly Riddle

Here is a riddle: "I am thinking of two numbers that add up to 5.678. The difference between them is 9.876. What are the two numbers?"

1. Name any pair of numbers whose sum is 5.678.

2. Name any pair of numbers whose difference is 9.876.

3. The riddle can be represented with two equations. Write the equations.

4. Solve the riddle. Explain or show your reasoning.

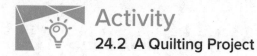

Activity
24.2 A Quilting Project

To make a quilt, a quilter is buying fabric in two colors, light and dark. He needs at least 9.5 yards of fabric in total.

The light color costs $9 a yard. The dark color costs $13 a yard.The quilter can spend up to $110 on fabric.

Here are two graphs that represent the two constraints.

A

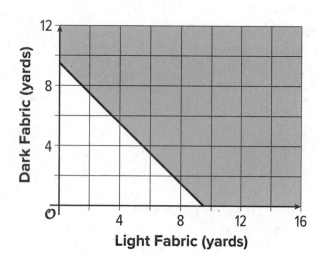

B

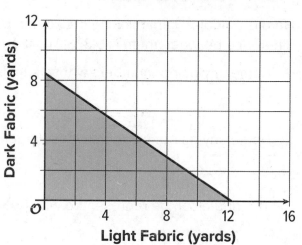

1. Write an inequality to represent the length constraint. Let *x* represent the yards of light fabric and *y* represent the yards of dark fabric.

2. Select **all** the pairs that satisfy the length constraint.

   (5, 5)   (2.5, 4.5)   (7.5, 3.5)   (12, 10)

3. Write an inequality to represent the cost constraint.

4. Select **all** the pairs that satisfy the cost constraint.

   (1, 1)   (4, 5)   (8, 3)   (10, 1)

NAME _____ DATE _____ PERIOD _____

5. Explain why (2, 2) satisfies the cost constraint, but not the length constraint.

6. Find at least one pair of numbers that satisfies *both* constraints. Be prepared to explain how you know.

7. What does the pair of numbers represent in this situation?

 Activity

24.3 Remember These Situations?

Here are some situations you have seen before. Answer the questions for one situation.

Bank Accounts

- A customer opens a checking account and a savings account at a bank. They will deposit a maximum of $600, some in the checking account and some in the savings account. (They might not deposit all of it and keep some of the money as cash.)

- The bank requires a minimum balance of $50 in the savings account. It does not matter how much money is kept in the checking account.

Concert Tickets

- Two kinds of tickets to an outdoor concert were sold: lawn tickets and seat tickets. Fewer than 400 tickets in total were sold.

- Lawn tickets cost $30 each and seat tickets cost $50 each. The organizers want to make at least $14,000 from ticket sales.

Advertising Packages

- An advertising agency offers two packages for small businesses who need advertising services. A basic package includes only design services. A premium package includes design and promotion. The agency's goal is to sell at least 60 packages in total.

- The basic advertising package has a value of $1,000 and the premium package has a value of $2,500. The goal of the agency is to sell more than $60,000 worth of small-business advertising packages.

1. Write a **system of inequalities** to represent the constraints. Specify what each variable represents.

2. Use technology to graph the inequalities and sketch the solution regions. Include labels and scales for the axes.

NAME _____ DATE _____ PERIOD _____

3. Identify **a solution to the system**. Explain what the numbers mean in the situation.

Members of a high school math club are doing a scavenger hunt. Three items are hidden in the park, which is a rectangle that measures 50 meters by 20 meters.

- The clues are written as systems of inequalities. One system has no solutions.

- The locations of the items can be narrowed down by solving the systems. A coordinate plane can be used to describe the solutions.

Can you find the hidden items? Sketch a graph to show where each item could be hidden.

Clue 1: $y > 14$

            $x < 10$

Clue 2: $x + y < 20$

            $x > 6$

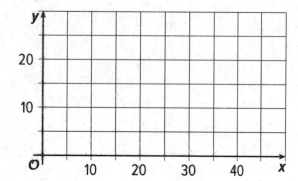

Clue 3: $y < -2x + 20$

            $y < -2x + 10$

Clue 4: $y \geq x + 10$

            $x > y$

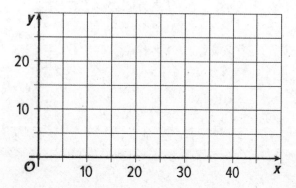

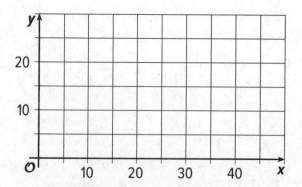

NAME _____ DATE _____ PERIOD _____

Are you ready for more?

Two non-negative numbers x and y satisfy $x + y \leq 1$.

1. Find a second inequality, also using x and y values greater than or equal to zero, to make a system of inequalities with exactly one solution.

2. Find as many ways to answer this question as you can.

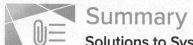

Summary

Solutions to Systems of Linear Inequalities in Two Variables

In this lesson, we used two linear inequalities in two variables to represent the constraints in a situation. Each pair of inequalities forms a **system of inequalities**.

A **solution to the system** is any (x, y) pair that makes both inequalities true, or any pair of values that simultaneously meet both constraints in the situation. The solution to the system is often best represented by a region on a graph.

Suppose there are two numbers, x and y, and there are two things we know about them:

- The value of one number is more than double the value of the other.

- The sum of the two numbers is less than 10.

We can represent these constraints with a system of inequalities.

$$\begin{cases} y > 2x \\ x + y < 10 \end{cases}$$

There are many possible pairs of numbers that meet the first constraint, for example: 1 and 3, or 4 and 9.

The same can be said about the second constraint, for example: 1 and 3, or 2.4 and 7.5.

The pair $x = 1$ and $y = 3$ meets both constraints, so it is a solution to the system.

The pair $x = 4$ and $y = 9$ meets the first constraint but not the second $(9 > 2(4)$ is a true statement, but $4 + 9 < 10$ is not true.)

Remember that graphing is a great way to show all the possible solutions to an inequality, so let's graph the solution region for each inequality.

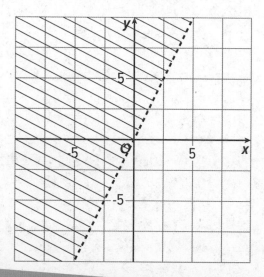
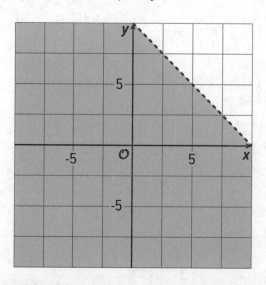

NAME _____ DATE _____ PERIOD _____

Because we are looking for a pair of numbers that meet both constraints or make both inequalities true at the same time, we want to find points that are in the solution regions of both graphs.

To do that, we can graph both inequalities on the same coordinate plane.

The solution set to the system of inequalities is represented by the region where the two graphs overlap.

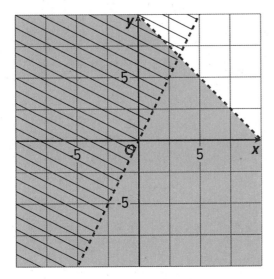

1. Two inequalities are graphed on the same coordinate plane.

   Which region represents the solution to the system of the two inequalities?

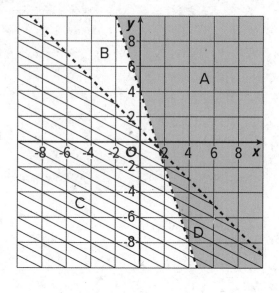

2. Select **all** the pairs of x and y that are solutions to the system of inequalities:

$$\begin{cases} y \le -2x + 6 \\ \quad x - y < 6 \end{cases}$$

   (A.) $x = 0, y = 0$

   (B.) $x = -5, y = -15$

   (C.) $x = 4, y = -2$

   (D.) $x = 3, y = 0$

   (E.) $x = 10, y = 0$

3. Jada has $200 to spend on flowers for a school celebration. She decides that the only flowers that she wants to buy are roses and carnations. Roses cost $1.45 each and carnations cost $0.65 each. Jada buys enough roses so that each of the 75 people attending the event can take home at least one rose.

   a. Write an inequality to represent the constraint that every person takes home at least one rose.

   b. Write an inequality to represent the cost constraint.

NAME _____ DATE _____ PERIOD _____

4. Here are the graphs of the equations
   $3x + y = 9$ and $3x - y = 9$ on the same
   coordinate plane.

   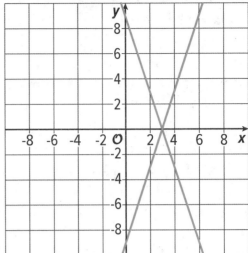

   a. Label each graph with the equation it
      represents.

   b. Identify the region that represents the
      solution set to $3x + y < 9$. Is the
      boundary line a part of the solution? Use
      a colored pencil or cross-hatching to
      shade the region.

   c. Identify the region that represents the
      solution set to $3x - y < 9$. Is the
      boundary line a part of the solution? Use
      a different colored pencil or cross-hatching to shade the region.

   d. Identify a point that is a solution to both $3x + y < 9$ and $3x - y < 9$.

5. Which coordinate pair is a solution to the inequality
   $4x - 2y < 22$? (Lesson 2-21)

   Ⓐ (4, -3)

   Ⓑ (4, 3)

   Ⓒ (8, -3)

   Ⓓ (8, 3)

6. Consider the linear equation $9x - 3y = 12$. **(Lesson 2-21)**

    a. The pair (3, 5) is a solution to the equation. Find another pair (x, y) that is a solution to the equation.

    b. Are (3, 5) and (2, -10) solutions to the inequality $9x - 3y \leq 12$? Explain how you know.

7. Elena is considering buying bracelets and necklaces as gifts for her friends. Bracelets cost $3, and necklaces cost $5. She can spend no more than $30 on the gifts. **(Lesson 2-22)**

    a. Write an inequality to represent the number of bracelets, b, and the number of necklaces n, she could buy while sticking to her budget.

NAME _____ DATE _____ PERIOD _____

b. Graph the solutions to the inequality on the coordinate plane.

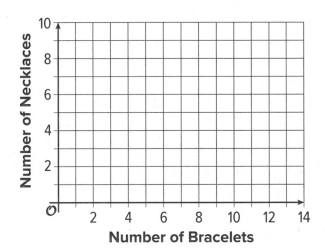

c. Explain how we could check if the boundary is included or excluded from the solution set.

8. In physical education class, Mai takes 10 free throws and 10 jump shots. She earns 1 point for each free throw she makes and 2 points for each jump shot she makes. The greatest number of points that she can earn is 30. (Lesson 2-23)

   a. Write an inequality to describe the constraints. Specify what each variable represents.

   b. Name one solution to the inequality and explain what it represents in that situation.

9. A rectangle with a width of w and a length of l has a perimeter greater than 100. (Lesson 2-23)

   Here is a graph that represents this situation.

   a. Write an inequality that represents this situation.

   b. Can the rectangle have width of 45 and a length of 10? Explain your reasoning.

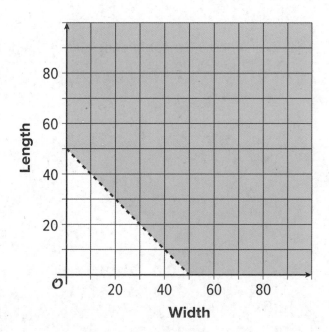

   c. Can the rectangle have a width of 30 and a length of 20? Explain your reasoning.

Lesson 2-25

Solving Problems with Systems of Linear Inequalities in Two Variables

NAME _____ DATE _____ PERIOD _____

Learning Goal Let's use systems of inequalities to solve some problems.

 ## Warm Up
25.1 Which One Doesn't Belong: Graphs of Solutions

Which one doesn't belong?

A

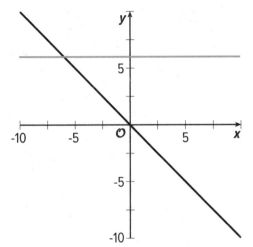

B

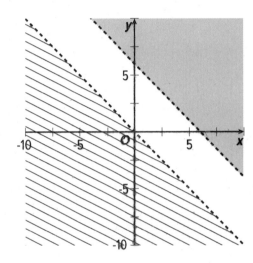

C

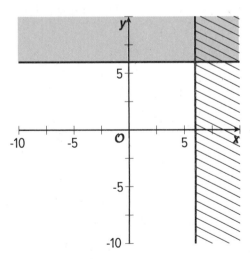

D

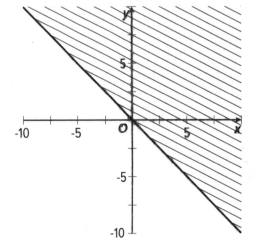

Activity

25.2 Focusing on the Details

Here are the graphs of the inequalities in this system:

$$\begin{cases} x < y \\ y \geq -2x - 6 \end{cases}$$

Decide whether each point is a solution to the system. Be prepared to explain how you know.

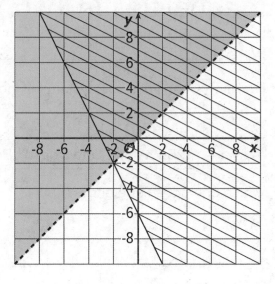

1. (3, -5)

2. 0, 5

3. (-6, 6)

4. (3, 3)

5. (-2, -2)

NAME _____ DATE _____ PERIOD _____

Are you ready for more?

Find a system of inequalities with this triangle as its set of solutions.

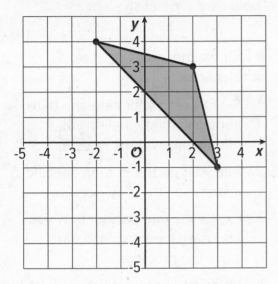

Activity

25.3 Info Gap: Terms of A Team

Your teacher will give you either a problem card or a data card. Do not show or read your card to your partner.

If your teacher gives you the data card:

1. Silently read the information on your card.

2. Ask your partner "What specific information do you need?" and wait for your partner to ask for information. Only give information that is on your card. (Do not figure out anything for your partner!)

3. Before telling your partner the information, ask "Why do you need to know (that piece of information)?"

4. Read the problem card, and solve the problem independently.

5. Share the data card, and discuss your reasoning.

If your teacher gives you the problem card:

1. Silently read your card and think about what information you need to answer the question.

2. Ask your partner for the specific information that you need.

3. Explain to your partner how you are using the information to solve the problem.

4. When you have enough information, share the problem card with your partner, and solve the problem independently.

5. Read the data card, and discuss your reasoning.

Pause here so your teacher can review your work. Ask your teacher for a new set of cards and repeat the activity, trading roles with your partner.

NAME _____ DATE _____ PERIOD _____

The blank coordinate planes are provided here in case they are useful.

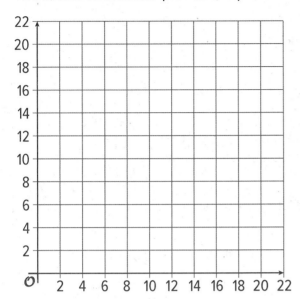
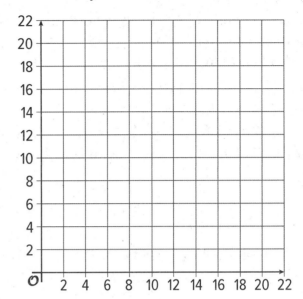

A family has at most $25 to spend on activities at Fun Zone. It costs $10 an hour to use the trampolines and $5 an hour to use the pool. The family can stay less than 4 hours.

What are some combinations of trampoline time and pool time that the family could choose given their constraints?

We could find some combinations by trial and error, but writing a system of inequalities and graphing the solution would allow us to see all the possible combinations.

Let t represent the time, in hours, on the trampolines and p represent the time, in hours, in the pool.

The constraints can be represented with the system of inequalities:

$$\begin{cases} 10t + 5p \leq 25 \\ t + p < 4 \end{cases}$$

Here are graphs of the inequalities in the system.

The solution set to the system is represented by the region where shaded parts of the two graphs overlap. Any point in that region is a pair of times that meet both the time and budget constraints.

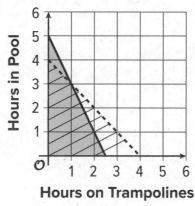

Hours in Pool (vertical axis)
Hours on Trampolines (horizontal axis)

The graphs give us a complete picture of the possible solutions.

- Can the family spend 1 hour on the trampolines and 3 hours in the pool?

  No. We can reason that it is because those times add up to 4 hours, and the family wants to spend *less than* 4 hours. But we can also see that the point (1, 3) lies on the dashed line of one graph, so it is not a solution.

- Can the family spend 2 hours on the trampolines and 1.5 hours in the pool?

  No. We know that these two times add up to less than 4 hours, but to find out the cost, we need to calculate $10(2) + 5(1.5)$, which is 27.5 and is more than the budget.

  It may be easier to know that this combination is not an option by noticing that the point (2, 1.5) is in the region with line shading, but not in the region with solid shading. This means it meets one constraint but not the other.

NAME _____ DATE _____ PERIOD _____

Practice
Solving Problems with Systems of Linear Inequalities in Two Variables

1. Jada has p pennies and n nickels that add up to more than 40 cents. She has fewer than 20 coins altogether.

   a. Write a system of inequalities that represents how many pennies and nickels that Jada could have.

   b. Is it possible that Jada has each of the following combinations of coins? If so, explain or show how you know. If not, state which constraint—the amount of money or the number of coins—it does not meet.

      i. 15 pennies and 5 nickels

      ii. 16 pennies and 2 nickels

      iii. 10 pennies and 8 nickels

2. A triathlon athlete swims at an average rate of 2.4 miles per hour, and bikes at an average rate of 16.1 miles per hour. At the end of one training session, she has swum and biked more than 20 miles in total.

   The inequality $2.4s + 16.1b > 20$ and the graph in part **b** represent the relationship between the hours of swimming, s, the hours of biking, h, and the total distance the athlete could have traveled in miles.

   Mai said, "I'm not sure the graph is right. For example, the point (10, 3) is in the shaded region, but it's not realistic for an athlete to swim for 10 hours and bike for 3 hours in a training session! I think triathlon athletes generally train for no more than 2 hours a day."

   a. Write an inequality to represent Mai's last statement.

b. Graph the solution set to your inequality.

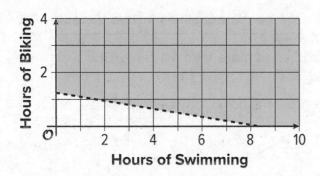

c. Determine a possible combination of swimming and biking times that meet both the distance and the time constraints in this situation.

3. Elena is considering buying bracelets and necklaces as gifts for her friends. Bracelets cost $3, and necklaces cost $5. She can spend no more than $30 on the gifts. Elena needs at least 7 gift items.

The graph below represents the inequality $3b + 5n \leq 30$, which describes the cost constraint in this situation.

Let b represent the number of bracelets and n the number of necklaces.

a. Write an inequality that represents the number of gift items that Elena needs.

b. On the same coordinate plane, graph the solution set to the inequality you wrote.

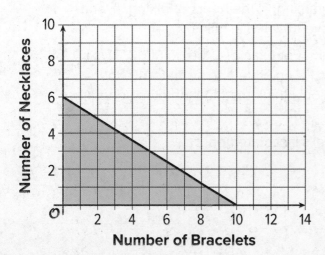

NAME _____ DATE _____ PERIOD _____

c. Use the graphs to find at least two possible combinations of bracelets and necklaces Elena could buy.

d. Explain how the graphs show that the combination of 2 bracelets and 5 necklaces meet one constraint in the situation but not the other constraint.

4. A gardener is buying some topsoil and compost to fill his garden. His budget is $70. Topsoil costs $1.89 per cubic foot, and compost costs $4.59 per cubic foot.

Select **all** statements or representations that correctly describe the gardener's constraints in this situation. Let t represent the cubic feet of topsoil and c the cubic feet of compost.

(Lesson 2-22)

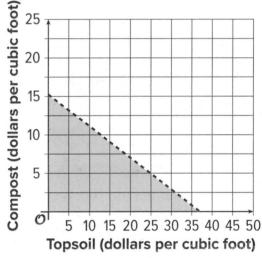

(A.) The combination of 7.5 cubic feet of topsoil and 12 cubic feet of compost is within the gardener's budget.

(B.) If the line represents the equation $1.89t + 4.59c = 70$, the top graph represents the solutions to the gardener's budget constraint.

(C.) $1.89t + 4.59c \geq 70$

(D.) The combination of 5 cubic feet of topsoil and 20 cubic feet of compost is within the gardener's budget.

(E.) $1.89t + 4.59c \leq 70$

(F.) If the line represents the equation $1.89t + 4.59c = 70$, the bottom graph represents the solutions to the gardener's budget constraint.

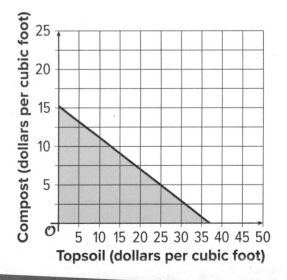

5. Priya writes the equation $y = -\frac{1}{2}x - 7$. Write an equation that has:
(Lesson 2-17)

    a. exactly one solution in common with Priya's equation

    b. no solutions in common with Priya's equation

    c. infinitely many solutions in common with Priya's equation, but looks different than hers

6. Two inequalities are graphed on the same coordinate plane.

Which region represents the solution to the system of the two inequalities? (Lesson 2-24)

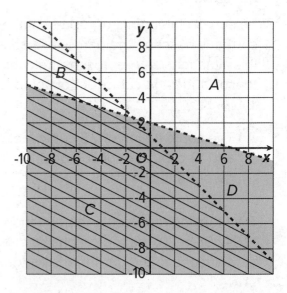

7. Here is a riddle: (Lesson 2-24)

    • The sum of two numbers is less than 10.

    • If we subtract the second number from the first, the difference is greater than 3.

Write a system of inequalities that represents this situation. Let *f* represent the first number and *s* represent the second number.

Lesson 2-26

Modeling with Systems of Inequalities in Two Variables

NAME _____ DATE _____ PERIOD _____

Learning Goal Let's create mathematical models using systems of inequalities.

 ## Warm Up
26.1 A Solution to Which Inequalities?

Is the ordered pair (5.43, 0) a solution to all, some, or none of these inequalities? Be prepared to explain your reasoning.

$x > 0$ $\qquad$ $y > 0$ $\qquad$ $x \geq 0$ $\qquad$ $y \geq 0$

Activity

26.2 Custom Trail Mix

Here is the nutrition information for some trail mix ingredients:

	Calories Per Gram (kcal)	Protein Per Gram (g)	Sugar Per Gram (g)	Fat Per Gram (g)	Fiber Per Gram (g)
Peanuts	5.36	0.21	0.04	0.46	0.07
Almonds	5.71	0.18	0.21	0.46	0.07
Raisins	3.00	0.03	0.60	0.00	0.05
Chocolate Pieces	4.76	0.05	0.67	0.19	0.02
Shredded Coconut	6.67	0.07	0.07	0.67	0.13
Sunflower Seeds	5.50	0.20	0.03	0.47	0.10
Dried Cherries	3.25	0.03	0.68	0.00	0.03
Walnuts	6.43	0.14	0.04	0.61	0.07

Tyler and Jada each designed their own custom trail mix using two of these ingredients. They wrote inequalities and created graphs to represent their constraints.

NAME _____ DATE _____ PERIOD _____

Tyler

- $x + y > 50$
- $4.76x + 6.67y \leq 400$
- $0.67x + 0.07y < 30$
- $x > 0$
- $y > 0$

Jada

- $w + z > 50$
- $0.14w + 0.03z > 4$
- $0.61w + 0z \leq 15$
- $w > 0$
- $z > 0$

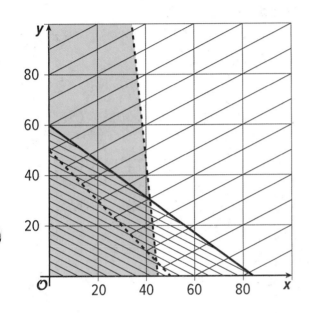

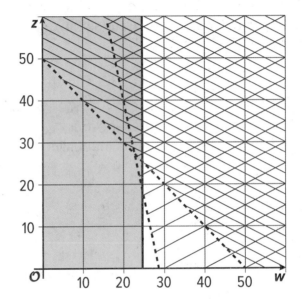

Use the inequalities and graphs to answer these questions about each student's trail mix. Be prepared to explain your reasoning.

1. Which two ingredients did they choose?

2. What do their variables represent?

3. What does each constraint mean?

4. Which graph represents which constraint?

5. Name one possible combination of ingredients for their trail mix.

It's time to design your own trail mix!

1. Choose two ingredients that you like to eat. (You can choose from the ingredients in the previous activity, or you can look up nutrition information for other ingredients.)

2. Think about the constraints for your trail mix. What do you want to be true about its calories, protein, sugar, fat, or fiber?

3. Write inequalities to represent your constraints. Then, graph the inequalities.

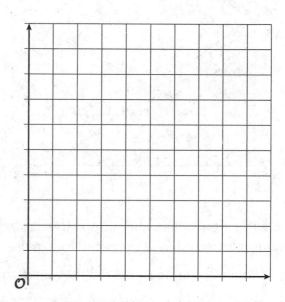

4. Is it possible to make trail mix that meets all your constraints using your ingredients? If not, make changes to your constraints or your ingredients and record them here.

5. Write a possible combination of ingredients for your trail mix.

Pause here so your teacher can review your work and give further instructions for displaying your work.

NAME _____ DATE _____ PERIOD _____

Practice
Modeling with Systems of Inequalities in Two Variables

The organizers of a conference needs to prepare at least 200 notepads for the event and have a budget of $160 for the notepads. A store sells notepads in packages of 24 and packages of 6. **(Lesson 2-25)**

1. This system of inequalities represent these constraints:

$$\begin{cases} 24x + 6y \geq 200 \\ 16x + 5.40y \leq 160 \end{cases}$$

   a. Explain what the second inequality in the system tells us about the situation.

   b. Here are incomplete graphs of the inequalities in the system, showing only the boundary lines of the solution regions.
   Which graph represents the boundary line of the second inequality?

   c. Complete the graphs to show the solution set to the system of inequalities.

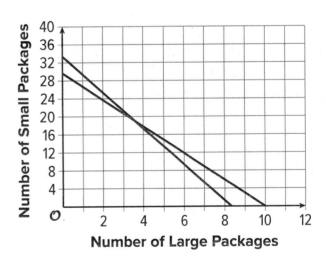

   d. Find a possible combination of large and small packages of notepads the organizer could order.

2. A certain stylist charges $15 for a haircut and $30 for hair coloring. A haircut takes on average 30 minutes, while coloring takes 2 hours. The stylist works up to 8 hours in a day, and she needs to make a minimum of $150 a day to pay for her expenses. **(Lesson 2-25)**

   a. Create a system of inequalities that describes the constraints in this situation. Be sure to specify what each variable represents.

   b. Graph the inequalities and show the solution set.

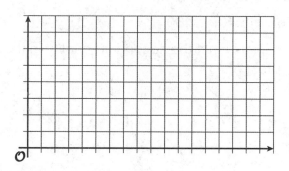

   c. Identify a point that represents a combination of haircuts and hair-coloring jobs that meets the stylist's requirements.

   d. Identify a point that is a solution to the system of inequalities but is not possible or not likely in the situation. Explain why this solution is impossible or unlikely.

NAME _____ DATE _____ PERIOD _____

3. Choose the graph that shows the solution to this system: $\begin{cases} y > 3x + 2 \\ -4x + 3y \le 12 \end{cases}$

(Lesson 2-24)

A.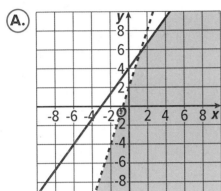

C.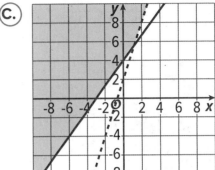

B.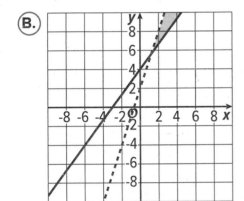

D.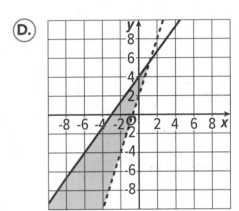

4. Match each inequality to the graph of its solution. (Lesson 2-23)

A.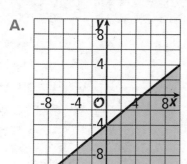

1. $2x - 5y \geq 20$

B.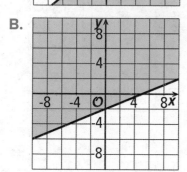

2. $5x + 2y \geq 20$

C.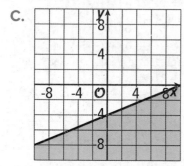

3. $4x - 10y \leq 20$

D.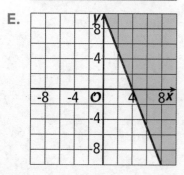

4. $4x - 5y \geq 20$

E.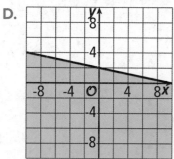

5. $2x + 10y \leq 20$

Learning Targets

Lesson	Learning Target(s)
2-1 Planning a Pizza Party	• I can explain the meaning of the term "constraints." • I can tell which quantities in a situation can vary and which ones cannot. • I can use letters and numbers to write expressions representing the quantities in a situation.
2-2 Writing Equations to Model Relationships (Part 1)	• I can tell which quantities in a situation can vary and which ones cannot. • I can use letters and numbers to write equations representing the relationships in a situation.
2-3 Writing Equations to Model Relationships (Part 2)	• I can use words and equations to describe the patterns I see in a table of values or in a set of calculations. • When given a description of a situation, I can use representations like diagrams and tables to help make sense of the situation and write equations for it.

(continued on the next page)

(continued from the previous page)

Lesson	Learning Target(s)
2-4 Equations and Their Solutions	• I can explain what it means for a value or pair of values to be a solution to an equation. • I can find solutions to equations by reasoning about a situation or by using algebra.
2-5 Equations and Their Graphs	• I can use graphing technology to graph linear equations and identify solutions to the equations. • I understand how the coordinates of the points on the graph of a linear equation are related to the equation. • When given the graph of a linear equation, I can explain the meaning of the points on the graph in terms of the situation it represents.
2-6 Equivalent Equations	• I can tell whether two expressions are equivalent and explain why or why not. • I know and can identify the moves that can be made to transform an equation into an equivalent one. • I understand what it means for two equations to be equivalent, and how equivalent equations can be used to describe the same situation in different ways.

(continued on the next page)

(continued from the previous page)

Lesson	Learning Target(s)
2-7 Explaining Steps for Rewriting Equations	• I can explain why some algebraic moves create equivalent equations but some do not. • I know how equivalent equations are related to the steps of solving equations. • I know what it means for an equation to have no solutions and can recognize such an equation.
2-8 Which Variable to Solve for? (Part 1)	• Given an equation, I can solve for a particular variable (like height, time, or length) when the equation would be more useful in that form. • I know the meaning of the phrase "to solve for a variable."
2-9 Which Variable to Solve for? (Part 2)	• I can write an equation to describe a situation that involves multiple quantities whose values are not known, and then solve the equation for a particular variable. • I know how solving for a variable can be used to quickly calculate the values of that variable.

(continued on the next page)

(continued from the previous page)

Lesson	Learning Target(s)
2-10 Connecting Equations to Graphs (Part 1)	• I can describe the connections between an equation of the form $ax + by = c$, the features of its graph, and the rate of change in the situation. • I can graph a linear equation of the form $ax + by = c$. • I understand that rewriting the equation for a line in different forms can make it easier to find certain kinds of information about the relationship and about the graph.
2-11 Connecting Equations to Graphs (Part 2)	• I can find the slope and vertical intercept of a line with equation $ax + by = c$. • I can take an equation of the form $ax + by = c$ and rearrange it into the equivalent form $y = mx + b$. • I can use a variety of strategies to find the slope and vertical intercept of the graph of a linear equation given in different forms.
2-12 Writing and Graphing Systems of Linear Equations	• I can explain what we mean by "the solution to a system of linear equations" and can explain how the solution is represented graphically. • I can explain what we mean when we refer to two equations as a system of equations. • I can use tables and graphs to solve systems of equations.

(continued on the next page)

(continued from the previous page)

Lesson	Learning Target(s)
2-13 Solving Systems by Substitution	• I can solve systems of equations by substituting a variable or an expression. • I know more than one way to perform substitution and can decide which way or what to substitute based on how the given equations are written.
2-14 Solving Systems by Elimination (Part 1)	• I can solve systems of equations by adding or subtracting them to eliminate a variable. • I know that adding or subtracting equations in a system creates a new equation, where one of the solutions to this equation is the solution to the system.
2-15 Solving Systems by Elimination (Part 2)	• I can explain why adding or subtracting two equations that share a solution results in a new equation that also shares the same solution.
2-16 Solving Systems by Elimination (Part 3)	• I can solve systems of equations by multiplying each side of one or both equations by a factor, then adding or subtracting the equations to eliminate a variable. • I understand that multiplying each side of an equation by a factor creates an equivalent equation whose graph and solutions are the same as that of the original equation.

(continued on the next page)

Lesson	Learning Target(s)
2-17 Systems of Linear Equations and Their Solutions	• I can tell how many solutions a system has by graphing the equations or by analyzing the parts of the equations and considering how they affect the features of the graphs. • I know the possibilities for the number of solutions a system of equations could have.
2-18 Representing Situations with Inequalities	• I can write inequalities that represent the constraints in a situation.
2-19 Solutions to Inequalities in One Variable	• I can graph the solution to an inequality in one variable. • I can solve one-variable inequalities and interpret the solutions in terms of the situation. • I understand that the solution to an inequality is a range of values (such as $x > 7$) that make the inequality true.
2-20 Writing and Solving Inequalities in One Variable	• I can analyze the structure of an inequality in one variable to help determine if the solution is greater or less than the solution to the related equation. • I can write and solve inequalities to answer questions about a situation.

(continued on the next page)

(continued from the previous page)

Lesson	Learning Target(s)
2-21 Graphing Linear Inequalities in Two Variables (Part 1)	• Given a two-variable inequality and the graph of the related equation, I can determine which side of the line the solutions to the inequality will fall. • I can describe the graph that represents the solutions to a linear inequality in two variables.
2-22 Graphing Linear Inequalities in Two Variables (Part 2)	• Given a two-variable inequality that represents a situation, I can interpret points in the coordinate plane and decide if they are solutions to the inequality. • I can find the solutions to a two-variable inequality by using the graph of a related two-variable equation. • I can write inequalities to describe the constraints in a situation.
2-23 Solving Problems with Inequalities in Two Variables	• I can use graphing technology to find the solution to a two-variable inequality. • When given inequalities, graphs, and descriptions that represent the constraints in a situation, I can connect the different representations and interpret them in terms of the situation.

(continued on the next page)

Lesson	Learning Target(s)
2-24 Solutions to Systems of Linear Inequalities in Two Variables	• I can write a system of inequalities to describe a situation, find the solution by graphing, and interpret points in the solution. • I know what is meant by "the solutions to a system of inequalities" and can describe the graphs that represent the solutions. • When given descriptions and graphs that represent two different constraints, I can find values that satisfy each constraint individually, and values that satisfy both constraints at once.
2-25 Solving Problems with Systems of Linear Inequalities in Two Variables	• I can explain how to tell if a point on the boundary of the graph of the solutions to a system of inequalities is a solution or not.
2-26 Modeling with Systems of Inequalities in Two Variables	• I can interpret inequalities and graphs in a mathematical model. • I know how to choose variables, specify the constraints, and write inequalities to create a mathematical model.

Notes

Two-variable Statistics

sauce7/123RF

Scientists collect data on bee populations. They can fit a model to their data to help them predict future needs. You will learn more about fitting models to data in this unit.

Topics
- Two-way Tables
- Scatterplots
- Correlation Coefficients
- Estimating Lengths

Unit 3

Two-variable Statistics

Lesson 3-1

Two-way Tables

NAME _____ DATE _____ PERIOD _____

Learning Goal Let's look at categorical data.

Warm Up

1.1 Utensils and Paper Preferences

Several students are surveyed about whether they prefer writing with a pen or a pencil and they are also asked whether they prefer lined paper or unlined paper. Some of the results are:

- The survey included 100 different students.

- 40 students said they prefer using pen more than pencil.

- 45 students said they prefer using unlined paper more than lined paper.

- 10 students said they prefer lined paper and pen.

- 45 students said they prefer pencil and lined paper.

For each part, explain or show your reasoning.

1. How many students prefer using pencil more than pen?

2. How many students prefer using pen and unlined paper?

3. How many students prefer using pencil and unlined paper?

Activity

1.2 Fruit Fly Mutations

A scientist is trying to determine the role of specific genes by looking at traits of fruit flies. The offspring of two fruit flies are examined to determine the color of their eyes and whether they have curled wings or standard wings. Eighty offspring are randomly selected, and the results are recorded in the **two-way table**.

	Curled Wings	Standard Wings
Red Eyes	17	45
White Eyes	5	13

1. Describe what the 17 in the table represents.

2. How many selected fly offspring had white eyes? Explain or show your reasoning.

3. How many selected fly offspring had standard wings? Explain or show your reasoning.

NAME _____ DATE _____ PERIOD _____

Are you ready for more?

1. Write 2 of your own survey questions that produce data which can be represented in a two-way table.

2. Give the survey to 20 or more students and record the results in a two-way table.

3. What questions can you answer with the information you found from your survey?

4. What does that tell you about the population of students who took your survey?

Activity

1.3 Info Gap: Running to the Dentist

Your teacher will give you either a problem card or a data card. Do not show or read your card to your partner.

If your teacher gives you the data card:

1. Silently read the information on your card.

2. Ask your partner "What specific information do you need?" and wait for your partner to ask for information. Only give information that is on your card. (Do not figure out anything for your partner!)

3. Before telling your partner the information, ask "Why do you need to know (that piece of information)?"

4. Read the problem card, and solve the problem independently.

5. Share the data card, and discuss your reasoning.

NAME _____ DATE _____ PERIOD _____

If your teacher gives you the problem card:

1. Silently read your card and think about what information you need to answer the question.

2. Ask your partner for the specific information that you need.

3. Explain to your partner how you are using the information to solve the problem.

4. When you have enough information, share the problem card with your partner, and solve the problem independently.

5. Read the data card, and discuss your reasoning.

Pause here so your teacher can review your work. Ask your teacher for a new set of cards and repeat the activity, trading roles with your partner.

In statistics, a **variable** is a characteristic that can take on different values. A **categorical variable** is a variable that takes on values which can be divided into groups or categories. Data from two categorical variables about a single subject can be organized using a **two-way table**.

For example, this two-way table shows the results from 170 responses to a survey asking people their age group and whether they have a cell phone or not.

	Has a Cell Phone	Does Not Have a Cell Phone
10–12 years old	25	35
13–15 years old	38	12
16–18 years old	52	8

The 38 in the table means that 38 of the 170 people surveyed are in both the 13–15 years old age category and have a cell phone. The two-way table also shows that 55 of the people surveyed do not have cell phones, since $35 + 12 + 8 = 55$.

The categories for a single variable should not overlap (a person cannot be 10–12 years old *and* 13–15 years old at the same time); each individual is included in only one of the cells in the table rather than in multiple places.

Glossary

categorical variable
two-way table
variable (statistics)

NAME _____ DATE _____ PERIOD _____

Practice
Two-way Tables

1. This two-way table shows the results of asking students if they prefer to have gym class in the morning or the afternoon.

	Morning	Afternoon	Total
Grade 6	15	8	23
Grade 8	18	21	39
Grade 10	12	26	38
Total	45	55	100

   a. How many students participated in the survey?

   b. How many students in grade 8 prefer to have gym in the morning?

   c. How many grade 10 students participated in the survey?

   d. How many students prefer to have gym in the afternoon?

2. A random sample of adults are asked about their preferences for a first dinner date with someone. Complete the two-way table so that it has the characteristics listed.

   • 122 people responded to the survey.

   • 50 of the people who said they order dessert said they also prefer to split the check.

   • 68 people prefer splitting the check.

   • 56 people prefer to skip dessert rather than ordering one.

	Order Dessert	No Dessert
Split the Check		
One Person Pays		

3. Students in the 7th, 8th, and 9th grade were asked whether they prefer to write in pen or pencil.

- 40 students prefer to write in pen.

- 60 students prefer to write in pencil.

Create values that could represent the number of students in the 7th, 8th, and 9th grade that responded to the survey.

4. A recent study observed the number of bike riders ages 0 to 20 that wear helmets. The results are represented in the table.

	Wear Helmet	Did Not Wear Helmet
0-5 years	21	4
6-10 years	34	14
11-15 years	14	18
16-20 years	3	12

Make an observation based on a data value that is not in the table. Explain your reasoning.

Lesson 3-2

Relative Frequency Tables

NAME _____ DATE _____ PERIOD _____

Learning Goal Let's find relative frequencies of categorical data.

 Warm Up

2.1 Notice and Wonder: Teacher Degrees

Several adults in a school building were asked about their highest degree completed and whether they were a teacher.

What do you notice? What do you wonder?

	Teacher	Not a Teacher
Associate Degree	4%	16%
Bachelor's Degree	52%	64%
Master's Degree or Higher	44%	20%

200 people were asked if they prefer dogs or cats, and whether they live in a rural or urban setting.

The actual values collected from the survey are in the first table.

	Urban	Rural	Total
Cat	54	42	96
Dog	80	24	104
Total	134	66	200

The next table shows what percentage of the 200 total people included are represented by each combination of categories. The segmented bar graph represents the same information graphically.

	Urban	Rural
Cat	27%	21%
Dog	40%	12%

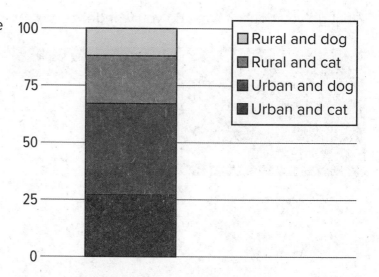

The next table shows the percentage of each column that had a certain pet preference in a column relative frequency table. The segmented bar graph represents the same information graphically.

	Urban	Rural
Cat	40%	64%
Dog	60%	36%

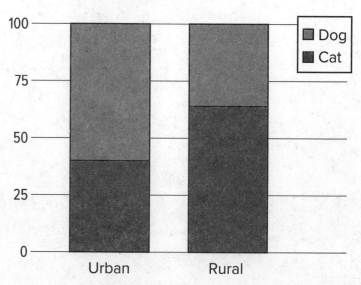

NAME _____ DATE _____ PERIOD _____

The last table shows the percentage of each row that live in a certain area in a row relative frequency table. The segmented bar graph represents the same information graphically.

	Urban	Rural
Cat	56%	44%
Dog	77%	23%

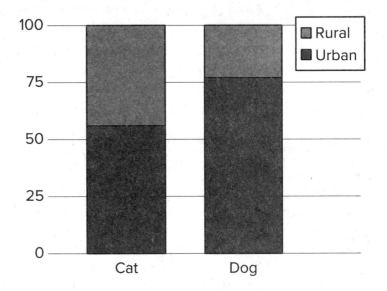

1. For each relative frequency table, select a percentage and explain how numbers from the original table were used to get the percentage.

2. What percentage of those surveyed live in an urban area and prefer dogs?

3. Among the people surveyed who prefer dogs, what percentage of them live in an urban setting?

4. What percentage of people surveyed who live in an urban setting prefer dogs?

5. How many of the people responded that they prefer dogs and live in an urban setting?

6. Among the people surveyed, are there more people who prefer dogs or cats?

7. Your pet food company has access to a billboard in a rural setting. Would you recommend advertising dog food or cat food on this billboard? Which table did you use to make this decision? Explain your reasoning.

In an experiment to test the effectiveness of vitamin C on the length of colds, two groups of people with colds are given a pill to take once a day. The pill for one of the groups contains 1,000 mg of vitamin C, while the other group takes a placebo pill. The researchers record the results in a table.

	Group A	Group B
Cold Lasts Less Than a Week	16	27
Cold Lasts a Week or More	17	53

1. First, the researchers want to know what percentage (to the nearest whole percent) of people are in each combination of categories. Fourteen percent of all the participants had a cold that lasted less than a week and were in group A. What percentage of all the participants had a cold that lasted less than a week and were in group B? Complete the rest of the relative frequency table with the corresponding percentages.

	Group A		Group B
Cold Lasts Less Than a Week	$14\%\left(\dfrac{16}{16 + 27 + 17 + 53} \approx 0.14\right)$		
Cold Lasts a Week or More			

2. Next, the researchers notice that, among participants who had colds that lasted less than a week, 37% were in group A. Among participants who had colds that lasted a week or more, what percentage were in group B? Complete the table with the corresponding percentages.

	Group A	Group B
Cold Lasts Less Than a Week	$37\%\left(\dfrac{16}{16 + 27} \approx 0.37\right)$	
Cold Lasts a Week or More		

NAME _____ DATE _____ PERIOD _____

3. Finally, the researchers notice that, among the participants in group A, 48% had colds that lasted less than one week. Among the participants in group B, how many had colds that lasted a week or more? Complete the table with the corresponding percentages.

	Group A	Group B
Cold Lasts Less Than a Week	$48\%\left(\dfrac{16}{16+17} \approx 0.48\right)$	
Cold Lasts a Week or More		

4. To understand the results, the researchers want to know: Among people whose colds lasted less than a week, what percentage are in group B? Explain your reasoning.

5. If the researchers believe that vitamin C has a small effect on the length of a cold, which group most likely got the pills containing vitamin C? Explain your reasoning.

A teacher surveyed a group of 25 8th graders and a group of 20 12th graders who indicated they knew a computer programming language. Python and Scratch are programming languages. The results from the 8th-grade survey are displayed in the two-way table.

	I Know Python the Best	I Know Scratch the Best	I Know a Different Programming Language the Best
I Have Been Taught a Programming Language at School	8	6	1
I Have Not Been Taught a Programming Language at School	1	7	2

The results from the 12th-grade survey are displayed in the two-way table.

	I Know Python the Best	I Know Scratch the Best	I Know a Different Programming Language the Best
I Have Been Taught a Programming Language at School	25%	35%	0%
I Have Not Been Taught a Programming Language at School	30%	5%	5%

1. Which programming language did a majority of 8th graders surveyed know best?

2. Which programming language did a majority of 12th graders surveyed know best?

NAME _____ DATE _____ PERIOD _____

3. How many of 12th graders surveyed reported that they were taught a programming language at school?

4. What percentage of 8th graders surveyed reported that they knew Python the best and were not taught a programming language at school?

5. Why is it difficult to decide if 12th graders or 8th graders use Python more with the way the information is given in the tables?

Summary
Relative Frequency Tables

Converting two-way tables to *relative frequency tables* can help reveal patterns in paired categorical variables. **Relative frequency tables** are created by dividing the value in each cell in a two-way table by the total number of responses in the entire table, or the total responses in a row or a column. Depending on what patterns are important, different types of relative frequency tables are used. To examine how individual combinations of the categorical variables relate to the whole group, divide each value in a two-way table by the total number of responses in the entire table to find the relative frequency.

For example, this two-way table displays the condition of a certain textbook and its price for 120 of the books at a college bookstore.

	$10 or Less	More Than $10 but Less Than $30	$30 or More
New	3	9	27
Used	33	36	12

A two-way relative frequency table is created by dividing each number in the two-way table by 120, because there are 120 values (3 + 9 + 27 + 33 + 36 + 12) in this data set. The resulting two-way relative frequency table can be represented using fractions or decimals.

	$10 or Less	More Than $10 but Less Than $30	$30 or More
New	0.025	0.075	0.225
Used	0.275	0.300	0.100

This two-way relative frequency table allows you to see what proportion of the total is represented by each number in the two-way table. The number 33 in the original two-way table represents the number of used books that also sell for $10 or less, which is 27.5% of all the books in the data set. Using this two-way relative frequency table, you can see that there are very few (2.5%) new books that are also inexpensive and that 10% of the books in the bookstore are both expensive and in used condition.

NAME _____ DATE _____ PERIOD _____

In other situations, it makes sense to examine row or column proportions in a relative frequency table. For example, to convert the original two-way table to a column relative frequency table using column proportions, divide each value by the sum of the column.

	$10 or Less	More Than $10 but Less Than $30	$30 or More
New	0.08	0.2	0.692
Used	0.917	0.8	0.308

This shows that about 91.7% $\left(\dfrac{33}{3 + 33} \approx 0.917\right)$ of the books that are sold for $10 or less are in used condition. Notice that each column of this column relative frequency table reveals the proportions of the books in each price category that are in each condition and the relative frequencies in each column sum to 1. In particular, this shows that most of the inexpensive and moderately priced books are used, and most of the expensive books are new.

Glossary

categorical variable

relative frequency table

variable (statistics)

1. A teacher asks their students whether they studied for a quiz, then scores the quiz. A relative frequency table displays some of the information they collected.

	Studied	Did Not Study
Passed Quiz	86%	14%
Failed Quiz	46%	54%

What does the 86% represent?

2. The accessory choices of 143 people are recorded in the table.

	Wearing a Watch	No Watch
Wearing a Belt	62	32
No Belt	29	20

Create a relative frequency table that could be used to show the percentages of belt wearers who wear a watch or not, as well as the percentages of people without belts who wear a watch or not.

	Wearing a Watch	No Watch
Wearing a Belt		
No Belt		

NAME _____ DATE _____ PERIOD _____

3. Scientists give two different treatments to people who have the flu and determine if their health improves. The results for the test are in the two-way table.

	Treatment 1	Treatment 2
Improved Health	23	25
No Improvement	17	35

a. What percentage of people receiving treatment 1 had improved health?

b. What percentage of people receiving treatment 2 had improved health?

4. A group of people are surveyed about whether they have any brothers or sisters or are an only child, and whether they have any pets.

	Have Sibling	Only Child
Have Pets	82	105
No Pets	141	

Which value could go in the blank cell so that the percentage of only children that have no pets is 37.5%?

(A.) 63

(B.) 82

(C.) 175

(D.) 205

5. Many adults are selected at random to respond to a survey about their favorite season and whether they have allergies or not. The two-way table summarizes the results from the survey. (Lesson 3-1)

	Allergies	No Allergies
Winter	43	21
Spring	12	13
Summer	35	33
Fall	33	35

a. Which season is the least popular in this group?

b. How many more people have allergies than people without allergies in this group?

c. How many people were surveyed in this group?

6. A random sample of people are asked about their preferences regarding home decoration and their interest in fashion. Complete the two-way table so it has the characteristics listed. (Lesson 3-1)

- 150 people responded to the survey.

- 70% of the responders do not pay attention to fashion.

- 33 of the responders who prefer neutral decorations also do not pay attention to fashion.

- 20 of the responders who pay attention to fashion also prefer colorful decorations in their home.

	Prefer Colorful Decorations	Prefer Neutral Decorations
Pay Attention to Fashion		
Do Not Pay Attention to Fashion		

Lesson 3-3

Associations in Categorical Data

NAME _____ DATE _____ PERIOD _____

Learning Goal Let's look for relationships between categorical variables.

 Warm Up
3.1 **Cake or Pie**

The table displays the dessert preference and dominant hand (left- or right-handed) for a sample of 300 people.

	Prefers Cake	Prefers Pie	Total
Left-Handed	10	20	30
Right-Handed	90	180	270
Total	100	200	300

For each of the calculations, describe the interpretation of the percentage in terms of the situation.

1. 10% from $\frac{10}{100} = 0.1$

2. 67% from $\frac{180}{270} \approx 0.67$

3. 30% from $\frac{90}{300} = 0.3$

3.2 Associations in Categorical Data

1. The two-way table displays data about 55 different locations. Scientists have a list of possible chemicals that may influence the health of the coral. They first look at how nitrate concentration might be related to coral health. The table displays the health of the coral (healthy or unhealthy) and nitrate concentration (low or high).

	Low Nitrate Concentration	High Nitrate Concentration	Total
Healthy	20	5	25
Unhealthy	8	22	30
Total	28	27	55

a. Complete the two-way relative frequency table for the data in the two-way table in which the relative frequencies are based on the total for each column.

	Low Nitrate Concentration	High Nitrate Concentration
Healthy		
Unhealthy		
Total	100%	100%

b. When there is a low nitrate concentration, which had a higher relative frequency, healthy or unhealthy coral?

c. When there is a high nitrate concentration, is there a higher relative frequency of healthy or unhealthy coral?

NAME _____ DATE _____ PERIOD _____

d. Based on this data, is there a possible **association** between coral health and the level of nitrate concentration? Explain your reasoning.

e. The scientists next look at how silicon dioxide concentration might be related to coral health. The relative frequencies based on the total for each column are shown in the table. Based on this data, is there a possible association between coral health and the level of silicon dioxide concentration? Explain your reasoning.

	Low Silicon Dioxide Concentration	High Silicon Dioxide Concentration
Healthy	44%	46%
Unhealthy	56%	54%
Total	100%	100%

2. Jada surveyed 300 people from various age groups about their shoe preference. The two-way table summarizes the results of the survey.

	Prefers Sneakers Without Laces	Prefers Sneakers With Laces	Prefers Shoes that are Not Sneakers	Total
4–10 Years Old	21	12	3	36
11–17 Years Old	21	48	39	108
18–24 Years Old	15	54	87	156
Total	57	114	129	300

Jada concludes that there is a possible association between age and shoe preference. Is Jada's conclusion reasonable? Explain your reasoning.

3. The two-way table summarizes data on writing utensil preference and the dominant hand for a sample of 100 people.

	Left-Handed	Right-Handed	Total
Prefers Pen	7	82	89
Prefers Pencil	6	5	11
Total	13	87	100

Is there a possible association between dominant hand and writing utensil preference? Explain your reasoning.

NAME _____ DATE _____ PERIOD _____

Are you ready for more?

The incomplete two-way table displays the results of a survey about the type of sports medicine treatment and recovery time for 33 student athletes who visited the athletic trainer.

	Returned to Playing in Less Than 2 Days	Returned to Playing in 2 or More Days
Treated With Ice	8	4
Treated With Heat		

1. What 2 values could you use to complete the two-way table to show that there is an association between returning to playing in less than 2 days and the treatment (ice or heat)? Explain your reasoning.

2. What 2 values could you use to complete the two-way table to show that there is no association between returning to playing in less than 2 days and the treatment (ice or heat)? Explain your reasoning.

3. Which 2 values were easier to choose, the 2 values showing an association, or the 2 values showing no association? Explain your reasoning.

Activity
3.3 Associating Your Own Variables

1. Work with your group to identify a pair of categorical variables you think might be associated and another pair you think would not be associated.

2. Imagine your group collected data for each pair of categorical variables. Create a two-way table that could represent each set of data. Invent some data with 100 total values to complete each table. Remember that one table shows a possible association, and the other table shows no association.

3. Explain or show why there appears to be an association for the first pair of variables and why there appears to be no association for the other pair of variables.

4. Prepare a display of your work to share.

NAME _____ DATE _____ PERIOD _____

Summary
Associations in Categorical Data

An **association** between two variables means that the two variables are statistically related to each other. For example, we might expect that ice cream sales would be higher on sunny days than on snowy days. If sales were higher on sunny days than on snowy days, then we would say that there is a possible association between ice cream sales and whether or not it is sunny or snowing. When dealing with categorical variables, row or column relative frequency tables are often used to look for associations in the data.

Here is a two-way table displaying ice cream sales and weather conditions for 41 days for a particular creamery.

	Sunny Day	Snowy Day	Total
Sold Fewer Than 50 Cones	8	7	15
Sold 50 Cones or More	22	4	26
Total	30	11	41

Noticing a pattern in the raw data can be difficult, especially when the row or column totals are not the same for different categories, so the data should be converted into a row or column relative frequency table to better compare the categories. For the creamery, notice that the number of days with low sales is about the same for the two weather types, which contradicts our intuition. In this case, it makes sense to look at the percentage of days that sold well under each weather condition separately. That is, consider the column relative frequencies.

	Sunny Day	Snowy Day
Sold Fewer Than 50 Cones	27%	64%
Sold 50 Cones or More	73%	36%
Total	100%	100%

From the column relative frequency table, it is clear that most of the sunny days resulted in sales of at least 50 cones (73%), while most of the snowy days resulted in fewer than 50 cones sold (64%). Because these percentages are quite different, this suggests there is an association between the weather condition and the number of cone sales. A bakery might wonder if the weather conditions impact their muffin sales as well.

	Sunny Day	Snowy Day
Sold Fewer Than 50 Muffins	32%	35%
Sold 50 Muffins or More	68%	65%
Total	100%	100%

For the bakery, it seems there is not an association between weather conditions and muffin sales, since the percentage of days with low sales are very similar under the different weather conditions, and the percentages are also close on days when they sold many muffins.

Using row or column relative frequency tables helps organize data so that columns (or rows) can be easily compared between different categories for a variable. This comparison can be accomplished using a two-way table or a two-way relative frequency table, but it requires you to account for the differences in the number of data values in a given category.

Glossary

association

NAME _____ DATE _____ PERIOD _____

Practice
Associations in Categorical Data

1. Which value would best fit in the missing cell to suggest there is no evidence of an association between the variables?

	Digital Watch	Analog Watch
Displays the Date	54	27
No Date Display	18	

(A.) 9

(B.) 18

(C.) 27

(D.) 54

2. The relative frequency table shows the percentage of each type of art (painting or sculpture) in a museum that would classify in the different styles (modern or classical). Based on these percentages, is there evidence to suggest an association between the variables? Explain your reasoning.

	Modern	Classical
Paintings	41%	59%
Sculptures	38%	62%

3. An automobile dealership keeps track of the number of cars and trucks they have for sale, as well as whether they are new or used. Based on the data, does there appear to be an association between the type of automobile and whether it is new or used? Explain your reasoning.

	Car	Truck
New	812	233
Used	422	51

4. A survey is given to 1,432 people about whether they take daily supplemental vitamins and whether they eat breakfast on a regular basis. The results are shown in the table. (Lesson 3-2)

	Take Daily Vitamins	No Daily Vitamins
Eat Breakfast	384	476
No Breakfast	268	304

Create a relative frequency table that shows the percentage of the entire group that is in each cell.

5. Several college students are surveyed about their college location and preferred locations for a spring break trip. (Lesson 3-2)

	College Near the Coast	College Away from the Coast
Beach Break	37	54
Ski Break	24	36

a. What percentage of people who prefer to spend spring break at the beach go to a college away from the coast?

b. What percentage of people who prefer to spend spring break skiing go to a college away from the coast?

6. A group of people are surveyed about whether they prefer to bike or run to exercise, and whether they prefer summer or winter weather. The results are in the table. (Lesson 3-2)

	Bike	Run
Summer	108	212
Winter		98

What value could go in the blank cell so that the percentage of people who like to bike and also prefer winter weather is 10%?

Lesson 3-4

Linear Models

NAME _____ DATE _____ PERIOD _____

Learning Goal Let's explore relationships between two numerical variables.

 Warm Up
4.1 Notice and Wonder: Crowd Noise

What do you notice? What do you wonder?

$y = 1.5x + 22.7$

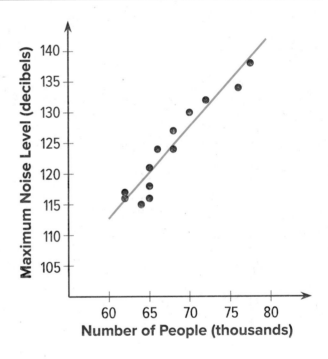

Activity

4.2 Orange You Glad We're Boxing Fruit?

1. Your teacher will show you a video. Record the weight for the number of oranges in the box.

Number of Oranges	Weight in Kilograms
3	
4	
5	
6	
7	
8	
9	
10	

2. Create a scatter plot of the data.

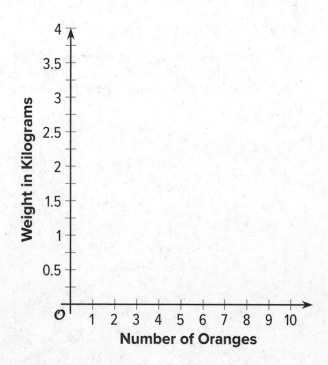

NAME _____ DATE _____ PERIOD _____

3. Draw a line through the data that fits the data well.

4. Estimate a value for the slope of the line that you drew. What does the value of the slope represent?

5. Estimate the weight of a box containing 11 oranges. Will this estimate be close to the actual value? Explain your reasoning.

6. Estimate the weight of a box containing 50 oranges. Will this estimate be close to the actual value? Explain your reasoning.

7. Estimate the coordinates for the vertical intercept of the line you drew. What might the *y*-coordinate for this point represent?

8. Which point(s) are best fit by your linear model? How did you decide?

9. Which point(s) are fit the least well by your linear model? How did you decide?

Activity

4.3 Food Markup

The scatter plot shows the sale price of several food items, y, and the cost of the ingredients used to produce those items, x, as well as a line that models the data. The line is also represented by the equation $y = 3.48x + 0.76$.

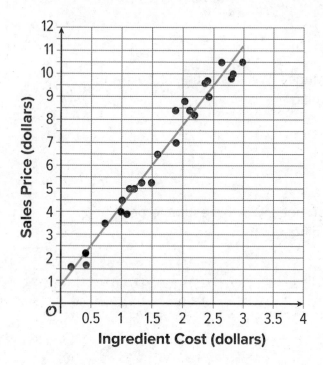

1. What is the predicted sale price of an item that has ingredients that cost $1.50? Explain or show your reasoning.

2. What is the predicted ingredient cost of an item that has a sale price of $7? Explain or show your reasoning.

3. What is the slope of the linear model? What does that mean in this situation?

4. What is the y-intercept of the linear model? What does this mean in this situation? Does this make sense?

NAME _____ DATE _____ PERIOD _____

Activity
4.4 The Slope is the Thing

1. Here are several scatter plots.

A. $y = -9.25x + 400$

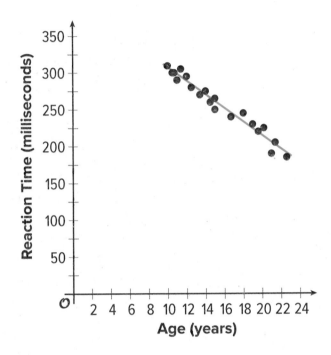

B. $y = 0.44x + 0.04$

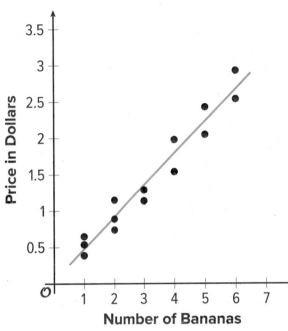

C. $y = 4x + 87$

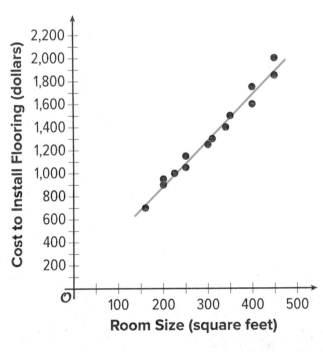

D. $y = -2.4x + 25.0$

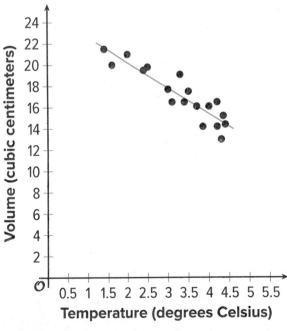

a. Using the horizontal axis for x and the vertical axis for y, interpret the slope of each linear model in the situations shown in the scatter plots.

b. If the linear relationship continues to hold for each of these situations, interpret the y-intercept of each linear model in the situations provided.

NAME _____ DATE _____ PERIOD _____

Are you ready for more?

Clare, Diego, and Elena collect data on the mass and fuel economy of cars at different dealerships. Clare finds the line of best fit for data she collected for 12 used cars at a used car dealership. The line of best fit is $y = \frac{-9}{1000}x + 34.3$ where x is the car's mass, in kilograms, and y is the fuel economy, in miles per gallon.

Diego made a scatter plot for the data he collected for 10 new cars at a different dealership.

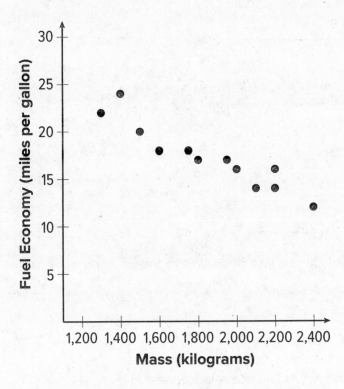

Elena made a table for data she collected on 11 hybrid cars at another dealership.

Mass (kilograms)	Fuel Economy (miles per gallon)
1,100	38
1,200	39
1,250	35
1,300	36
1,400	31
1,600	27
1,650	28
1,700	26
1,800	28
2,000	24
2,050	22

1. Interpret the slope and *y*-intercept of Clare's line of best fit in this situation.

2. Diego looks at the data for new cars and used cars. He claims that the fuel economy of new cars decreases as the mass increases. He also claims that the fuel economy of used cars increases as the mass increases. Do you agree with Diego's claims? Explain your reasoning.

3. Elena looks at the data for hybrid cars and correctly claims that the fuel economy decreases as the mass increases. How could Elena compare the decrease in fuel economy as mass increases for hybrid cars to the decrease in fuel economy as mass increases for new cars? Explain your reasoning.

NAME _____ DATE _____ PERIOD _____

Summary
Linear Models

While working in math class, it can be easy to forget that reality is somewhat messy. Not all oranges weigh exactly the same amount, beans have different lengths, and even the same person running a race multiple times will probably have different finishing times. We can approximate these messy situations with more precise mathematical tools to better understand what is happening. We can also predict or estimate additional results as long as we continue to keep in mind that reality will vary a little bit from what our mathematical model predicts.

For example, the data in this scatter plot represents the price of a package of broccoli and its weight. The data can be modeled by a line given by the equation $y = 0.46x + 0.92$. The data does not all fall on the line because there may be factors other than weight that go into the price, such as: the quality of the broccoli, the region where the package is sold, and any discounts happening in the store.

$y = 0.46x + 0.92$

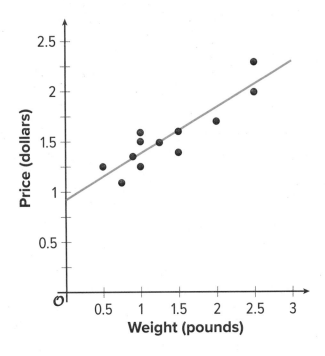

We can interpret the y-intercept of the line as the price for the package without any broccoli (which might include the cost of things like preparing the package and shipping costs for getting the vegetable to the store). In many situations, the data may not follow the same linear model farther away from the given data, especially as one variable gets close to zero. For this reason, the interpretation of the *y*-intercept should always be considered in context to determine if it is reasonable to make sense of the value in that way.

We can also interpret the slope as the approximate increase in price of the package for the addition of 1 pound of broccoli to the package.

The equation also allows us to predict additional values for the price of a package of broccoli for packages that have weights near the weights observed in the data set. For example, even though the data does not include the price of a package that contains 1.7 pounds of broccoli, we can predict the price to be about $1.70 based on the equation of the line, since $0.46 \cdot 1.7 + 0.92 \approx 1.70$.

On the other hand, it does not make sense to predict the price of 1,000 pounds of broccoli with this data, because there may be many more factors that will influence the pricing of packages that far away from the data presented here.

NAME _____ DATE _____ PERIOD _____

Practice
Linear Models

1. The scatter plot shows the number of times a player came to bat and the number of hits they had.

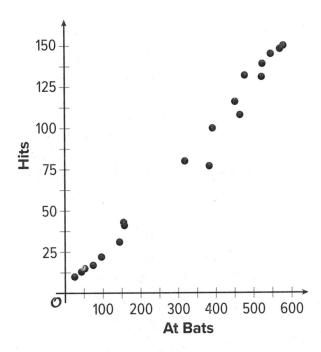

The scatter plot includes a point at (318, 80). Describe the meaning of this point in this situation.

2. The scatter plot shows the number of minutes people had to wait for service at a restaurant and the number of staff available at the time.

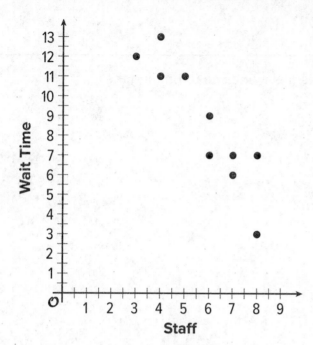

A line that models the data is given by the equation $y = -1.62x + 18$, where y represents the wait time, and x represents the number of staff available.

a. The slope of the line is -1.62. What does this mean in this situation? Is it realistic?

b. The y-intercept is (0, 18). What does this mean in this situation? Is it realistic?

NAME _____ DATE _____ PERIOD _____

3. A taxi driver records the time required to complete various trips and the distance for each trip.

The best fit line is given by the equation $y = 0.467x + 0.417$, where y represents the distance in miles, and x represents the time for the trip in minutes.

a. Use the best fit line to estimate the distance for a trip that takes 20 minutes. Show your reasoning.

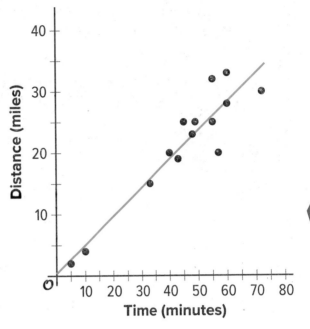

b. Use the best fit line to estimate the time for a trip that is 6 miles long. Show your reasoning.

4. Data is collected about the number of wins and losses by a random sample of teams with an animal mascot and those with another kind of mascot. The column relative frequencies are shown in the table. Based on the information in the table, is there an association between the variables? Explain your reasoning. **(Lesson 3-3)**

	Animal Mascot	Other Type of Mascot
Wins	74%	49%
Losses	26%	51%

5. A random selection of indoor and outdoor pool managers are surveyed about the number of people in each age group that swim there. Results from the survey are displayed in the two-way table. Based on the data, does there appear to be an association between pool type and age group? Explain your reasoning. (Lesson 3-3)

	Outdoor Pool	Indoor Pool
Younger Than 18	317	41
18 or Older	352	163

6. Data from a random sample of people are collected about how they watch movies in the genres of action or mystery. Which value would best fit in the missing cell to suggest there is no association between the genre and how the movies are watched? (Lesson 3-3)

	Streaming Movies	Disc Rental
Action	526	147
Mystery	317	

(A.) 19

(B.) 89

(C.) 147

(D.) 320

Lesson 3-5

Fitting Lines

NAME _____ DATE _____ PERIOD _____

Learning Goal Let's find the best linear model for some data.

 Warm Up

5.1 Selecting the Best Line

Which of the lines is the best fit for the data in each scatter plot? Explain your reasoning.

1.

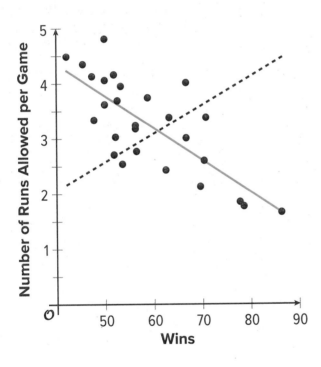

2.

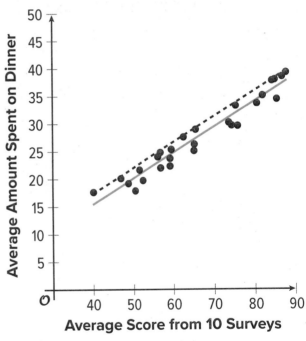

3.

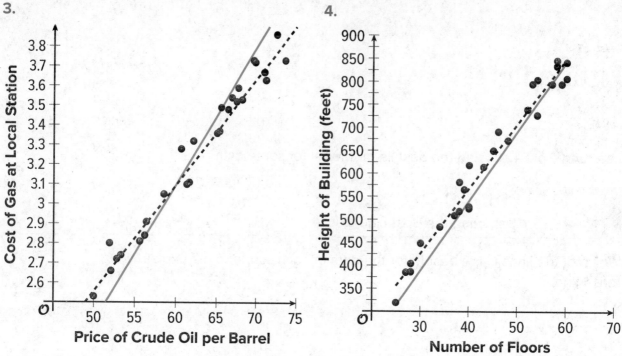

4.

NAME _____ DATE _____ PERIOD _____

Activity

5.2 Card Sort: Data Patterns

Your teacher will give you a set of cards that show scatter plots.

1. Arrange all the cards in three different ways. Ensure that you and your partner agree on the arrangement before moving on to the next one. Sort all the cards in order from:

   a. best to worst for representing with a linear model

   b. least to greatest slope of a linear model that fits the data well

   c. least to greatest vertical intercept of a linear model that fits the data well

2. For each card, write a sentence that describes how *y* changes as *x* increases and whether the linear model is a good fit for the data or not.

The weight of ice cream sold at a small store in pounds (*x*) and the average temperature outside in degrees Celsius (*y*) are recorded in the table.

x	20	18	21	17	21.5	19.5	21	18
y	6	4.5	6.5	3.5	7.5	6.5	7	5

1. For this data, create a scatter plot and sketch a line that fits the data well.

2. Use technology to compute the best fit line. Round any numbers to 2 decimal places.

3. What are the values for the slope and *y*-intercept for the best fit line? What do these values mean in this situation?

hdere/Getty Images

NAME _____ DATE _____ PERIOD _____

4. Use the best fit line to predict the *y* value when *x* is 10. Is this a good estimate for the data? Explain your reasoning.

5. Your teacher will give you a data table for one of the other scatter plots from the previous activity. Use technology and this table of data to create a scatter plot that also shows the line of best fit, then interpret the slope and *y*-intercept.

Priya uses several different ride services to get around her city. The table shows the distance, in miles, she traveled during her last 10 trips and the price of each trip, in dollars.

Distance (miles)	Price ($)
3.1	12.5
4.2	14.75
5	16
3.5	13.25
2.5	12
1	9
0.8	8.75
1.6	9.75
4.3	12
3.3	14

1. Priya creates a scatter plot of the data using the distance, x, and the price, y. She determines that a linear model is appropriate to use with the data. Use technology to find the equation of a line of best fit.

2. Interpret the slope and the y-intercept of the equation of the line of best fit in this situation.

NAME _____ DATE _____ PERIOD _____

3. Use the line of best fit to estimate the cost of a 3.6-mile trip. Will this estimate be close to the actual value? Explain your reasoning.

4. On her next trip, Priya tries a new ride service and travels 3.6 miles, but pays only $4.00 because she receives a discount. Include this trip in the table and calculate the equation of the line of best fit for the 11 trips. Did the slope of the equation of the line of best fit increase, decrease, or stay the same? Why? Explain your reasoning.

5. Priya uses the new ride service for her 12th trip. She travels 4.1 miles and is charged $24.75. How do you think the slope of the equation of the line of best fit will change when this 12th trip is added to the table?

Some data appear to have a linear relationship, so finding an equation for a line that fits the data can help you understand the relationship between the variables.

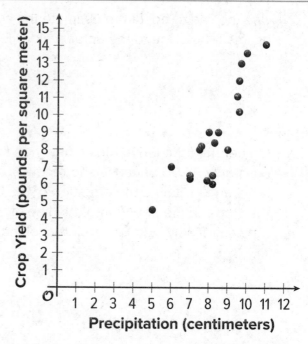

Other data may follow non-linear trends or not have an apparent trend at all.

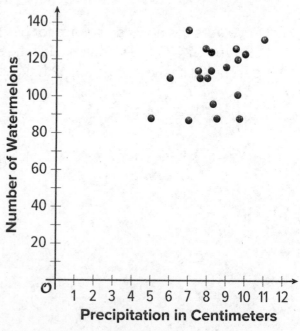

When modeling data with a linear function seems useful, it is important to find a linear function that is close to the data. The line should have a *y*-intercept and slope to follow the shape of the data in the scatter plot as much as possible.

Technology can be used to quickly find a line of best fit for the data and provide the equation of the line that we can use to analyze the situation.

NAME _____ DATE _____ PERIOD _____

Practice
Fitting Lines

1. *Technology required.*

x	y
83	102
87	115
91	107
93	122
97	125
97	127
101	120
104	127

a. Use graphing technology to create a scatter plot and find the best fit line.

b. What does the best fit line estimate for the *y* value when *x* is 100?

2. *Technology required.*

x	y
2.3	6.2
2.8	5.7
3.1	4.7
3	3.2
3.5	3
3.8	2.8

a. What is the equation of the line of best fit? Round numbers to 2 decimal places.

b. What does the equation estimate for y when x is 2.3? Round to 3 decimal places.

c. How does the estimated value compare to the actual value from the table when x is 2.3?

d. How does the estimated value compare to the actual value from the table when x is 3?

NAME _____ DATE _____ PERIOD _____

3. Which of these scatter plots are best fit by the shown linear model?

(A.)

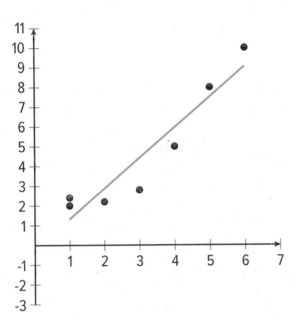

(B.)

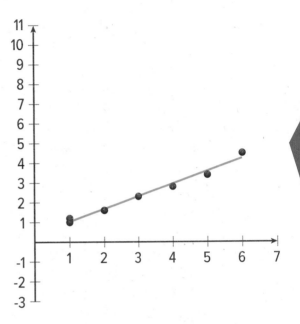

(C.)

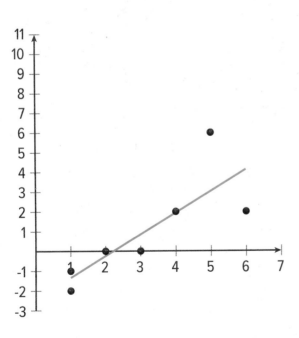

(D.)

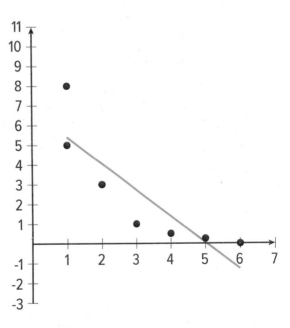

4. A seed is planted in a glass pot and its height is measured in centimeters every day. (Lesson 3-4)

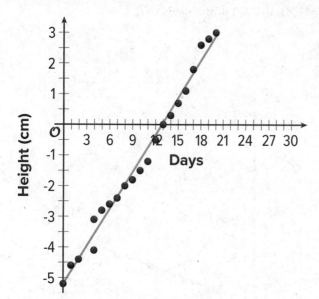

The best fit line is given by the equation $y = 0.404x - 5.18$, where y represents the height of the plant above ground level, and x represents the number of days since it was planted.

a. What is the slope of the best fit line? What does the slope of the line mean in this situation? Is it reasonable?

b. What is the y-intercept of the best fit line? What does the y-intercept of the line mean in this situation? Is it reasonable?

NAME _____ DATE _____ PERIOD _____

5. At a restaurant, the total bill and the percentage of the bill left as a tip is represented in the scatter plot. (Lesson 3-4)

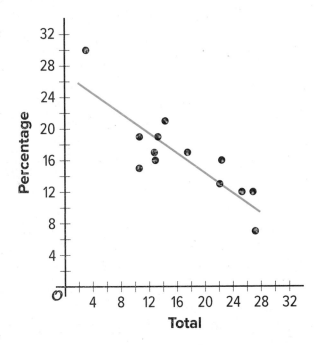

The best fit line is represented by the equation $y = -0.632x + 27.1$, where x represents the total bill in dollars, and y represents the percentage of the bill left as a tip.

a. What does the best fit line estimate for the percentage of the bill left as a tip when the bill is $15? Is this reasonable?

b. What does the best fit line predict for the percentage of the bill left as a tip when the bill is $50? Is this reasonable?

6. A recent study investigated the amount of battery life remaining in alkaline batteries of different ages. The scatter-plot shows this relationship between the different alkaline batteries tested. (Lesson 3-4)

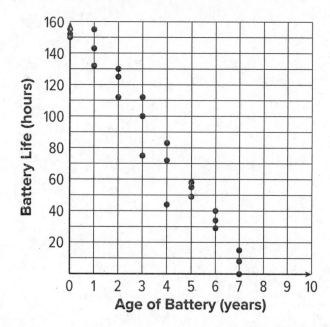

The scatter plot includes a point at (7, 15). Describe the meaning of this point in this situation.

Lesson 3-6

Residuals

NAME _____ DATE _____ PERIOD _____

Learning Goal Let's examine how close data is to linear models.

 ## Warm Up

6.1 Math Talk: Differences in Expectations

Mentally calculate how close the estimate is to the actual value using the difference: actual value − estimated value.

Actual value: 24.8 grams. Estimated value: 19.6 grams

Actual value: $112.11. Estimated value: $109.30

Actual value: 41.5 centimeters. Estimated value: 45.90 centimeters

Actual value: -1.34 degrees Celsius. Estimated value: -2.45 degrees Celsius

1. For the scatter plot of orange weights from a previous lesson, use technology to find the line of best fit.

2. What level of accuracy makes sense for the slope and intercept values? Explain your reasoning.

3. What does the linear model estimate for the weight of the box of oranges for each of the number of oranges?

Number of Oranges	Actual Weight in Kilograms	Linear Estimate Weight in Kilograms
3	1.027	
4	1.162	
5	1.502	
6	1.617	
7	1.761	
8	2.115	
9	2.233	
10	2.569	

NAME _____ DATE _____ PERIOD _____

4. Compare the weights of the box with 3 oranges in it to the estimated weight of the box with 3 oranges in it. Explain or show your reasoning.

5. How many oranges are in the box when the linear model estimates the weight best? Explain or show your reasoning.

6. How many oranges are in the box when the linear model estimates the weight least well? Explain or show your reasoning.

7. The difference between the actual value and the value estimated by a linear model is called the **residual.** If the actual value is greater than the estimated value, the residual is positive. If the actual value is less than the estimated value, the residual is negative. For the orange weight data set, what is the residual for the best fit line when there are 3 oranges? On the same axes as the scatter plot, plot this residual at the point where $x = 3$ and y has the value of the residual.

8. Find the residuals for each of the other points in the scatter plot and graph them

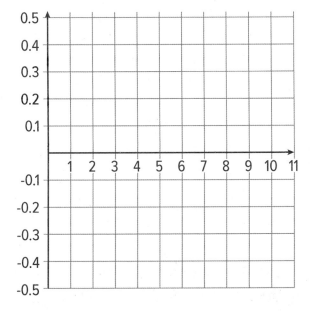

9. Which point on the scatter plot has the residual closest to zero? What does this mean about the weight of the box with that many oranges in it?

10. How can you use the residuals to decide how well a line fits the data?

 Activity

6.3 Best Residuals

1. Match the scatter plots and given linear models to the graph of the residuals.

2. Turn the scatter plots over so that only the residuals are visible. Based on the residuals, which line would produce the most accurate estimates? Which line fits its data worst?

NAME _____ DATE _____ PERIOD _____

Are you ready for more?

1. Tyler estimates a line of best fit for some linear data about the mass, in grams, of different numbers of apples. Here is the graph of the residuals.

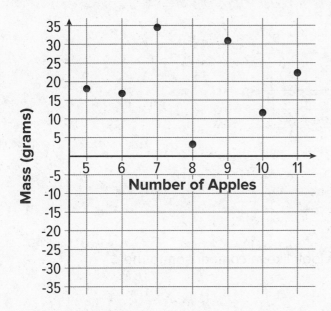

a. What does Tyler's line of best fit look like according to the graph of the residuals?

b. How well does Tyler's line of best fit model the data? Explain your reasoning.

2. Lin estimates a line of best fit for the same data. The graph shows the residuals.

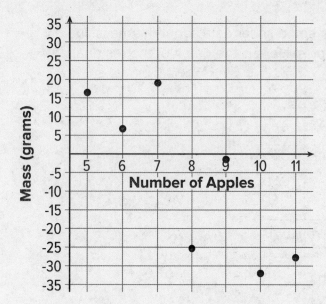

a. What does Lin's line of best fit look like in comparison to the data?

b. How well does Lin's line of best fit model the data? Explain your reasoning.

NAME _____ DATE _____ PERIOD _____

3. Kiran also estimates a line of best fit for the same data. The graph shows the residuals.

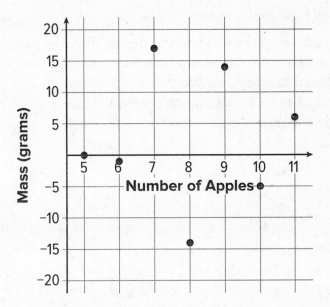

a. What does Kiran's line of best fit look like in comparison to the data?

b. How well does Kiran's line of best fit model the data? Explain your reasoning.

4. Who has the best estimate of the line of best fit—Tyler, Lin, or Kiran? Explain your reasoning.

When fitting a linear model to data, it can be useful to look at the residuals. **Residuals** are the difference between the y-value for a point in a scatter plot and the value predicted by the linear model for that x value.

For example, in the scatter plot showing the length of fish and the age of the fish, the residual for the fish who is 2 years old and 100 mm long is 8.06 mm, because the point is at (2, 100) and the linear function has the value 91.94 mm ($34.08 \cdot 2 + 23.78$) when x is 2. The residual of 8.06 mm means that the actual fish is about 8 millimeters longer than the linear model estimates for a fish of that same age.

$y = 34.08x + 23.78$

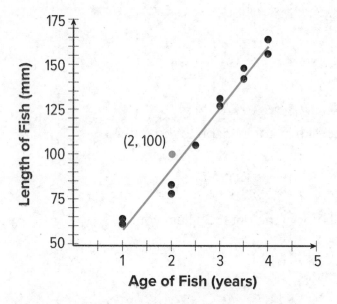

When the point on the scatter plot is above the line, it has a positive residual. When the point on the scatter plot is below the line, the residual is a negative value. A line that has smaller residuals would be more likely to produce estimates that are close to the actual value.

Glossary

residual

NAME _____ DATE _____ PERIOD _____

Practice
Residuals

1. Han creates a scatter plot that displays the relationship between the number of items sold, x, and the total revenue, y, in dollars. Han creates a line of best fit and finds that the residual for the point (12, 1,000) is 75. The point (13, 930) has a residual of -40. Interpret the meaning of -40 in the context of the problem.

2. The line of best fit for a data set is $y = 1.1x + 3.4$. Find the residual for each of the coordinate pairs, (x, y).

   a. (5, 8.8)

   b. (2.5, 5.95)

   c. (0, 3.72)

   d. (1.5, 5.05)

   e. (-3, 0)

   f. (-5, -4.86)

3. Plots of the residuals for four different models of the same data set are displayed. Which of the following represents the plot of the residuals from a model that fits its data best?

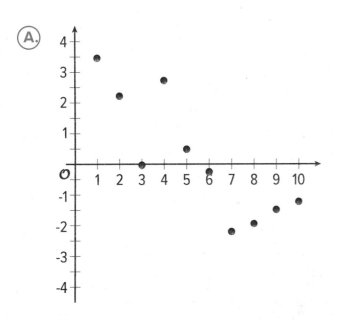

(A.)

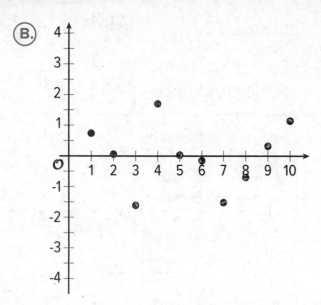

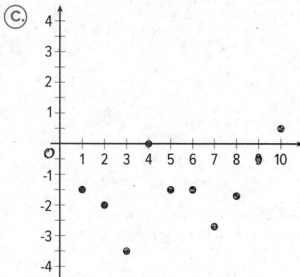

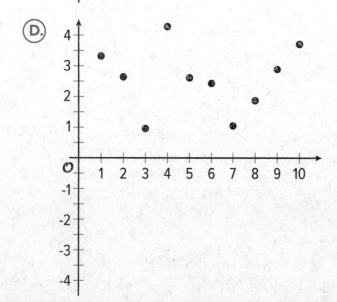

NAME _____ DATE _____ PERIOD _____

4. A local car salesperson created a scatter plot to display the relationship between a car's sale price in dollars, *y*, and the age of the car in years, *x*. The scatter plot and the line of best fit are displayed in the graph.

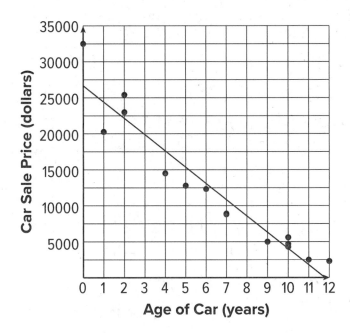

The car salesperson looks at the residuals for the car sales.

a. For a car that is 4 years old, does the salesperson sell above or below her average selling price? Explain your reasoning.

b. For a car that is 12 years old, does the salesperson sell above or below her average selling price? Explain your reasoning.

5. *Technology required.*

Data about the outside temperature and gas used for heating a building are given in the table.

Use a graphing calculator to create the line of best fit for the data. **(Lesson 3-5)**

a. What is the equation of the line of best fit for this data? Round numbers to the nearest whole number.

b. What is the slope of the line of best fit? What does it mean in this situation?

c. What does the line of best fit estimate for gas usage when the outside temperature is 59 degrees Fahrenheit?

d. How does the actual gas usage compare to the estimated gas usage when the outside temperature is 59 degrees Fahrenheit?

Temperature (deg F) x	Gas Usage (therms) y
58	5,686
62	7,373
64	5,805
67	5,636
70	3,782
73	3,976
74	3,351
74	3,396
75	2,936
73	3,078
65	4,549
59	7,022
58	6,106
62	4,566
64	4,608
67	5,790
70	6,501
73	3,843

Lesson 3-7

The Correlation Coefficient

NAME _____ DATE _____ PERIOD _____

Learning Goal Let's see how good a linear model is for some data.

 Warm Up
7.1 Which One Doesn't Belong: Linear Models

Which one doesn't belong?

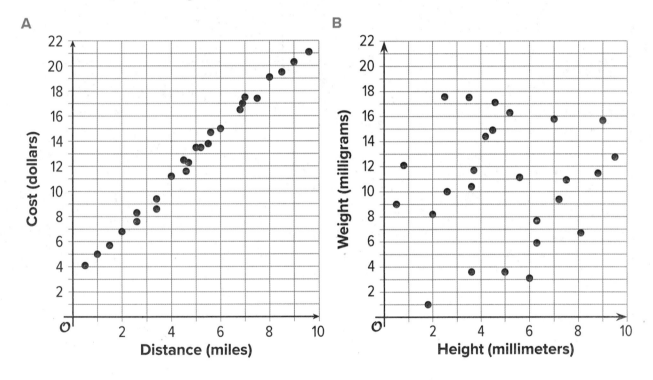

A

B

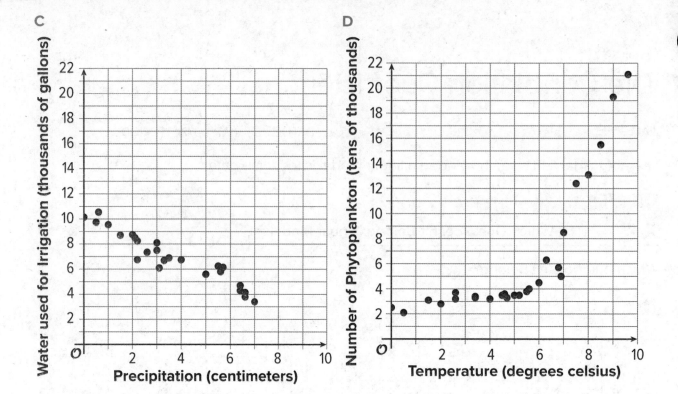

C Water used for Irrigation (thousands of gallons) vs. Precipitation (centimeters)

D Number of Phytoplankton (tens of thousands) vs. Temperature (degrees celsius)

Activity

7.2 Card Sort: Scatter Plot Fit

Your teacher will give you a set of cards that show scatter plots of data. Sort the cards into 2 categories of your choosing. Be prepared to explain the meaning of your categories. Then, sort the cards into 2 categories in a different way. Be prepared to explain the meaning of your new categories.

NAME _____ DATE _____ PERIOD _____

Activity
7.3 Matching Correlation Coefficients

1. Take turns with your partner to match a scatter plot with a **correlation coefficient**.

2. For each match you find, explain to your partner how you know it's a match.

3. For each match your partner finds, listen carefully to their explanation. If you disagree, discuss your thinking and work to reach an agreement.

A

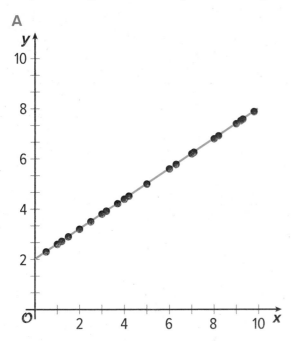

1. $r = -1$
2. $r = -0.95$
3. $r = -0.74$
4. $r = -0.06$
5. $r = 0.48$
6. $r = 0.65$
7. $r = 0.9$
8. $r = 1$

B

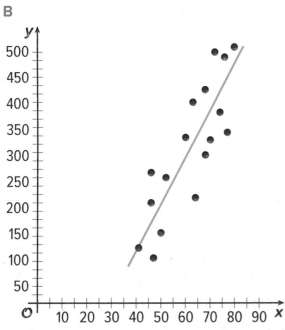

C

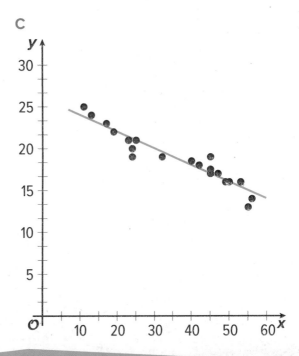

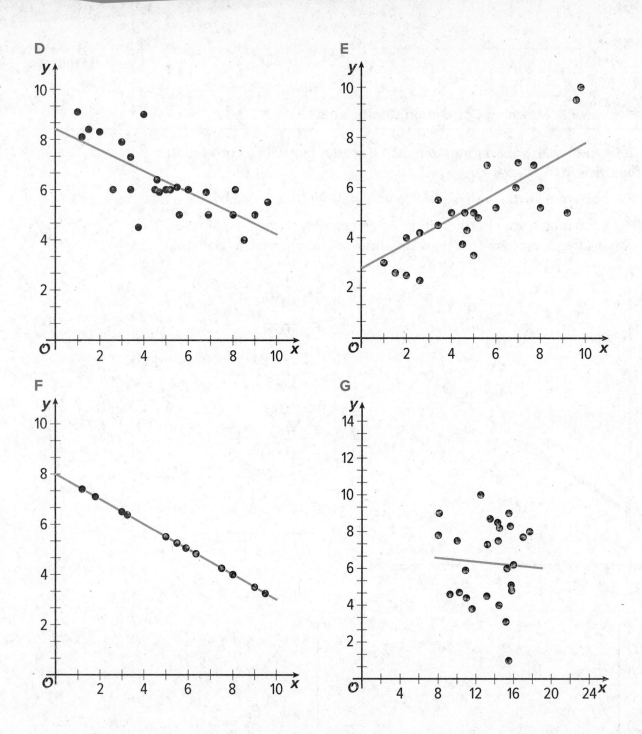

NAME _____ DATE _____ PERIOD _____

H

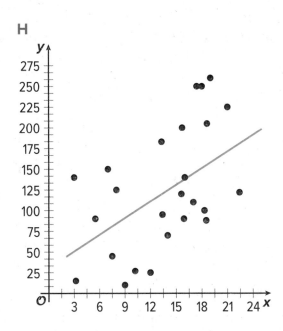

Are you ready for more?

Jada wants to know if the speed that people walk is correlated with their texting speed. To investigate this, she measured the distance, in feet, that 5 of her friends walked in 30 seconds and the number of characters they texted during that time. Each of the 5 friends took 4 walks for a total of 20 walks. Here are the results of the first 20 walks.

Distance (feet)	Number of Characters Texted	Distance (feet)	Number of Characters Texted
105	142	95	138
125	110	125	110
115	120	160	80
140	98	175	64
145	102	130	106
160	89	140	95
170	72	150	95
140	100	155	90
130	107	160	74
105	113	135	108

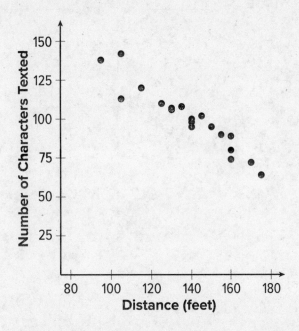

Over the next few days, the same 5 friends practiced walking and texting to see if they could walk faster and text more characters. They did not record any more data while practicing. After practicing, each of the 5 friends took another 4 walks. Here are the results of the final 20 walks.

Distance (feet)	Number of Characters Texted	Distance (feet)	Number of Characters Texted
140	140	165	151
150	155	170	136
160	151	190	143
155	170	205	132
180	125	205	128
205	130	210	140
225	95	215	109
175	161	220	105
195	108	230	126
155	142	225	138

NAME _____ DATE _____ PERIOD _____

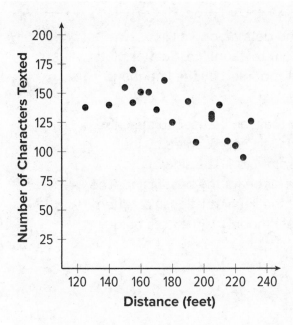

1. What do you notice about the 2 scatter plots?

2. Jada noticed that her friends walked further and texted faster during the last 20 walks than they did during the first 20 walks. Since both were faster, she predicts that the correlation coefficient of the line of best fit for the last 20 walks will be closer to -1 than the correlation coefficient of the line of best fit for the first 20 walks. Do you agree with Jada? Explain your reasoning.

3. Use technology to find an equation of the line of best fit and the correlation coefficient for each data set. Was your answer to the previous question correct?

4. Why do you think the correlation coefficients for the 2 data sets are so different? Explain your reasoning.

While residuals can help pick the best line to fit the data among all lines, we still need a way to determine the strength of a linear relationship. Scatter plots of data that are close to the best fit line are better modeled by the line than scatter plots of data that are farther from the line.

The **correlation coefficient** is a convenient number that can be used to describe the strength and direction of a linear relationship. Usually represented by the letter r, the correlation coefficient can take values from -1 to 1. The sign of the correlation coefficient is the same as the sign of the slope for the best fit line. The closer the correlation coefficient is to 0, the weaker the linear relationship. When the correlation coefficient is closer to 1 or -1, the linear model fits the data better.

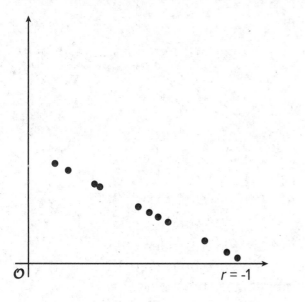

$r = -1$

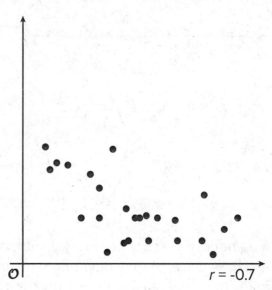

$r = -0.7$

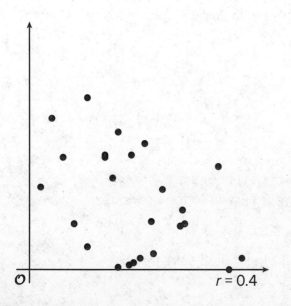

$r = 0.4$

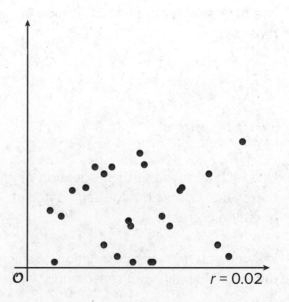

$r = 0.02$

NAME _____ DATE _____ PERIOD _____

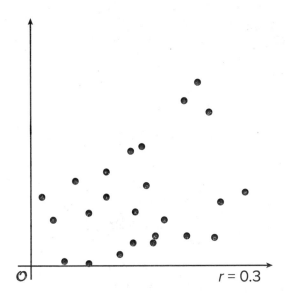

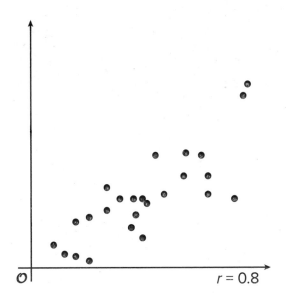

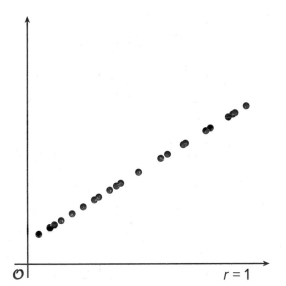

While it is possible to try to fit a linear model to any data, you should always look at the scatter plot to see if there is a possible linear trend. The correlation coefficient and residuals can also help determine whether the linear model makes sense to use to estimate the situation. In some cases, another type of function might be a better fit for the data, or the two variables you are examining may be uncorrelated, and you should look for other connections using other variables.

Glossary

correlation coefficient

Practice
The Correlation Coefficient

1. Select **all** the values for *r* that indicate a positive slope for the line of best fit.

   (A.) 1

   (B.) -1

   (C.) 0.5

   (D.) -0.5

   (E.) 0

   (F.) 0.8

   (G.) -0.8

2. The correlation coefficient, *r*, is given for several different data sets. Which value for *r* indicates the strongest correlation?

   (A.) 0.01

   (B.) -0.34

   (C.) -0.82

   (D.) -0.95

3. Which of the values is the best estimate of the correlation coefficient for the line of best fit shown in the scatter plot?

   (A.) -0.9

   (B.) -0.4

   (C.) 0.4

   (D.) 0.9

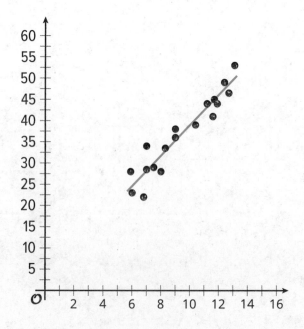

NAME _____ DATE _____ PERIOD _____

4. *Technology required.*

A study investigated the relationship between the amount of daily food waste measured in pounds and the number of people in a household. The data in the table displays the results of the study. **(Lesson 3-5)**

Number of People in Household, x	Food Waste (pounds), y
2	3.4
3	2.5
4	8.9
4	4.7
4	3.5
4	4
5	5.3
5	4.6
5	7.8
6	3.2
8	12

Use graphing technology to create the line of best fit for the data in the table.

a. What is the equation of the line of best fit for this data? Round numbers to two decimal places.

b. What is the slope of the line of best fit? What does it mean in this situation? Is this realistic?

c. What is the *y*-intercept of the line of best fit? What does it mean in this situation? Is this realistic?

5. A table of values and the plot of the residuals for the line of best fit are shown. (Lesson 3-6)

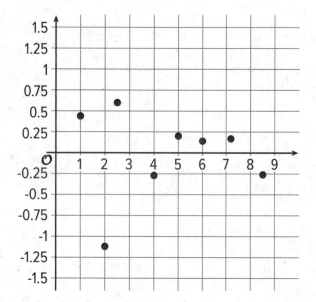

x	y
1	10
2	8
2.5	9.5
4	8
5	8
6	7.5
7.2	7
8.5	6

a. Which point does the line estimate the best?

b. Which point does the line estimate the worst?

6. Tyler creates a scatter plot that displays the relationship between the grams of food a hamster eats, *x*, and the total number of rotations that the hamster's wheel makes, *y*. Tyler creates a line of best fit and finds that the residual for the point (1.4, 1250) is -132. The point (1.2, 1364) has a residual of 117. Interpret the meaning of 117 in the context of the problem. (Lesson 3-6)

Lesson 3-8

Using the Correlation Coefficient

NAME _____ DATE _____ PERIOD _____

Learning Goal Let's look closer at correlation coefficients.

 ## Warm Up

8.1 Putting the Numbers in Context

Match the variables to the scatter plot you think they best fit. Be prepared to explain your reasoning.

	x Variable	*y* Variable
1.	daily low temperature in Celsius for Denver, CO	boxes of cereal in stock at a grocery in Miami, FL
2.	average number of free throws shot in a season	basketball team score per game
3.	measured student height in feet	measured student height in inches
4.	average number of minutes spent in a waiting room	hospital satisfaction rating

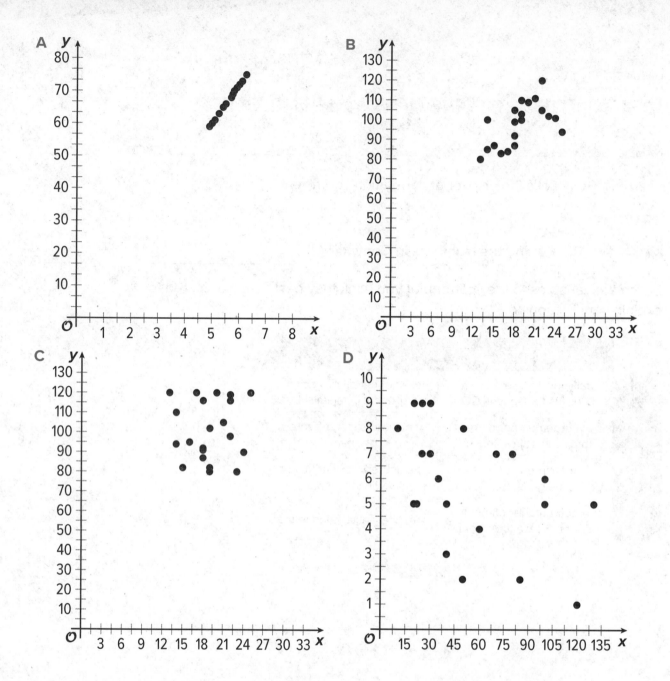

NAME _____ DATE _____ PERIOD _____

Activity
8.2 Never Know How Far You'll Go

Priya takes note of the distance the car drives and the time it takes to get to the destination for many trips.

1. Distance is one factor that influences the travel time of Priya's car trips. What are some other factors?

2. Which of these factors (including distance) most likely has the most consistent influence for all the car trips? Explain your reasoning.

3. Use technology to create a scatter plot of the data and add the best fit line to the graph.

Distance (mi) (x)	Travel Time (min) (y)
2	4
5	7
10	11
10	15
12	16
15	22
20	23
25	25
26	28
30	36
32	35
40	37
50	51
65	70
78	72

4. What do the slope and *y*-intercept for the line of best fit mean in this situation?

5. Use technology to find the correlation coefficient for this data. Based on the value, how would you describe the strength of the linear relationship?

6. How long do you think it would take Priya to make a trip of 90 miles if the linear relationship continues? If she drives 90 miles, do you think the prediction you made will be close to the actual value? Explain your reasoning.

NAME _____ DATE _____ PERIOD _____

Activity

8.3 Correlation Zoo

For each situation, describe the relationship between the variables, based on the correlation coefficient. Make sure to mention whether there is a **strong relationship** or not as well as whether it is a **positive relationship** or **negative relationship**.

1. Number of steps taken per day and number of kilometers walked per day.
    $r = 0.92$

2. Temperature of a rubber band and distance the rubber band can stretch.
    $r = 0.84$

3. Car weight and distance traveled using a full tank of gas. $r = -0.86$

4. Average fat intake per citizen of a country and average cancer rate of a country. $r = 0.73$

5. Score on science exam and number of words written on the essay question.
    $r = 0.28$

6. Average time spent listening to music per day and average time spent watching TV per day. $r = -0.17$

A biologist is trying to determine if a group of dolphins is a new species of dolphin or if it is a new group of individuals within the same species of dolphin. The biologist measures the width (in millimeters) of the largest part of the skull, zygomatic width, and the length (in millimeters) of the snout, rostral length, of 10 dolphins from the same group of individuals.

x, Rostral Length (mm)	y, Zygomatic Width (mm)
288	147
247	147
268	171
278	177
258	168
272	184
272	161
258	159
273	168
277	166

The data appears to be linear and the equation of the line of best fit is $y = 0.201x + 110.806$ and the r-value is 0.201.

1. After checking the data, the biologist realizes that the first zygomatic width listed as 147 mm is an error. It is supposed to be 180 mm. Use technology to find the equation of a line of best fit and the correlation coefficient for the corrected data. What is the equation of the line of best fit and the correlation coefficient?

2. Compare the new equation of the line of best fit with the original. What impact did changing one data point have on the slope, y-intercept, and correlation coefficient on the line of best fit?

3. Why do you think that weak positive association became a moderately strong association? Explain your reasoning.

4. Use technology to change the y-value for the first and second entries in the table.

   a. How does changing each point's y-value impact the correlation coefficient?

NAME _____ DATE _____ PERIOD _____

b. Can you change two values to get the correlation coefficient closer to 1? Use data to support your answer.

c. By leaving (288, 180), can you change a value to get the relationship to change from a positive one to a negative one? Use data to support or refute your answer.

Summary
Using the Correlation Coefficient

The value for the correlation coefficient can be used to determine the strength of the relationship between the two variables represented in the data.

In general, when the variables increase together, we can say they have a **positive relationship**. If an increase in one variable's data tends to be paired with a decrease in the other variable's data, the variables have a **negative relationship**. When the data is tightly clustered around the best fit line, we say there is a **strong relationship**. When the data is loosely spread around the best fit line, we say there is a **weak relationship**.

A correlation coefficient with a value near 1 suggests a strong, positive relationship between the variables. This means that most of the data tends to be tightly clustered around a line, and that when one of the variables increases in value, the other does as well. The number of schools in a community and the population of the community is an example of variables that have a strong, positive correlation. When there is a large population, there is usually a large number of schools, and small communities tend to have fewer schools, so the correlation is positive. These variables are closely tied together, so the correlation is strong.

Similarly, a correlation coefficient near -1 suggests a strong, negative relationship between the variables. Again, most of the data tends to be tightly clustered around a line, but now, when one value increases, the other decreases. The time since you left home and the distance left to reach school has a strong, negative correlation. As the travel time increases, the distance to school tends to decrease, so this is a negative correlation. The variables are again closely, linearly related, so this is a strong correlation.

Weaker correlations mean there may be other reasons the data is variable other than the connection between the two variables. For example, number of pets and number of siblings has a weak correlation. There may be some relationship, but there are many other factors that account for the variability in the number of pets other than the number of siblings.

The context of the situation should be considered when determining whether the correlation value is strong or weak. In physics, measuring with precise instruments, a correlation coefficient of 0.8 may not be considered strong. In social sciences, collecting data through surveys, a correlation coefficient of 0.8 may be very strong.

Glossary

negative relationship
positive relationship
strong relationship
weak relationship

NAME _____ DATE _____ PERIOD _____

 Practice

Using the Correlation Coefficient

1. The number of hours worked, x, and the total dollars earned, y, have a strong positive relationship.

   Explain what it means to have a strong positive relationship in this situation.

2. The number of minutes on the phone and the customer satisfaction rating have a weak negative relationship.

   Explain what it means to have a weak negative relationship in this context.

3. *Technology required.* Use a graphing calculator to answer the questions.

   a. What is an equation of the line of best fit for the data in the table?

   b. What is the value of the correlation coefficient?

4. Elena collects data to investigate the relationship between the number of bananas she buys at the store, x, and the total cost of the bananas, y. Which value for the correlation coefficient is most likely to match a line of best fit of the form $y = mx + b$ for this situation?

   (A.) -0.9   (C.) 0.4

   (B.) -0.4   (D.) 0.9

x	y
5	2
6	4.6
7.5	7.2
8	8.4
8.3	8.2
9	9.1
10.2	10.3
11.4	9.9
11.4	11
12	12.5

5. A researcher creates a scatter plot that displays the relationship between the number of years in business, x, and the percentage of company business that is fair trade, y. The researcher creates a line of best fit, $y = 0.091x + 0.060$, and wants to finds the residuals for the companies that have been in business for 3 years, (Lesson 3-6)

   a. Find the residuals for the two points representing companies that have been in business for 3 years, (3,0.42) and (3,0.3).

   b. Compare the residuals for the two companies who have been in business for 3 years. How are they different? How are they similar? What does the information about the residuals for the two companies tell you about their fair trade business?

6. The correlation coefficient, r, is given for several different data sets. Which value for r indicates the weakest correlation? (Lesson 3-7)

   (A.) 0.01

   (B.) 0.5

   (C.) -0.99

   (D.) 1

7. Which of the following is the best estimate of the correlation coefficient for the data shown in the scatter plot? (Lesson 3-7)

   (A.) -0.9

   (B.) -0.4

   (C.) 0.4

   (D.) 0.9

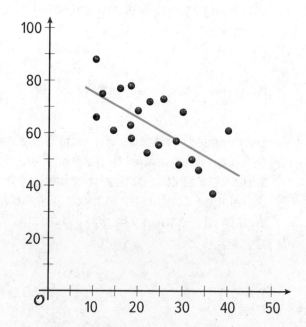

Lesson 3-9

Causal Relationships

NAME _____ DATE _____ PERIOD _____

Learning Goal Let's get a closer look at related variables.

Warm Up
9.1 Used Car Relationships

Describe the strength and sign of the relationship you expect for each pair of variables. Explain your reasoning.

1. Used car price and original sale price of the car.

2. Used car price and number of cup holders in the car.

3. Used car price and number of oil changes the car has had.

4. Used car price and number of miles the car has been driven.

Activity

9.2 Cause or Effect?

Each of the scatter plots show a strong relationship. Write a sentence or two describing how you think the variables are related.

1. During the month of April, Elena keeps track of the number of inches of rain recorded for the day and the percentage of people who come to school with rain jackets.

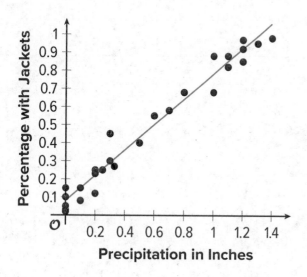

2. A school book club has a list of 100 books for its members to read. They keep track of the number of pages in the books the members read from the list and the amount of time it took to read the book.

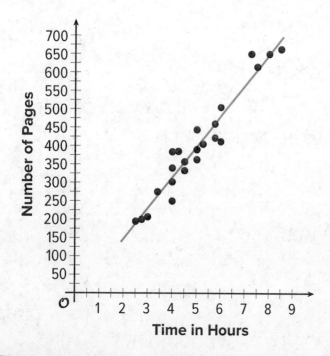

NAME _____ DATE _____ PERIOD _____

3. Number of tickets left for holiday parties at a venue and noise level at the party.

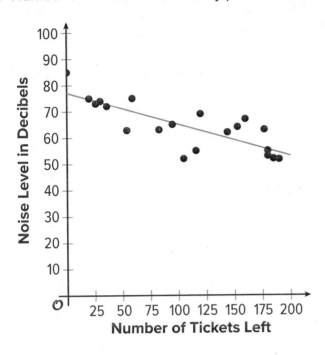

4. The height and score on a test of vocabulary for several children ages 6 to 13.

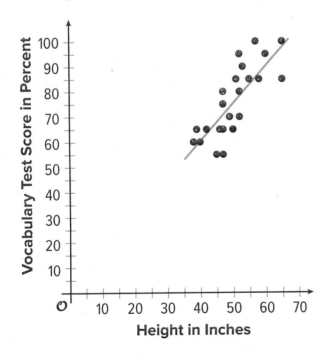

Activity

9.3 Find Your Cause

Describe a pair of variables with each condition. Explain your reasoning.

1. Two variables with a **causal relationship**.

2. The variables are strongly related, but a third factor might be the cause for the changes in the variables.

3. The variables are only weakly related.

NAME _____ DATE _____ PERIOD _____

Are you ready for more?

1. Look through news articles or advertisement for claims of causation or correlation. Find 2 or 3 claims and read or watch the articles or the advertisement. Answer these questions for each of the claims.

   a. What is the claim?

   b. What evidence is provided for the claim?

   c. Does there appear to be evidence for causation or correlation? Explain your thinking.

2. Choose the claim with the least or no evidence. Describe an experiment or other way that you could collect data to show correlation or causation.

NAME _____ DATE _____ PERIOD _____

Summary
Causal Relationships

Humans are wired to look for connections and then use those connections to learn about the world around them. One way to notice connections is by looking for a pair of variables with a relationship. In order to learn about how the variables are related, we want to control one of the variables and see if there are changes in the other variable. For example, if we notice that people who tend to eat many calories also have a higher chance of having a heart attack, we might wonder if lowering our calorie intake would improve our health.

One common mistake people tend to make using statistics is to think that all relationships between variables are causal. Scatter plots can only show a relationship between the two variables. To determine if change in one of the variables actually causes a change in the other variable, or has a **causal relationship**, the context must be better understood and other options ruled out.

For example, we might expect to see a strong, positive relationship between the number of snowboard rentals and sales of hot chocolate during the months of September through January. This does not mean that an increase in snowboard rentals causes people to purchase more hot chocolate. Nor does it mean that increased sales of hot chocolate cause people to rent snowboards more. More likely there is a third variable, such as colder weather, that might be causing both variables to increase at the same time.

On the other hand, sometimes there is a causal relationship. A strong, positive relationship between hot chocolate sales and small marshmallow sales may be linked, because people buying hot chocolate may want to add small marshmallows to the drink, so an increase in the sales of hot chocolate are actually causing the marshmallow sale increase.

Finding relationships with the help of the correlation coefficient is a very good way to notice that there is a connection between variables. To determine whether the relationship is causal, the next step is usually to carefully design an experiment that isolates and precisely controls only one of the variables to determine how it affects the other variable.

Glossary

causal relationship

Practice
Causal Relationships

1. Priya creates a scatter plot showing the relationship between the number of steps she takes and her heart rate. The correlation coefficient for the data is 0.88.

   a. Are they correlated? Explain your reasoning.

   b. Do either of the variables cause the other to change? Explain your reasoning.

2. Kiran creates a scatter plot showing the relationship between the number of students attending drama club and the number of students attending poetry club each week. The correlation coefficient for the data is -0.36.

   a. Are they correlated? Explain your reasoning.

   b. Do either of the variables cause the other to change? Explain your reasoning.

NAME _____ DATE _____ PERIOD _____

3. A news website shows a scatter plot with a negative relationship between the amount of sugar eaten and happiness levels. The headline reads, "Eating sugar causes happiness to decrease!"

   a. What is wrong with this claim?

   b. What is a better headline for this information?

4. A group of 125 college students are surveyed about their note taking and study habits. Some results are represented in the table. (Lesson 3-1)

	Prefer Writing Notes by Hand	Prefer Typing Notes	Don't Take Notes
Study for Less Than 1 Hour	22		8
Study for 1 Hour or More	38	28	3

   How many students prefer typing notes and study for less than 1 hour?

5. The number of miles driven, x, and the number of gallons remaining in the gas tank, y, have a strong negative relationship.

   Explain what it means to have a strong negative relationship in this context. (Lesson 3-8)

6. *Technology required. Use a graphing calculator to answer the questions.* **(Lesson 3-8)**

x	y
10.2	31
10.4	27
10.5	29
10.5	30
10.5	31
10.6	26
10.8	25
10.8	26
10.9	27
11	24
11.2	22

a. What is an equation of the line of best fit?

b. What is the value of the correlation coefficient?

Lesson 3-10

Fossils and Flags

NAME _____ DATE _____ PERIOD _____

Learning Goal Let's collect some data and analyze it.

 Warm Up

10.1 A Fossil Puzzle

An anthropologist finds a fossilized humerus bone of an ancient human ancestor. The humerus is an arm bone running from the shoulder to the elbow. It is 24 centimeters in length. Use data from your classmates to estimate the height of this ancient human.

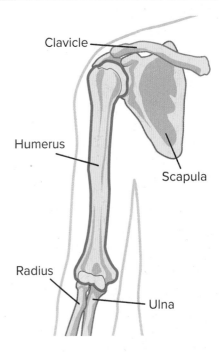

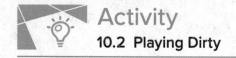

1. Before researching data, do you think the number of penalties a team gets in a season is correlated to the number of wins the team gets in that season? Do you think it is a positive or negative relationship? Do you think the relationship is weak or strong? Explain your reasoning.

2. Is there a relationship between the number of penalties a team gets in a season and the number of wins in that season? Show any mathematical work that leads to your answer.

3. Do penalties cause a change in wins or wins cause a change in penalties or neither? Explain your reasoning.

NAME _____ DATE _____ PERIOD _____

Summary
Fossils and Flags

Scatter plots can be useful to display possible relationships between two variables. Once a pattern is recognized, fitting a function to the data and then recognizing how well the function represents the relationship in the data can quantify your intuition about relationships between variables. The best fit line, for example, can be used to predict the value of one variable based on the value of a second variable. Although technology can aid in finding best fit lines and evaluating the strength of the line's fit, human understanding of the variables and of how the data were collected is still required to determine whether the relationship is merely a relationship or whether there is a causal relationship.

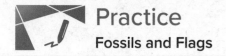

Practice
Fossils and Flags

1. A random sampling of people are asked whether they like comedies or dramatic movies, and whether they prefer to watch them in the theater or at home. The results are in the table.

	Comedy	Drama
Theater	57	36
Home	78	53

Create a relative frequency table that shows the percentage of comedy lovers that prefer each location and the percentage of drama lovers that prefer each location. (Lesson 3-2)

2. Two types of soil are used to grow corn to see if there is an association between the type of soil and the time it takes to reach harvest readiness. Due to the difficulty in making soil B, there were only 100 corn plants grown in that soil type. Complete the table so that it suggests no association between soil type and time to reach harvest readiness. Explain your reasoning. (Lesson 3-3)

	Soil A	Soil B
Less Than 80 Days	279	
80 Days or More	131	

NAME _____ DATE _____ PERIOD _____

3. A recent survey investigated the relationship between the number of traffic tickets a person received and the cost of the person's car insurance. The scatter plot displays the relationship.

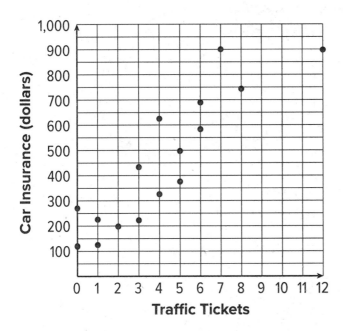

The line that models the data is given by the equation $y = 73x + 146.53$, where x represents the number of traffic tickets, and y represents the cost of car insurance. **(Lesson 3-4)**

a. The slope of the line is 73. What does this mean in this situation? Is it realistic?

b. The y-intercept is (0, 146.53). What does this mean in this situation?

4. *Technology required.*

A survey wanted to determine if there was a relationship between the number of joggers who used a local park for exercise and the temperature outside. The data in the table display their findings.

Temperature in Fahrenheit, x	Number of Joggers, y
15	4
30	8
30	8
41	4
42	16
49	20
49	14
55	16
66	34
72	44
85	40
94	15

Use graphing technology to create a scatter plot of the data. **(Lesson 3-5)**

a. Is a linear model appropriate for this data? Explain your reasoning.

b. If the data seems appropriate, create the line of best fit. Round to two decimal places.

c. What is the slope of the line of best fit and what does it mean in this context? Is it realistic?

d. What is the *y*-intercept of the line of best fit and what does it mean in this context? Is it realistic?

5. Data for a local hospital is displayed in the scatter plot. The graph shows the relationship between the length of a person's stay in the hospital in days, *x*, and the amount owed for the hospital bill. The line of best fit for the data is $y = 1,899.66x + 852.81$. **(Lesson 3-6)**

   a. Find the residuals for the three patients who were in the hospital for 6 days, (6, 14320), (6, 7900), and (6, 13998).

   b. Compare the residuals for the three patients. How are they similar? How are they different? What does the information about the residuals for the three patients tell you about their hospital bills?

6. Select **all** the values for *r* that indicate a strong, negative relationship for the line of best fit. (Lesson 3-7)

(A.) 1

(B.) -0.97

(C.) -0.45

(D.) 0.53

(E.) 0.9

(F.) -0.8

(G.) -1

7. Noah collects data to investigate the relationship between the number of runs scored by his favorite baseball team, *x*, and the number of runs scored by his high school baseball team, *y*. Which value for the correlation coefficient is most likely to match a line of best fit of the form $y = mx + b$ for this situation? (Lesson 3-8)

(A.) -1

(B.) 0

(C.) 0.7

(D.) 1

NAME _____ DATE _____ PERIOD _____

8. Noah creates a scatter plot showing the relationship between number of free throws made in a basketball game and the number of points scored. The correlation coefficient for the line of best fit is 0.76. **(Lesson 3-9)**

   a. Are they correlated? Explain your reasoning.

   b. Do either of the variables cause the other to change? Explain your reasoning.

9. A news headline claims that "Essential Oils Cause Hormone Levels to Drop." They show a scatter plot displaying a weak negative relationship ($r = -0.13$) between essential oil use and hormone levels. **(Lesson 3-9)**

   a. What is wrong with this claim?

   b. What is a better headline for this information?

10. *Technology required.*

Data in the table shows the relationship between average number of social network notifications a student receives during one class, x, and average test scores, y.

Average Number of Social Network Notifications, x	Average Test Score, y
12	92
26	84
17	87
43	65
51	57
29	75
13	83
4	100
16	86
12	73
25	67
22	77
12	89
8	91
34	98

a. What conclusions, if any, can you draw from the information provided? Justify your thinking with mathematics learned from this unit.

b. What conclusions can you not draw from the information provided? Justify your thinking with mathematics learned from this unit.

Learning Targets

Lesson	Learning Target(s)
3-1 Two-way Tables	• I can calculate missing values in a two-way table. • I can create a two-way table for categorical data given information in everyday language. • I can describe what the values in a two-way table mean in everyday language.
3-2 Relative Frequency Tables	• I can calculate values in a relative frequency table and describe what the values mean in everyday language.
3-3 Associations in Categorical Data	• I can look for patterns in two-way tables and relative frequency tables to see if there is a possible association between two variables.

(continued on the next page)

Lesson	Learning Target(s)
3-4 Linear Models	• I can describe the rate of change and *y*-intercept for a linear model in everyday language. • I can draw a linear model that fits the data well and use the linear model to estimate values I want to find.
3-5 Fitting Lines	• I can describe the rate of change and *y*-intercept for a linear model in everyday language. • I can use technology to find the line of best fit.
3-6 Residuals	• I can plot and calculate residuals for a data set and use the information to judge whether a linear model is a good fit.
3-7 The Correlation Coefficient	• I can describe the goodness of fit of a linear model using the correlation coefficient. • I can match the correlation coefficient with a scatter plot and linear model.

(continued on the next page)

(continued from the previous page)

Lesson	Learning Target(s)
3-8 Using the Correlation Coefficient	• I can describe the strength of a relationship between two variables. • I can use technology to find the correlation coefficient and explain what the value tells me about a linear model in everyday language.
3-9 Causal Relationships	• I can look for connections between two variables to analyze whether or not there is a causal relationship.
3-10 Fossils and Flags	• I can collect data, create a linear model to fit the data, determine if the linear model is a good fit, and use the information from my linear model to answer questions.

Notes

_____

_____

_____

_____

_____

_____

_____

Unit 4
Functions

Phoenixns/Shutterstock

The total distance traveled by a drone can be found using an absolute value function. You will learn more about absolute value functions in this unit.

Topics
- Functions and Their Representations
- Analyzing and Creating Graphs of Functions
- A Closer Look at Inputs and Outputs
- Inverse Functions
- Putting it All Together

Functions

Lesson 4-1

Describing and Graphing Situations

NAME _____ DATE _____ PERIOD _____

Learning Goal Let's look at some fun functions around us and try to describe them!

Warm Up
1.1 Bagel Shop

FRESH BAGELS!

1 bagel	*$ 1.25*
6 bagels	*$ 6.00*
9 bagels	*$ 8.00*
12 bagels	*$ 10.00*

Illustrative Math

A customer at a bagel shop is buying 13 bagels. The shopkeeper says, "That would be $16.25."

Jada, Priya, and Han, who are in the shop, all think it is a mistake.

- Jada says to her friends, "Shouldn't the total be $13.25?"
- Priya says, "I think it should be $13.00."
- Han says, "No, I think it should be $11.25."

Explain how the shopkeeper, Jada, Priya, and Han could all be right.

Your teacher will give you instructions for completing the table.

Number of Bagels	
1	
2	
3	
4	
5	
6	
7	
8	
9	
10	
11	
12	
13	

NAME _____ DATE _____ PERIOD _____

Activity

1.2 Be Right Back!

Three days in a row, a dog owner tied his dog's 5-foot-long leash to a post outside a store while he ran into the store to get a drink. Each time, the owner returned within minutes.

The dog's movement each day is described here.

- Day 1: The dog walked around the entire time while waiting for its owner.

- Day 2: The dog walked around for the first minute, and then laid down until its owner returned.

- Day 3: The dog tried to follow its owner into the store but was stopped by the leash. Then, it started walking around the post in one direction. It kept walking until its leash was completely wound up around the post. The dog stayed there until its owner returned.

- Each day, the dog was 1.5 feet away from the post when the owner left.

- Each day, 60 seconds after the owner left, the dog was 4 feet from the post.

Your teacher will assign one of the days for you to analyze.

Sketch a graph that could represent the dog's distance from the post, in feet, as a function of time, in seconds, since the owner left.

Day _____

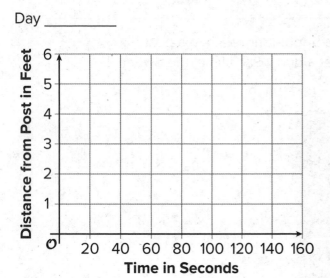

Are you ready for more?

From the graph, is it possible to tell how many times the dog changed directions while walking around? Explain your reasoning.

NAME _____ DATE _____ PERIOD _____

Activity
1.3 Talk about a Function

The following pair of quantities has a relationship that can be defined as a function.

- time, in seconds, since the dog owner left and the total number of times the dog has barked

Express this relationship as a function.

1. In that function, which variable is independent? Which one is dependent?

2. Write a sentence of the form "_____ is a function of _____."

3. Sketch a possible graph of the relationship on the coordinate plane. Be sure to label and indicate a scale on each axis, and be prepared to explain your reasoning.

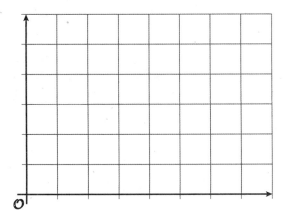

The following pair of quantities has a relationship that can be defined as a function.

- time, in seconds, since the owner left and the total distance, in feet, that the dog has walked while waiting.

Express this relationship as a function.

1. In that function, which variable is independent? Which one is dependent?

2. Write a sentence of the form "_____ is a function of _____."

3. Sketch a possible graph of the relationship on the coordinate plane. Be sure to label and indicate a scale on each axis, and be prepared to explain your reasoning.

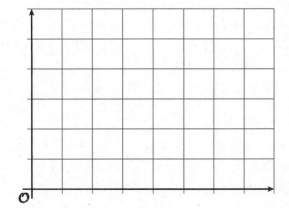

NAME _____ DATE _____ PERIOD _____

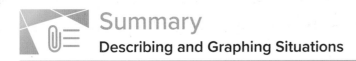

Summary
Describing and Graphing Situations

A relationship between two quantities is a *function* if there is exactly one output for each input. We call the input the *independent variable* and the output the *dependent variable*.

Let's look at the relationship between the amount of time since a plane takes off, in seconds, and the plane's height above the ground, in feet.

- These two quantities form a function if time is the independent variable (the input) and height is the dependent variable (the output). This is because at any amount of time since takeoff, the plane could only be at one height above the ground.

  For example, 50 seconds after takeoff, the plane might have a height of 180 feet. At that moment, it cannot be simultaneously 180 feet and 95 feet above the ground.

  For any input, there is only one possible output, so the height of the plane is *a function of* the time since takeoff.

- The two quantities do not form a function, however, if we consider height as the input and time as the output. This is because the plane can be at the same height for multiple lengths of times since takeoff.

  For instance, when the plane is 1,500 feet above ground, it is possible that 300 seconds have passed. It is also possible that 425 seconds, 275 seconds, or some other amounts of time have passed.

  For any input, there are multiple possible outputs, so the time since takeoff is *not a function of* the height of the plane.

Functions can be represented in many ways—with a verbal description, a table of values, a graph, an expression or an equation, or a set of ordered pairs.

When a function is represented with a graph, each point on the graph is a specific pair of input and output.

Here is a graph that shows the height of a plane as a function of time since takeoff.

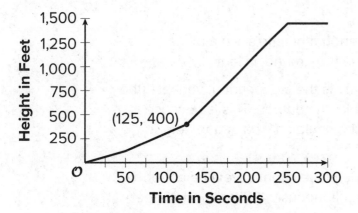

The point (125, 400) tells us that 125 seconds after takeoff, the height of the plane is 400 feet.

NAME _____ DATE _____ PERIOD _____

Practice
Describing and Graphing Situations

1. The relationship between the amount of time a car is parked, in hours, and the cost of parking, in dollars, can be described with a function.

   a. Identify the independent variable and the dependent variable in this function.

   b. Describe the function with a sentence of the form "_____is a function of_____."

   c. Suppose it costs $3 per hour to park, with a maximum cost of $12.

   Sketch a possible graph of the function. Be sure to label the axes.

   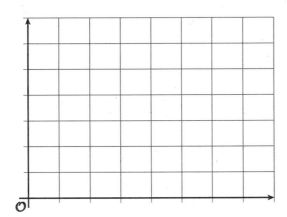

   d. Identify one point on the graph and explain its meaning in this situation.

2. The prices of different burgers are shown on this sign.

Based on the information from the menu, is the price of a burger a function of the number of patties? Explain your reasoning.

BURGER MENU

Served Anytime

Cheeseburger $3.49
1 patty, 1 cheese slice

Just the Patties................... $4.09
2 patties, no cheese

Double Cheeseburger....... $4.59
2 patties, 2 cheese slices

Big Island........................... $6.79
4 patties, 4 cheese slices

3. The distance a person walks, *d*, in kilometers, is a function of time, *t*, in minutes, since the walk begins.

Select **all** true statements about the input variable of this function.

A. Distance is the input.

B. Time of day is the input.

C. Time since the person starts walking is the input.

D. *t* represents the input.

E. *d* represents the input.

F. The input is not measured in any particular unit.

G. The input is measured in hours.

H. For each input, there are sometimes two outputs.

4. It costs $3 per hour to park in a parking lot, with a maximum cost of $12.

Explain why the amount of time a car is parked is *not* a function of the parking cost.

5. Here are clues for a puzzle involving two numbers. **(Lesson 2-12)**

- Seven times the first number plus six times the second number equals 31.

- Three times the first number minus ten times the second number is 29.

What are the two numbers? Explain or show your reasoning.

6. To keep some privacy about the students, a professor releases only summary statistics about student scores on a difficult quiz.

Mean	Standard Deviation	Minimum	Q1	Median	Q3	Maximum
66.91	12.74	12	57	66	76	100

Based on this information, what can you know about outliers in the student scores? **(Lesson 1-14)**

(A.) There is an outlier at the upper end of the data.

(B.) There is an outlier at the lower end of the data.

(C.) There are outliers on both ends of the data.

(D.) There is not enough information to determine whether there are any outliers.

7. An airline company creates a scatter plot showing the relationship between the number of flights an airport offers and the average distance in miles travelers must drive to reach the airport. The correlation coefficient of the line of best fit is -0.52. (Lesson 3-9)

   a. Are they correlated? Explain your reasoning.

   b. Do either of the variables cause the other to change? Explain your reasoning.

Lesson 4-2

Function Notation

NAME _____ DATE _____ PERIOD _____

Learning Goal Let's learn about a handy way to refer to and talk about a function.

 ## Warm Up
2.1 Back to the Post!

Here are the graphs of some situations you saw before. Each graph represents the distance of a dog from a post as a function of time since the dog owner left to purchase something from a store. Distance is measured in feet and time is measured in seconds.

Day 1

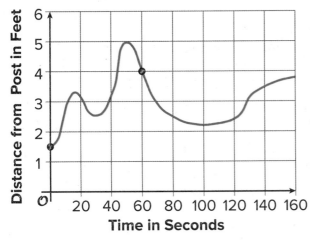

Day 2

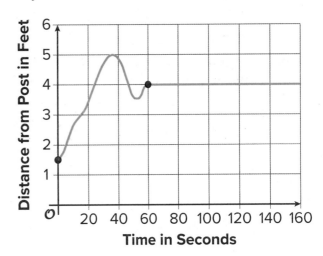

Day 3

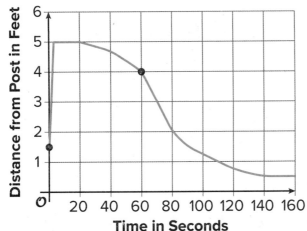

1. Use the given graphs to answer these questions about each of the three days:

   a. How far away was the dog from the post 60 seconds after the owner left?

      Day 1:

      Day 2:

      Day 3:

   b. How far away was the dog from the post when the owner left?

      Day 1:

      Day 2:

      Day 3:

   c. The owner returned 160 seconds after he left. How far away was the dog from the post at that time?

      Day 1:

      Day 2:

      Day 3:

   d. How many seconds passed before the dog reached the farthest point it could reach from the post?

      Day 1:

      Day 2:

      Day 3:

2. Consider the statement, "The dog was 2 feet away from the post after 80 seconds." Do you agree with the statement?

3. What was the distance of the dog from the post 100 seconds after the owner left?

NAME _____ DATE _____ PERIOD _____

Activity

2.2 A Handy Notation

Let's name the functions that relate the dog's distance from the post and the time since its owner left: function *f* for Day 1, function *g* for Day 2, function *h* for Day 3. The input of each function is time in seconds, *t*.

1. Use function notation to complete the table.

	Day 1	Day 2	Day 3
a. distance from post 60 seconds after the owner left			
b. distance from post when the owner left			
c. distance from post 150 seconds after the owner left			

2. Describe what each expression represents in this context:

    a. $f(15)$

    b. $g(48)$

    c. $h(t)$.

3. The equation $g(120) = 4$ can be interpreted to mean: "On Day 2, 120 seconds after the dog owner left, the dog was 4 feet from the post."

    What does each equation mean in this situation?

    a. $h(40) = 4.6$

    b. $f(t) = 5$

    c. $g(t) = d$

Activity

2.3 Birthdays

Rule *B* takes a person's name as its input, and gives their birthday as the output.

Rule *P* takes a date as its input and gives a person with that birthday as the output.

Input	Output
Abraham Lincoln	February 12

Input	Output
August 26	Katherine Johnson

1. Complete each table with three more examples of input-output pairs.

2. If you use your name as the input to *B*, how many outputs are possible? Explain how you know.

3. If you use your birthday as the input to *P*, how many outputs are possible? Explain how you know.

4. Only one of the two relationships is a function. The other is not a function. Which one is which? Explain how you know.

5. For the relationship that is a function, write two input-output pairs from the table using function notation.

NAME _____ DATE _____ PERIOD _____

Are you ready for more?

1. Write a rule that describes these input-output pairs:

    F(ONE) = 3
    F(TWO) = 3
    F(THREE) = 5
    F(FOUR) = 4

2. Here are some input-output pairs with the same inputs but different outputs:

    v(ONE) = 2
    v(TWO) = 1
    v(THREE) = 2
    v(FOUR) = 2

    What rule could define function v?

Here are graphs of two functions, each representing the cost of riding in a taxi from two companies—Friendly Rides and Great Cabs.

For each taxi, the cost of a ride is a function of the distance traveled. The input is distance in miles, and the output is cost in dollars.

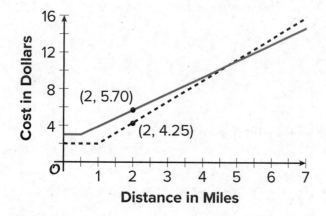

Distance in Miles

- The point (2, 5.70) on one graph tells us the cost of riding a Friendly Rides taxi for 2 miles.

- The point (2, 4.25) on the other graph tells us the cost of riding a Great Cabs taxi for 2 miles.

We can convey the same information much more efficiently by naming each function and using **function notation** to specify the input and the output.

- Let's name the function for Friendly Rides function f.

- Let's name the function for Great Cabs function g.

- To refer to the cost of riding each taxi for 2 miles, we can write: $f(2)$ and $g(2)$.

- To say that a 2-mile trip with Friendly Rides will cost $5.70, we can write $f(2) = 5.70$.

- To say that a 2-mile trip with Great Cabs will cost $4.25, we can write $g(2) = 4.25$.

NAME _____ DATE _____ PERIOD _____

In general, function notation has this form:

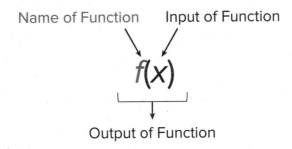

Name of Function   Input of Function

$f(x)$

Output of Function

It is read "f of x" and can be interpreted to mean: $f(x)$ is the output of a function f when x is the input.

The function notation is a concise way to refer to a function and describe its input and output, which can be very useful. Throughout this unit and the course, we will use function notation to talk about functions.

Glossary

function notation

Practice
Function Notation

1. The height of water in a bathtub, w, is a function of time, t. Let P represent this function. Height is measured in inches and time in minutes.

   Match each statement in function notation with a description.

   A $P(0) = 0$

   B $P(4) = 10$

   C $P(10) = 4$

   D $P(20) = 0$

   1 After 20 minutes, the bathtub is empty.

   2 The bathtub starts out with no water.

   3 After 10 minutes, the height of the water is 4 inches.

   4 The height of the water is 10 inches after 4 minutes.

2. Function C takes time for its input and gives a student's Monday class for its output.

   a. Use function notation to represent: A student has English at 10:00.

   b. Write a statement to describe the meaning of $C(11:15) = $ chemistry.

3. Function f gives the distance of a dog from a post, in feet, as a function of time, in seconds, since its owner left.

   Find the value of $f(20)$ and of $f(140)$.

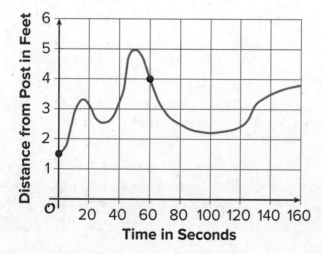

NAME _____ DATE _____ PERIOD _____

4. Function *C* gives the cost, in dollars, of buying *n* apples. What does each expression or equation represent in this situation?

   a. $C(5) = 4.50$

   b. $C(2)$

5. A number of identical cups are stacked up. The number of cups in a stack and the height of the stack in centimeters are related. **(Lesson 4-1)**

   a. Can we say that the height of the stack is a function of the number of cups in the stack? Explain your reasoning.

   b. Can we say that the number of cups in a stack is a function of the height of the stack? Explain your reasoning.

6. In a function, the number of cups in a stack is a function of the height of the stack in centimeters. **(Lesson 4-1)**

   a. Sketch a possible graph of the function on the coordinate plane. Be sure to label the axes.

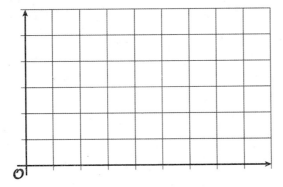

   b. Identify one point on the graph and explain the meaning of the point in the situation.

7. Solve each system of equations without graphing. Show your reasoning. (Lesson 2-16)

a. $\begin{cases} -5x + 3y = -8 \\ 3x - 7y = -3 \end{cases}$

b. $\begin{cases} -8x - 2y = 24 \\ 5x - 3y = 2 \end{cases}$

Lesson 4-3

Interpreting & Using Function Notation

NAME _____ DATE _____ PERIOD _____

Learning Goal Let's use function notation to talk about functions.

Warm Up

3.1 Observing a Drone

Here is a graph that represents function *f*, which gives the height of a drone, in meters, *t* seconds after it leaves the ground.

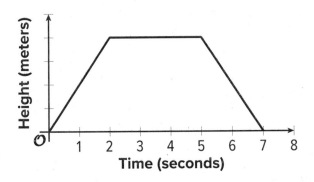

Decide which function value is greater.

1. $f(0)$ or $f(4)$

2. $f(2)$ or $f(5)$

3. $f(3)$ or $f(7)$

4. $f(t)$ or $f(t + 1)$

Activity

3.2 Smartphones

The function *P* gives the number of people, in millions, who own a smartphone, *t* years after year 2000.

1. What does each equation tell us about smartphone ownership?

   a. $P(17) = 2{,}320$

   b. $P(-10) = 0$

2. Use function notation to represent each statement.

   a. In 2010, the number of people who owned a smartphone was 296,600,000.

   b. In 2015, about 1.86 billion people owned a smartphone.

3. Mai is curious about the value of *t* in $P(t) = 1{,}000$.

   a. What would the value of *t* tell Mai about the situation?

   b. Is 4 a possible value of *t* here?

NAME _____ DATE _____ PERIOD _____

4. Use the information you have so far to sketch a graph of the function.

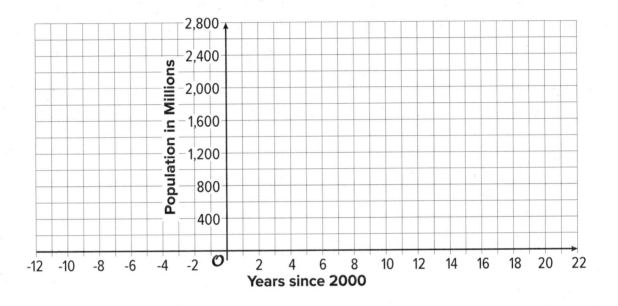

Are you ready for more?

What can you say about the value or values of *t* when $P(t) = 1,000$?

The function W gives the temperature, in degrees Fahrenheit, of a pot of water on a stove, t minutes after the stove is turned on.

1. Take turns with your partner to explain the meaning of each statement in this situation. When it's your partner's turn, listen carefully to their interpretation. If you disagree, discuss your thinking and work to reach an agreement.

   a. $W(0) = 72$

   b. $W(5) > W(2)$

   c. $W(10) = 212$

   d. $W(12) = W(10)$

   e. $W(15) > W(30)$

   f. $W(0) < W(30)$

NAME _____ DATE _____ PERIOD _____

2. If all statements in the previous question represent the situation, sketch a possible graph of function *W*.

Be prepared to show where each statement can be seen on your graph.

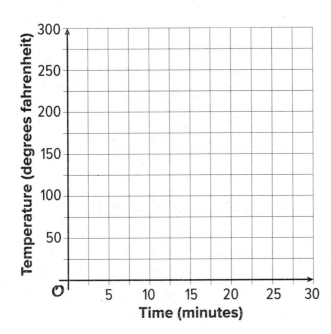

What does a statement like $p(3) = 12$ mean?

On its own, $p(3) = 12$ only tells us that when p takes 3 as its input, its output is 12.

If we know what quantities the input and output represent, however, we can learn much more about the situation that the function represents.

- If function p gives the perimeter of a square whose side length is x and both measurements are in inches, then we can interpret $p(3) = 12$ to mean "a square whose side length is 3 inches has a perimeter of 12 inches."

  We can also interpret statements like $p(x) = 32$ to mean "a square with side length x has a perimeter of 32 inches," which then allows us to reason that x must be 8 inches and to write $p(8) = 32$.

- If function p gives the number of blog subscribers, in thousands, x months after a blogger started publishing online, then $p(3) = 12$ means "3 months after a blogger started publishing online, the blog has 12,000 subscribers."

It is important to pay attention to the units of measurement when analyzing a function. Otherwise, we might mistake what is happening in the situation. If we miss that $p(x)$ is measured in thousands, we might misinterpret $p(x) = 36$ to mean "there are 36 blog subscribers after x months," while it actually means "there are 36,000 subscribers after x months."

A graph of a function can likewise help us interpret statements in function notation.

Function f gives the depth, in inches, of water in a tub as a function of time, t, in minutes, since the tub started being drained.

Here is a graph of f.

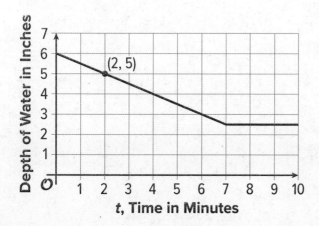

Each point on the graph has the coordinates $(t, f(t))$, where the first value is the input of the function and the second value is the output.

- $f(2)$ represents the depth of water 2 minutes after the tub started being drained. The graph passes through $(2,5)$, so the depth of water is 5 inches when $t = 2$. The equation $f(2) = 5$ captures this information.

- $f(0)$ gives the depth of the water when the draining began, when $t = 0$. The graph shows the depth of water to be 6 inches at that time, so we can write $f(0) = 6$.

- $f(t) = 3$ tells us that t minutes after the tub started draining, the depth of the water is 3 inches. The graph shows that this happens when t is 6.

1. Function *f* gives the temperature, in degrees Celsius, *t* hours after midnight.

   Choose the equation that represents the statement: "At 1:30 p.m., the temperature was 20 degrees Celsius."

   (A.) $f(1{:}30) = 20$

   (B.) $f(1.5) = 20$

   (C.) $f(13{:}30) = 20$

   (D.) $f(13.5) = 20$

2. Tyler filled up his bathtub, took a bath, and then drained the tub. The function *B* gives the depth of the water, in inches, *t* minutes after Tyler began to fill the bathtub.

   Explain the meaning of each statement in this situation.

   a. $B(0) = 0$

   b. $B(1) < B(7)$

   c. $B(9) = 11$

   d. $B(10) = B(22)$

   e. $B(20) > B(40)$

NAME _____ DATE _____ PERIOD _____

3. Function *f* gives the temperature, in degrees Celsius, *t* hours after midnight.
   Use function notation to write an equation or expression for each statement.

   a. The temperature at 12 p.m.

   b. The temperature was the same at 9 a.m. and at 4 p.m.

   c. It was warmer at 9 a.m. than at 6 a.m.

   d. Some time after midnight, the temperature was 24 degrees Celsius.

4. Select **all** points that are on the graph of *f* if we know that $f(2) = -4$ and $f(5) = 3.4$.

   (A.) (-4, 2)   (D.) (5, 3.4)

   (B.) (2, -4)   (E.) (2, 5)

   (C.) (3.4, 5)

5. Write three statements that are true about this situation. Use function notation.

   Function *f* gives the distance of a dog from a post, in feet, as a function of time, *t*, in seconds, since its owner left.

   Use the = sign in at least one statement and the < sign in another statement.

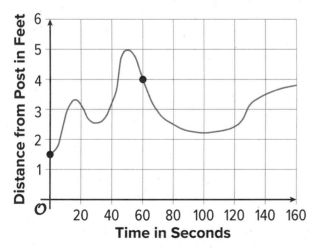

6. Elena writes the equation $6x + 2y = 12$. Write a new equation that has: (Lesson 2-17)

a. exactly one solution in common with Elena's equation

b. no solutions in common with Elena's equation

c. infinitely many solutions in common with Elena's equation

7. A restaurant owner wants to see if there is a relationship between the amount of sugar in some food items on her menu and how popular the items are.

She creates a scatter plot to show the relationship between amount of sugar in menu items and the number of orders for those items. The correlation coefficient for the line of best fit is 0.58. (Lesson 3-9)

a. Are the two variables correlated? Explain your reasoning.

b. Does either of the variables cause the other to change? Explain your reasoning.

Lesson 4-4

Using Function Notation to Describe Rules (Part 1)

NAME _____ DATE _____ PERIOD _____

Learning Goal Let's look at some rules that describe functions and write some, too.

 ## Warm Up
4.1 Notice and Wonder: Two Functions

What do you notice? What do you wonder?

x	$f(x) = 10 - 2x$
1	8
1.5	7
5	0
-2	14

x	$g(x) = x^3$
-2	-8
0	0
1	1
3	27

Here are descriptions and equations that represent four functions.

1. Match each equation with a verbal description that represents the same function. Record your results.

   $f(x) = 3x - 7$

   A To get the output, subtract 7 from the input, then divide the result by 3.

   $g(x) = 3(x - 7)$

   B To get the output, subtract 7 from the input, then multiply the result by 3.

   $h(x) = \frac{x}{3} - 7$

   C To get the output, multiply the input by 3, then subtract 7 from the result.

   $k(x) = \frac{x - 7}{3}$

   D To get the output, divide the input by 3, and then subtract 7 from the result.

2. For one of the functions, when the input is 6, the output is -3. Which is that function: f, g, h, or k? Explain how you know.

3. Which function value—$f(x)$, $g(x)$, $h(x)$, or $k(x)$—is the greatest when the input is 0? What about when the input is 10?

Are you ready for more?

Mai says $f(x)$ is always greater than $g(x)$ for the same value of x. Is this true? Explain how you know.

NAME _____ DATE _____ PERIOD _____

Activity

4.3 Rules for Area and Perimeter

1. A square that has a side length of 9 cm has an area of 81 cm². The relationship between the side length and the area of the square is a function.

   a. Complete the table with the area for each given side length.

   Then, write a rule for a function, A, that gives the area of the square in cm² when the side length is s cm. Use function notation.

Side Length (cm)	Area (cm²)
1	
2	
4	
6	
s	

   b. What does $A(2)$ represent in this situation? What is its value?

   c. On the coordinate plane, sketch a graph of this function.

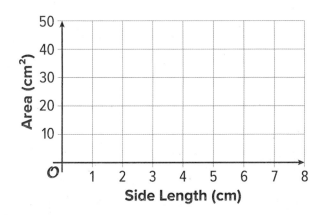

2. A roll of paper that is 3 feet wide can be cut to any length.

   a. If we cut a length of 2.5 feet, what is the perimeter of the paper?

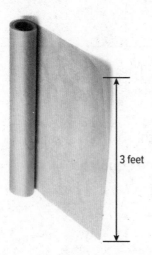

3 feet

   b. Complete the table with the perimeter for each given side length.

      Then, write a rule for a function, P, that gives the perimeter of the paper in feet when the side length in feet is ℓ. Use function notation.

Side Length (feet)	Perimeter (feet)
1	
2	
6.3	
11	
ℓ	

   c. What does $P(11)$ represent in this situation? What is its value?

   d. On the coordinate plane, sketch a graph of this function.

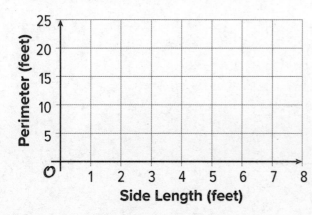

Shutterstock / Coprid

NAME _____ DATE _____ PERIOD _____

Summary
Using Function Notation to Describe Rules (Part 1)

Some functions are defined by rules that specify how to compute the output from the input. These rules can be verbal descriptions or expressions and equations. For example:

Rules in words:

- To get the output of function f, add 2 to the input, then multiply the result by 5.

- To get the output of function m, multiply the input by $\frac{1}{2}$ and subtract the result from 3.

Rules in function notation:

- $f(x) = (x + 2) \cdot 5$ or $5(x + 2)$

- $m(x) = 3 - \frac{1}{2}x$

Some functions that relate two quantities in a situation can also be defined by rules and can therefore be expressed algebraically, using function notation.

Suppose function c gives the cost of buying n pounds of apples at \$1.49 per pound. We can write the rule $c(n) = 1.49n$ to define function c.

To see how the cost changes when n changes, we can create a table of values.

Pounds of Apples, n	Cost in Dollars, $c(n)$
0	0
1	1.49
2	2.98
3	4.47
n	$1.49n$

Plotting the pairs of values in the table gives us a graphical representation of c.

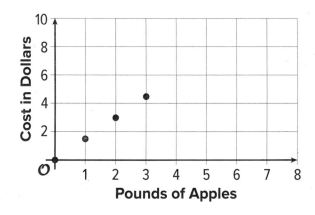

Practice

Using Function Notation to Describe Rules (Part 1)

1. Match each equation with a description of the function it represents.

   A $f(x) = 2x + 4$

   B $g(x) = 2(x + 4)$

   C $h(x) = 4x + 2$

   D $k(x) = 4(x + 2)$

   1 To get the output, add 4 to the input, then multiply the result by 2.

   2 To get the output, add 2 to the input, then multiply the result by 4.

   3 To get the output, multiply the input by 2, then add 4 to the result.

   4 To get the output, multiply the input by 4, then add 2 to the result.

2. Function P represents the perimeter, in inches, of a square with side length x inches.

   a. Complete the table.

x	0	1	2	3	4	5	6
$P(x)$							

   b. Write an equation to represent function P.

   c. Sketch a graph of function P.

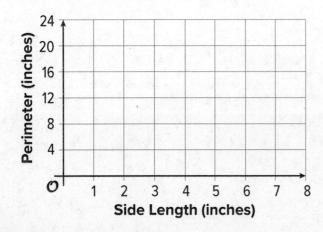

NAME _____ DATE _____ PERIOD _____

3. Functions f and A are defined by these equations.

$f(x) = 80 - 15x$

$A(x) = 25 + 10x$

Which function has a greater value when x is 2.5?

4. An equilateral triangle has three sides of equal length. Function P gives the perimeter of an equilateral triangle of side length s.

a. Find $P(2)$

b. Find $P(10)$

c. Find $P(s)$

5. Imagine a situation where a person is using a garden hose to fill a child's pool. Think of two quantities that are related in this situation and that can be seen as a function. **(Lesson 4-1)**

a. Define the function using a statement of the form " _____ is a function of _____. Be sure to consider the units of measurement.

b. Sketch a possible graph of the function. Be sure to label the axes.

Then, identify the coordinates of one point on the graph and explain its meaning.

6. Function C gives the cost, in dollars, of buying *n* apples.

   Which statement best represents the meaning of $C(10) = 9$? (Lesson 4-2)

   (A.) The cost of buying 9 apples

   (B.) The cost of 9 apples is $10.

   (C.) The cost of 10 apples

   (D.) Ten apples cost $9.

7. Diego is baking cookies for a fundraiser. He opens a 5-pound bag of flour and uses 1.5 pounds of flour to bake the cookies.

   Which equation or inequality represents *f*, the amount of flour left in the bag after Diego bakes the cookies? (Lesson 2-18)

   (A.) $f = 1.5$

   (B.) $f < 1.5$

   (C.) $f = 3.5$

   (D.) $f > 3.5$

Lesson 4-5

Using Function Notation to Describe Rules (Part 2)

NAME _____ DATE _____ PERIOD _____

Learning Goal Let's graph and find the values of some functions.

Warm Up
5.1 Make It True

Consider the equation $q = 4 + 0.8p$.

1. What value of q would make the equation true when:

   a. p is 7?

   b. p is 100?

2. What value of p would make the equation true when:

   a. q is 12?

   b. q is 60?

Be prepared to explain or show your reasoning.

A college student is choosing between two data plans for her new cell phone. Both plans include an allowance of 2 gigabytes of data per month. The monthly cost of each option can be seen as a function and represented with an equation:

- Option A: $A(x) = 60$

- Option B: $B(x) = 10x + 25$

In each function, the input, x, represents the gigabytes of data used *over* the monthly allowance.

1. The student decides to find the values of $A(1)$ and $B(1)$ and compare them. What are those values?

2. After looking at some of her past phone bills, she decided to compare $A(7.5)$ and $B(7.5)$. What are those values?

3. Describe each data plan in words.

4. Graph each function on the same coordinate plane. Then, explain which plan you think she should choose.

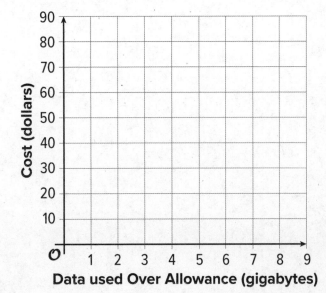

5. The student only budgeted $50 a month for her cell phone. She thought, "I wonder how many gigabytes of data I would have for $50 if I go with Option B?" and wrote $B(x) = 50$. What is the answer to her question?

Explain or show how you know.

Are you ready for more?

Describe a different data plan that, for any amount of data used, would cost no more than one of the given plans and no less than the other given plan. Explain or show how you know this data plan would meet these requirements.

Activity

5.3 Function Notation and Graphing Technology

The function B is defined by the equation $B(x) = 10x + 25$. Use graphing technology to:

1. Find the value of each expression:

   $B(6)$

   $B(2.75)$

   $B(1.482)$

2. Solve each equation:

   $B(x) = 93$

   $B(x) = 42.1$

   $B(x) = 116.25$

Summary

Using Function Notation to Describe Rules (Part 2)

Knowing the rule that defines a function can be very useful. It can help us to:

- Find the output when we know the input.

  - If the rule $f(x) = 5(x + 2)$ defines f, we can find $f(100)$ by evaluating $5(100 + 2)$.

  - If $m(x) = 3 - \frac{1}{2}x$ defines function m, we can find $m(10)$ by evaluating $3 - \frac{1}{2}(10)$.

- Create a table of values.

  Here are tables representing functions f and m:

x	$f(x) = 5(x + 2)$
0	10
1	15
2	20
3	25
4	30

x	$m(x) = 3 - \frac{1}{2}x$
0	3
1	$2\frac{1}{2}$
2	2
3	$1\frac{1}{2}$
4	1

NAME _____ DATE _____ PERIOD _____

- Graph the function. The horizontal values represent the input, and the vertical values represent the output.

  For function f, the values of $f(x)$ are the vertical values, which are often labeled y, so we can write $y = f(x)$. Because $f(x)$ is defined by the expression $5(x + 2)$, we can graph $y = 5(x + 2)$.

  For function m, we can write $y = m(x)$ and graph $y = 3 - \frac{1}{2}x$.

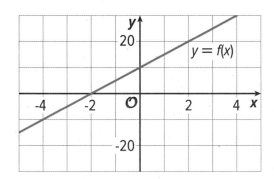 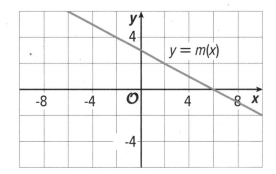

- Find the input when we know the output.

  Suppose the output of function f is 65 at some value of x, or $f(x) = 65$, and we want to find out what that value is. Because $f(x)$ is equal to $5(x + 2)$, we can write $5(x + 2) = 65$ and solve for x.

  $5(x + 2) = 65$

  $\quad x + 2 = 13$

  $\quad\quad x = 11$

Each function here is a **linear function** because the value of the function changes by a constant rate and its graph is a line.

Glossary

linear function

1. The cell phone plan from Company C costs $10 per month, plus $15 per gigabyte for data used. The plan from Company D costs $80 per month, with unlimited data.

   Rule C gives the monthly cost, in dollars, of using g gigabytes of data on Company C's plan. Rule D gives the monthly cost, in dollars, of using g gigabytes of data on Company D's plan.

   a. Write a sentence describing the meaning of the statement $C(2) = 40$.

   b. Which is less, $C(4)$ or $D(4)$? What does this mean for the two phone plans?

   c. Which is less, $C(5)$ or $D(5)$? Explain how you know.

   d. For what number g is $C(g) = 130$?

   e. Draw the graph of each function.

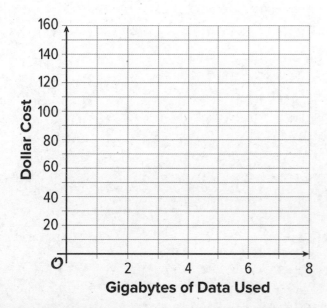

NAME _____ DATE _____ PERIOD _____

2. Function g is represented by the graph.
For what input value or values is $g(x) = 4$?

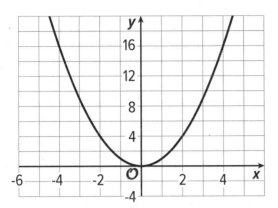

(A.) 2

(B.) -2 and 2

(C.) 16

(D.) none

3. Function P gives the perimeter of an equilateral triangle of side length s. It is represented by the equation $P(s) = 3s$.

a. What does $P(s) = 60$ mean in this situation?

b. Find a value of s to make the equation $P(s) = 60$ true.

4. Function G takes a student's first name for its input and gives the number of letters in the first name for its output. **(Lesson 4-2)**

a. Describe the meaning of $G(\text{Jada}) = 4$.

b. Find the value of $G(\text{Diego})$.

5. *W* gives the weight of a puppy, in pounds, as a function of its age, *t*, in months.

Describe the meaning of each statement in function notation. (Lesson 4-3)

a. $W(2) = 5$

b. $W(6) > W(4)$

c. $W(12) = W(15)$

6. Diego is building a fence for a rectangular garden. It needs to be at least 10 feet wide and at least 8 feet long. The fencing he uses costs $3 per foot. His budget is $120. (Lesson 2-18)

He wrote some inequalities to represent the constraints in this situation:

$f = 2x + 2y$ $\qquad$ $x \geq 10$ $\qquad$ $y \geq 8$ $\qquad$ $3f \leq 120$

a. Explain what each equation or inequality represents.

b. His mom says he should also include the inequality $f > 0$. Do you agree? Explain your reasoning.

Lesson 4-6

Features of Graphs

NAME _____ DATE _____ PERIOD _____

Learning Goal Let's use graphs of functions to learn about situations.

Warm Up
6.1 Walking Home

Diego is walking home from school at a constant rate. This graph represents function d, which gives his distance from home, in kilometers, m minutes since leaving the school.

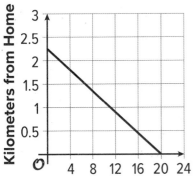

Use the graph to find or estimate:

1. $d(0)$

2. $d(12)$

3. the solution to $d(m) = 1$

4. the solution to $d(m) = 0$

Activity

6.2 A Toy Rocket and a Drone

A toy rocket and a drone were launched at the same time.

Here are the graphs that represent the heights of two objects as a function of time since they were launched.

Height is measured in meters above the ground and time is measured in seconds since launch.

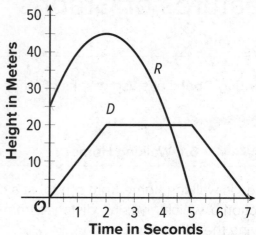

1. Analyze the graphs and describe—as precisely as you can—what was happening with each object. Your descriptions should be complete and precise enough that someone who is not looking at the graph could visualize how the objects were behaving.

NAME _____ DATE _____ PERIOD _____

2. Which parts or features of the graphs show important information about each object's movement? List the features or mark them on the graphs.

Activity

6.3 The Jump

In a bungee jump, the height of the jumper is a function of time since the jump begins.

Function h defines the height, in meters, of a jumper above a river, t seconds since leaving the platform.

Here is a graph of function h, followed by five expressions or equations and five graphical features.

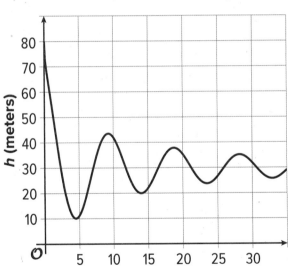

$h(0)$ • first dip in the graph

$h(t) = 0$ • vertical intercept

$h(4)$ • first peak in the graph

$h(t) = 80$ • horizontal intercept

$h(t) = 45$ • maximum

1. Match each description about the jump to a corresponding expression or equation and to a feature on the graph.

   One expression or equation does not have a matching verbal description. Its corresponding graphical feature is also not shown on the graph. Interpret that expression or equation in terms of the jump and in terms of the graph of the function. Record your interpretation in the last row of the table.

Description of Jump	Expression or Equation	Feature of graph
a. The greatest height that the jumper is from the river		
b. The height from which the jumper was jumping		
c. The time at which the jumper reached the highest point after the first bounce		
d. The lowest point that the jumper reached in the entire jump		
e.		

2. Use the graph to:

   a. estimate $h(0)$ and $h(4)$

   b. estimate the solutions to $h(t) = 45$ and $h(t) = 0$

NAME _____ DATE _____ PERIOD _____

Are you ready for more?

Based on the information available, how long do you think the bungee cord is? Make an estimate and explain your reasoning.

The graph of the function can give us useful information about the quantities in a situation. Some points and features of a graph are particularly informative, so we pay closer attention to them.

Let's look at the graph of function *h*, which gives the height, in meters, of a ball *t* seconds after it is tossed up in the air. From the graph, we can see that:

- The point (0, 20) is the *vertical intercept* of the graph, or the point where the graph intersects the vertical axis.

  This point tells us that the initial height of the ball is 20 meters, because when *t* is 0, the value of *h(t)* is 20.

  The statement $h(0) = 20$ captures this information.

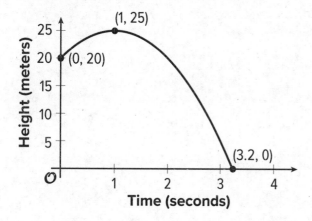

- The point (1, 25) is the highest point on the graph, so it is a **maximum** of the graph.

  The value 25 is also the maximum value of the function *h*. It tells us that the highest point the ball reaches is 25 feet, and that this happens 1 second after the ball is tossed.

- The point (3.2, 0) is a *horizontal intercept* of the graph, a point where the graph intersects the horizontal axis. This point is also the lowest point on the graph, so it represents a **minimum** of the graph.

  This point tells us that the ball hits the ground 3.2 seconds after being tossed up, so the height of the ball is 0 when *t* is 3.2, which we can write as $h(3.2) = 0$. Because *h* cannot have any lower value, 0 is also the minimum value of the function.

NAME _____ DATE _____ PERIOD _____

- The height of the graph increases when *t* is between 0 and 1. Then, the graph changes direction and the height decreases when *t* is between 1 and 3.2. Neither the increasing part nor the decreasing part is a straight line.

  This suggests that the ball increases in height in the first second after being tossed, and then starts falling between 1 second and 3.2 seconds. It also tells us that the height does not increase or decrease at a constant rate.

Because the intercepts of a graph are points on an axis, at least one of their coordinates is 0. The 0 corresponds to the input or the output of a function, or both.

- A vertical intercept is on the vertical axis, so its coordinates have the form (0, *b*), where the first coordinate is 0 and *b* can be any number. The 0 is the input.

- A horizontal intercept is on the horizontal axis, so its coordinates have the form (*a*, 0), where *a* can be any number and the second coordinate is 0. The 0 is an output.

- A graph that passes through (0, 0) intersects both axes, so that point is both a horizontal intercept and a vertical intercept. Both the input and output are 0.

Glossary

decreasing (function)

horizontal intercept

increasing (function)

maximum

minimum

vertical intercept

1. This graph represents Andre's distance from his bicycle as he walks in a park.

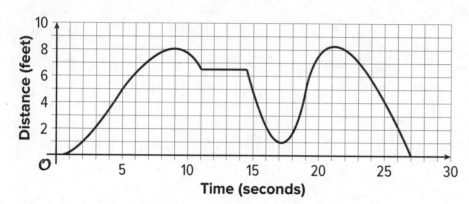

Decide whether the following statements are true or false.

a. The graph has multiple horizontal intercepts.

b. A horizontal intercept of the graph represents the time when Andre was with his bike.

c. A minimum of the graph is (17, 1).

d. The graph has two maximums.

e. About 21 seconds after he left his bike, he was the farthest away from it, at about 8.3 feet.

2. The graph represents the temperature in degrees Fahrenheit as a function of time.

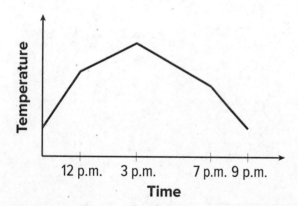

Tell the story of the temperature throughout the day.

NAME _____ DATE _____ PERIOD _____

Identify the maximum and minimum of the function and where the function is increasing and decreasing.

3. Match each feature of the situation with a corresponding statement in function notation.

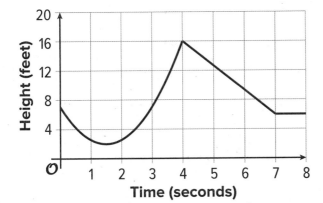

A   maximum height

B   minimum height

C   height staying the same

D   starting height

1   $h(0) = 7$

2   $h(1.5)$

3   $h(4)$

4   $h(t) = 6$ for $7 \leq t \leq 8$

4. Here are the equations that define three functions.

$f(x) = 4x - 5$ $g(x) = 4(x - 5)$ $h(x) = \frac{x}{4} - 5$

   a. Which function value is the largest: $f(100)$, $g(100)$, or $h(100)$?

   b. Which function value is the largest: $f(-100)$, $g(-100)$, or $h(-100)$?

   c. Which function value is the largest: $f\left(\frac{1}{100}\right)$, $g\left(\frac{1}{100}\right)$, or $h\left(\frac{1}{100}\right)$?

5. Function f is defined by the equation $f(x) = x^2$. (Lesson 4-4)

   a. What is $f(2)$?

   b. What is $f(3)$?

   c. Explain why $f(2) + f(3) \neq f(5)$.

6. Priya bought two plants for a science experiment. When she brought them home, the first plant was 5 cm tall and the second plant was 4 cm. Since then, the first plant has grown 0.5 cm a week and the second plant has grown 0.75 cm a week. (Lesson 2-19)

   a. Which plant is taller at the end of 2 weeks? Explain your reasoning.

   b. Which plant is taller at the end of 10 weeks? Explain your reasoning.

   c. Priya represents this situation with the equation $5 + 0.5w = 4 + 0.75w$, where w represents the end of week w. What does the solution to this equation, $w = 4$ represent in this situation?

   d. What does the solution to the inequality $5 + 0.5w > 4 + 0.75w$ represent in this situation?

Lesson 4-7

Using Graphs to Find Average Rate of Change

NAME _____ DATE _____ PERIOD _____

Learning Goal Let's measure how quickly the output of a function changes.

Warm Up
7.1 Temperature Drop

Here are the recorded temperatures at three different times on a winter evening.

Time	4 p.m.	6 p.m.	10 p.m.
Temperature	25°F	17°F	8°F

Tyler says the temperature dropped faster between 4 p.m. and 6 p.m. Mai says the temperature dropped faster between 6 p.m. and 10 p.m. Who do you agree with? Explain your reasoning.

Activity

7.2 Drop Some More

The table and graph show a more complete picture of the temperature changes on the same winter day. The function *T* gives the temperature in degrees Fahrenheit, *h* hours since noon.

h	T(h)
0	18
1	19
2	20
3	20
4	25
5	23
6	17
7	15
8	11
9	11
10	8
11	6
12	7

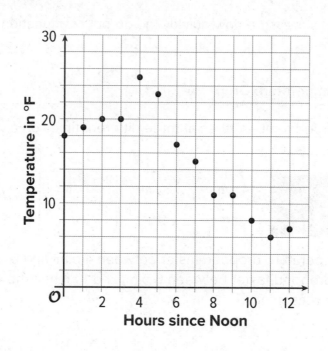

1. Find the **average rate of change** for the following intervals. Explain or show your reasoning.

    a. between noon and 1 p.m.

    b. between noon and 4 p.m.

    c. between noon and midnight

NAME _____ DATE _____ PERIOD _____

2. Remember Mai and Tyler's disagreement? Use average rate of change to show which time period—4 p.m. to 6 p.m. or 6 p.m. to 10 p.m.—experienced a faster temperature drop.

Are you ready for more?

1. Over what interval did the temperature decrease the most rapidly?

2. Over what interval did the temperature increase the most rapidly?

Activity

7.3 Populations of Two States

The graphs show the populations of California and Texas over time.

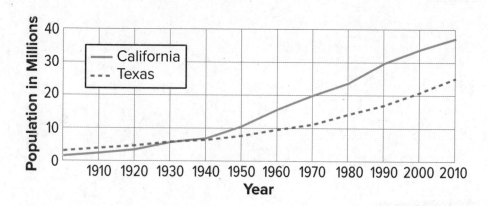

1. Respond to each question.

   a. Estimate the average rate of change in the population in each state between 1970 and 2010. Show your reasoning.

   b. In this situation, what does each rate of change mean?

2. Which state's population grew more quickly between 1900 and 2000? Show your reasoning.

NAME _____ DATE _____ PERIOD _____

Summary

Using Graphs to Find Average Rate of Change

Here is a graph of one day's temperature as a function of time.

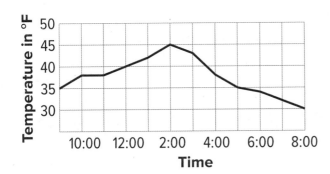

The temperature was 35°F at 9 a.m. and 45°F at 2 p.m., an increase of 10°F over those 5 hours.

The increase wasn't constant, however. The temperature rose from 9 a.m. and 10 a.m., stayed steady for an hour, then rose again.

On average, how fast was the temperature rising between 9 a.m. and 2 p.m.?

Let's calculate the **average rate of change** and measure the temperature change per hour. We do that by finding the difference in the temperature between 9 a.m. and 2 p.m. and dividing it by the number of hours in that interval.

$$\text{average rate of change} = \frac{45 - 35}{5} = \frac{10}{5} = 2$$

On average, the temperature between 9 a.m. and 2 p.m. increased 2°F per hour.

How quickly was the temperature falling between 2 p.m. and 8 p.m.?

$$\text{average rate of change} = \frac{30 - 45}{6} = \frac{-15}{6} = \text{-}2.5$$

On average, the temperature between 2 p.m. and 8 p.m. dropped by 2.5°F per hour.

In general, we can calculate the average rate of change of a function f, between input values a and b, by dividing the difference in the outputs by the difference in the inputs.

$$\text{average rate of change} = \frac{f(b) - f(a)}{b - a}$$

If the two points on the graph of the function are $(a, f(a))$ and $(b, f(b))$, the average rate of change is the slope of the line that connects the two points.

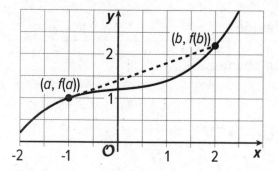

Glossary

average rate of change

NAME _____ DATE _____ PERIOD _____

Practice
Using Graphs to Find Average Rate of Change

1. The temperature was recorded at several times during the day. Function T gives the temperature in degrees Fahrenheit, n hours since midnight.

   Here is a graph for this function.

   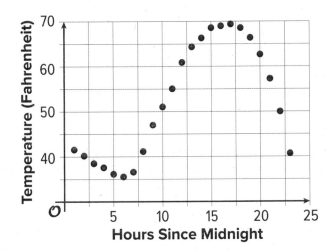

   For each time interval, decide if the average rate of change is positive, negative, or zero:

   a. From $n = 1$ to $n = 5$

   b. From $n = 5$ to $n = 7$

   c. From $n = 10$ to $n = 20$

   d. From $n = 15$ to $n = 18$

   e. From $n = 20$ to $n = 24$

2. The graph shows the total distance, in feet, walked by a person as a function of time, in seconds.

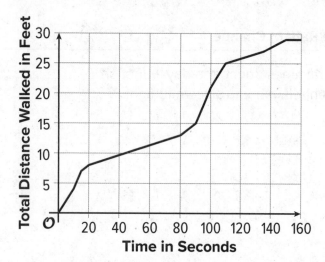

a. Was the person walking faster between 20 and 40 seconds or between 80 and 100 seconds?

b. Was the person walking faster between 0 and 40 seconds or between 40 and 100 seconds?

3. The height, in feet, of a squirrel running up and down a tree is a function of time, in seconds.

Here are statements describing the squirrel's movement during four intervals of time. Match each description with a statement about the average rate of change of the function for that interval.

A The squirrel runs up the tree very fast.

B The squirrel starts and ends at the same height.

C The squirrel runs down the tree.

D The squirrel runs up the tree slowly.

1 The average rate of change is negative.

2 The average rate of change is zero.

3 The average rate of change is small and positive.

4 The average rate of change is large and positive.

NAME _____ DATE _____ PERIOD _____

4. The percent of voters between the ages of 18 and 29 that participated in each United States presidential election between the years 1988 to 2016 are shown in the table.

Year	1988	1992	1996	2000	2004	2008	2012	2016
Percentage of Voters Ages 18-29	35.7	42.7	33.1	34.5	45.0	48.4	40.9	43.4

The function P gives the percent of voters between 18 and 29 years old that participated in the election in year t.

a. Determine the average rate of change for P between 1992 and 2000.

b. Pick two different values of t so that the function has a negative average rate of change between the two values. Determine the average rate of change.

c. Pick two values of t so that the function has a positive average rate of change between the two values. Determine the average rate of change.

5. Jada walks to school. The function D gives her distance from school, in meters, as a function of time, in minutes, since she left home.

What does $D(10) = 0$ represent in this situation? **(Lesson 4-2)**

6. Jada walks to school. The function D gives her distance from school, in meters, t minutes since she left home. **(Lesson 4-2)**

Which equation tells us "Jada is 600 meters from school after 5 minutes"?

(A.) $D(5) = 600$

(B.) $D(600) = 5$

(C.) $t(5) = 600$

(D.) $t(600) = 5$

7. A news website shows a scatter plot with a positive relationship between the number of vending machines in a school and the percentage of students who are absent from school on average. The headline reads, "Vending machines are causing our youth to miss school!" **(Lesson 3-9)**

   a. What is wrong with this claim?

   b. What is a better headline for this information?

Lesson 4-8

Interpreting and Creating Graphs

NAME _____ DATE _____ PERIOD _____

Learning Goal Let's sketch graphs to represent situations.

Warm Up
8.1 Which One Doesn't Belong: Temperature Over Time

Which graph doesn't belong?

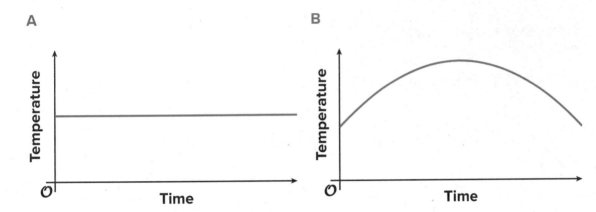

A

B

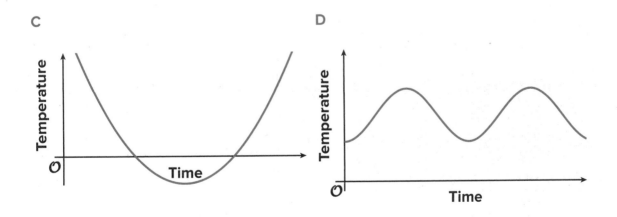

C

D

Activity

8.2 Flag Raising (Part 1)

A flag ceremony is held at a Fourth of July event. The height of the flag is a function of time. Here are some graphs that could each be a possible representation of the function.

A

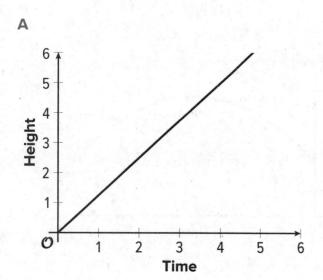

B

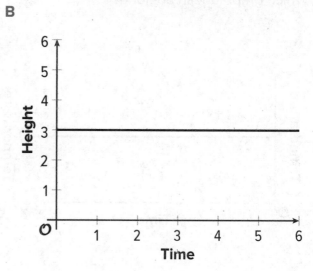

C

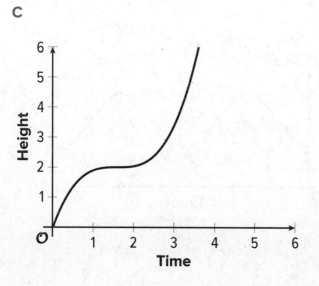

D

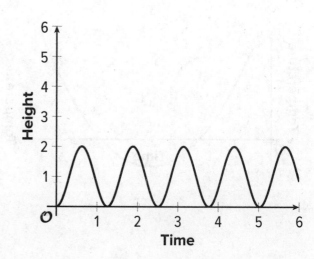

NAME _____ DATE _____ PERIOD _____

E

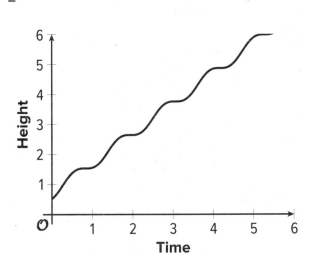

F

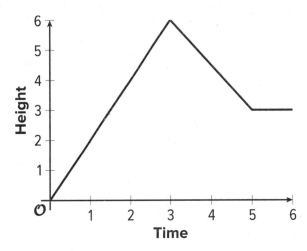

1. Respond to each question.

   a. For each graph assigned to you, explain what it tells us about the flag.

   b. Decide as a group which graph(s) appear to be most realistic and which ones least realistic.

2. Here is another graph that relates time and height.

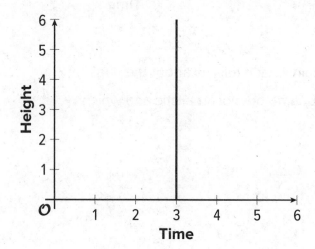

a. Can this graph represent the time and height of the flag? Explain your reasoning.

b. Is this a graph of a function? Explain your reasoning.

NAME _____ DATE _____ PERIOD _____

Are you ready for more?

Suppose an ant is moving at a rate of 1 millimeter per second and keeps going at that rate for a long time.

If time, x, is measured in seconds, then the distance the ant has traveled in millimeters, y, is $y = 1x$. If time, x, is measured in minutes, the distance in millimeters is $y = 60x$.

1. Explain why the equation $y = (365 \cdot 24 \cdot 3{,}600)x$ gives the distance the ant has traveled, in millimeters, as a function of time, x, in years.

2. Use graphing technology to graph the equation.

   a. Label the axes with appropriate quantities and units.

   b. Does the graph look like that of a function? Why do you think it looks this way?

3. Adjust the graphing window until the graph no longer looks this way. If you manage to do so, describe the graphing window that you use.

4. Do you think the last graph in the flag activity could represent a function relating time and height of flag? Explain your reasoning.

NAME _____ DATE _____ PERIOD _____

Activity
8.3 Flag Raising (Part 2)

Use the internet to find a video of a flag being raised.

1. On the coordinate plane, sketch a graph that could represent function *H*, the height of the flag over time. Be sure to include a label and a scale for each axis.

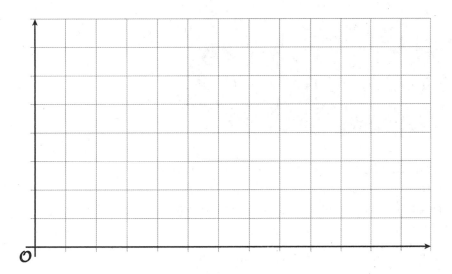

2. Use your graph to estimate the average rate of change from the time the flag starts moving to the time it stops. Be prepared to explain what the average rate of change tells us about the flag.

To prepare for a backyard party, a parent uses two identical hoses to fill a small pool that is 15 inches deep and a large pool that is 27 inches deep.

The height of the water in each pool is a function of time since the water is turned on.

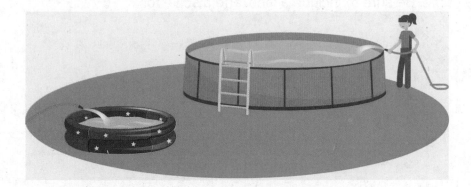

Here are descriptions of three situations. For each situation, sketch the graphs of the two functions on the same coordinate plane, so that $S(t)$ is the height of the water in the small pool after t minutes, and $L(t)$ is the height of the water in the large pool after t minutes.

In both functions, the height of the water is measured in inches.

Situation 1: Each hose fills one pool at a constant rate. When the small pool is full, the water for that hose is shut off. The other hose keeps filling the larger pool until it is full.

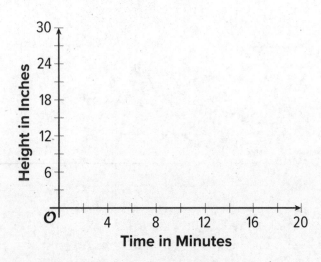

Situation 2: Each hose fills one pool at a constant rate. When the small pool is full, both hoses are shut off.

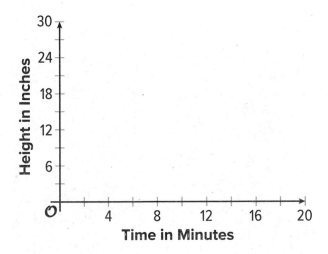

Situation 3: Each hose fills one pool at a constant rate. When the small pool is full, both hoses are used to fill the large pool until it is full.

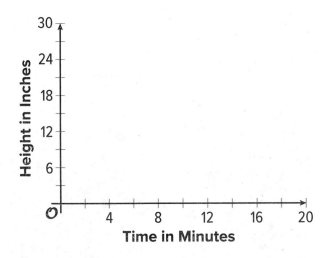

 Activity

8.5 The Bouncing Ball

Your teacher will show you one or more videos of a tennis ball being dropped. Here are some still images of the situation.

The height of the ball is a function of time. Suppose the height is h feet, t seconds after the ball is dropped.

1. Use the blank coordinate plane to sketch a graph of the height of the tennis ball as a function of time.

To help you get started, here are some pictures and a table. Complete the table with your estimates before sketching your graph.

0
seconds

0.28
seconds

0.54
seconds

0.74
seconds

1.03
seconds

1.48
seconds

1.88
seconds

2.25
seconds

NAME _____ DATE _____ PERIOD _____

Time (seconds)	Height (feet)
0	
0.28	
0.54	
0.74	
1.03	
1.48	
1.88	
2.25	

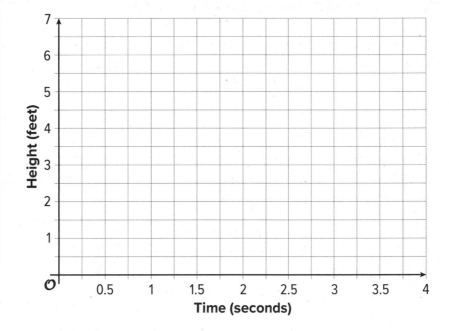

2. Identify horizontal and vertical intercepts of the graph. Explain what the coordinates tell us about the tennis ball.

3. Find the maximum and minimum values of the function. Explain what they tell us about the tennis ball.

Are you ready for more?

If you only see the still images of the ball and not the video of the ball bouncing, can you accurately graph the height of the ball as a function of time? Explain your reasoning.

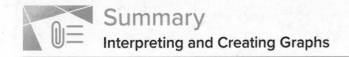
We can use graphs to help visualize the relationship between quantities in a situation, even if we only have a general description.

Here is a description of a hiker's journey on a trail:

A hiker walked briskly and steadily for about 30 minutes and then took a 10-minute break. Afterward, she jogged all the way to the end of the trail, which took about 20 minutes. There, she took a 15-minute break, and then started walking back leisurely, stopping twice to enjoy the scenery. Her return trip along the same trail took 105 minutes.

We can sketch a graph of the distance the hiker has traveled as a function of time based on this description.

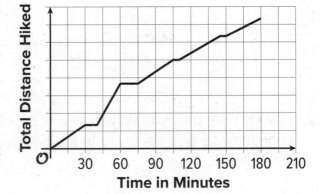

Even though we don't know the specific distances she has traveled or the length of the trail, we can show in the graph some important features of the situation. For example:

- the intervals in which the distance increased or stayed constant

- how quickly the distance was increasing

- the time when the hiker traveled the greatest distance on that hike

If we are looking at distance from the trailhead (the start of the trail) as a function of time, the graph of the function might look something like this:

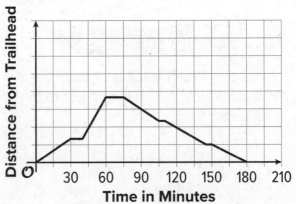

It shows the distance increasing as the hiker was walking away from the trailhead, and then decreasing as she was returning to the trailhead.

NAME _____ DATE _____ PERIOD _____

Practice
Interpreting and Creating Graphs

1. The graphs show the distance, *d*, traveled by two cars, A and B, over time, *t*. Distance is measured in miles and time is measured in hours.
Which car traveled slower? Explain how you know.

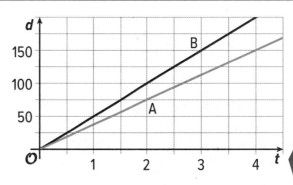

2. Here are descriptions of four situations in which the volume of water in a tank is a function of time. Match each description to a corresponding graph.

Graph 1

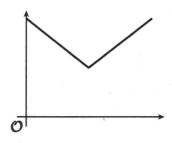

Graph 2

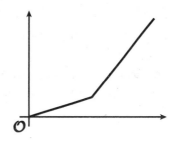

Graph 3

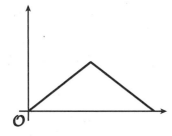

Graph 4

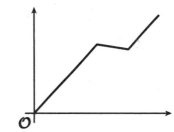

A. An empty 20-gallon water tank is filled at a constant rate for 3 minutes until it is half full. Then, it is emptied at a constant rate for 3 minutes.

B. A full 10-gallon water tank is drained for 30 seconds, until it is half full. Afterwards, it gets refilled.

C. A 2,000-gallon water tank starts out empty. It is being filled for 5 hours, slowly at first, and faster later.

D. An empty 100-gallon water tank is filled in 50 minutes. Then, a dog jumps in and splashes around for 10 minutes, letting 7 gallons of water out. The tank is refilled afterwards.

1. Graph 1
2. Graph 2
3. Graph 3
4. Graph 4

3. Clare describes her morning at school yesterday: "I entered the school on the first floor, then walked up to the third floor and stayed for my class for an hour. Afterwards, I had an hour-long class in the basement, and after that I went back to the ground level and sat outside to eat my lunch." Sketch a possible graph of her height from the ground floor as a function of time.

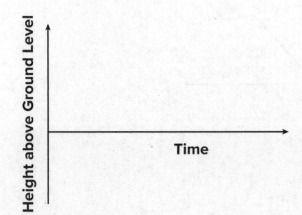

NAME _____ DATE _____ PERIOD _____

4. Tyler filled up his bathtub, took a bath, and then drained the tub. The function B gives the depth of the water, in inches, t minutes after Tyler began to fill the bathtub.

These statements describe how the water level in the tub was changing over time. Use the statements to sketch an approximate graph of function B.

- $B(0) = 0$
- $B(1) < B(7)$
- $B(9) = 11$
- $B(10) = B(23)$
- $B(20) > B(40)$

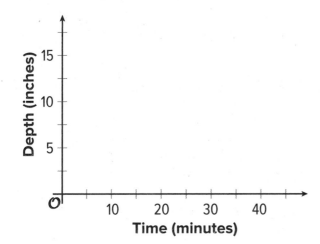

5. Two functions are defined by these equations: **(Lesson 4-5)**

$$f(x) = 5.1 + 0.8x \qquad\qquad g(x) = 3.4 + 1.2x$$

Which function has a greater value when x is 3.9? How much greater?

6. Function f is defined by the equation $f(x) = 3x - 7$. Find the value of c so that $f(c) = 20$ is true. **(Lesson 4-5)**

7. Function V gives the volume of water (liters) in a water cooler as a function of time, t (minutes). **(Lesson 4-6)**

This graph represents function V.

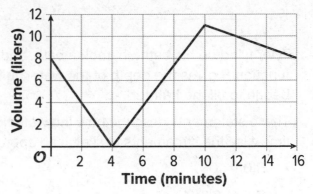

a. What is the greatest water volume in the cooler?

b. Find the value or values of t that make $V(t) = 4$ true. Explain what the value or values tell us about the volume of the water in the cooler.

c. Identify the horizontal intercept of the graph. What does it tell you about the situation?

8. Noah draws this box plot for data that has measure of variability 0.

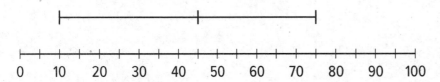

Explain why the box plot is complete even though there do not appear to be any boxes. **(Lesson 1-15)**

Lesson 4-9

Comparing Graphs

NAME _____ DATE _____ PERIOD _____

Learning Goal Let's compare graphs of functions to learn about the situations they represent.

Warm Up
9.1 Population Growth

This graph shows the populations of Baltimore and Cleveland in the 20th century. $B(t)$ is the population of Baltimore in year t. $C(t)$ is the population of Cleveland in year t.

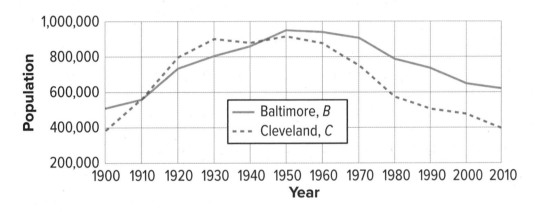

1. Estimate $B(1930)$ and explain what it means in this situation.

2. Here are pairs of statements about the two populations. In each pair, which statement is true? Be prepared to explain how you know.

   a. $B(2000) > C(2000)$ or $B(2000) < C(2000)$

   b. $B(1900) = C(1900)$ or $B(1900) > C(1900)$

3. Were the two cities' populations ever the same? If so, when?

$H(t)$ is the percentage of homes in the United States that have a landline phone in year t. $C(t)$ is the percentage of homes with *only* a cell phone. Here are the graphs of H and C.

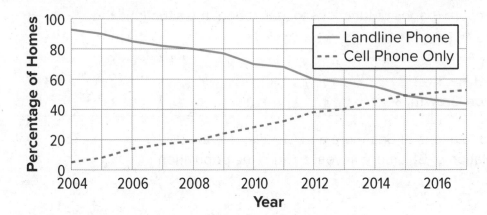

1. Estimate $H(2006)$ and $C(2006)$. Explain what each value tells us about the phones.

2. What is the approximate solution to $C(t) = 20$? Explain what the solution means in this situation.

NAME _____ DATE _____ PERIOD _____

3. Determine if each equation is true. Be prepared to explain how you know.

   a. $C(2011) = H(2011)$

   b. $C(2015) = H(2015)$

4. Between 2004 and 2015, did the percentage of homes with landlines decrease at the same rate at which the percentage of cell-phones-only homes increased? Explain or show your reasoning.

1. Explain why the statement $C(t) + H(t) \leq 100$ is true in this situation.

2. What value does $C(t) + H(t)$ appear to take between 2004 and 2017? How much does this value vary in that interval?

NAME _____ DATE _____ PERIOD _____

Activity
9.3 Audience of TV Shows

The number of people who watched a TV episode is a function of that show's episode number. Here are three graphs of three functions—A,B, and C—representing three different TV shows.

Show A

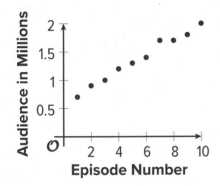

Show B

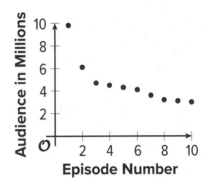

Show C

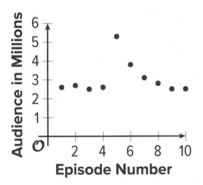

1. Match each description with a graph that could represent the situation described. One of the descriptions has no corresponding graph.

   a. This show has a good core audience. They had a guest star in the fifth episode that brought in some new viewers, but most of them stopped watching after that.

   b. This show is one of the most popular shows, and its audience keeps increasing.

   c. This show has a small audience, but it's improving, so more people are noticing.

   d. This show started out huge. Even though it feels like it crashed, it still has more viewers than another show.

2. Which is greatest, $A(7)$, $B(7)$, or $C(7)$? Explain what the answer tells us about the shows.

3. Sketch a graph of the viewership of the fourth TV show that did not have a matching graph.

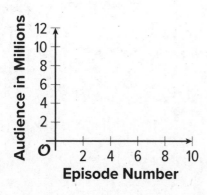

NAME _____ DATE _____ PERIOD _____

Activity
9.4 Functions *f* and *g*

1. Here are graphs that represent two functions, *f* and *g*.

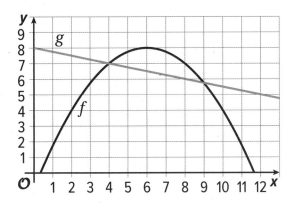

Decide which function value is greater for each given input. Be prepared to explain your reasoning.

   a. $f(2)$ or $g(2)$

   b. $f(4)$ or $g(4)$

   c. $f(6)$ or $g(6)$

   d. $f(8)$ or $g(8)$

2. Is there a value of x at which the equation $f(x) = g(x)$ is true? Explain your reasoning.

3. Identify at least two values of x at which the inequality $f(x) < g(x)$ is true.

Graphs are very useful for comparing two or more functions. Here are graphs of functions C and T, which give the populations (in millions) of California and Texas in year x.

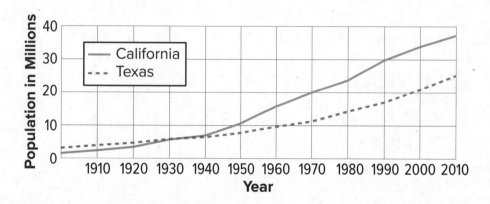

What can We Tell about the Populations?	How can We Tell?	How can We Convey This with Function Notation?
In the early 1900s, California had a smaller population than Texas.	The graph of C is below the graph of T when x is 1900.	$C(1900) < T(1900)$
Around 1935, the two states had the same population of about 5 million people.	The graphs intersect at about (1935, 5).	$C(1935) = 5$ and $T(1935) = 5$, and $C(1935) = T(1935)$
After 1935, California has had more people than Texas.	When x is greater than 1935, the graph of $C(x)$ is above that of $T(x)$.	$C(x) > T(x)$ for $x > 1935$
Both populations have increased over time, with no periods of decline.	Both graphs slant upward from left to right.	
From 1900 to 2010, the population of California has risen faster than that of Texas. California had a greater average rate of change.	If we draw a line to connect the points for 1900 and 2010 on each graph, the line for C has a greater slope than that for T.	$\dfrac{C(2010) - C(1900)}{2010 - 1900} > \dfrac{T(2010) - T(1900)}{2010 - 1900}$

NAME _____ DATE _____ PERIOD _____

Practice
Comparing Graphs

1. Functions *R* and *D* give the height, in feet, of a toy rocket and a drone, *t* seconds after they are released.

   Here are the graphs of *R* (for the rocket) and *D* (for the drone).

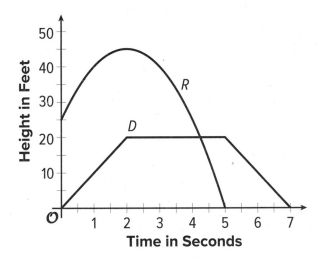

   Write each statement in function notation:

   a. At 3 seconds, the toy rocket is higher than the drone.

   b. At the start, the toy rocket is 25 feet above the drone.

2. *A(t)* is the average high temperature in Aspen, Colorado, *t* months after the start of the year. *M(t)* is the average high temperature in Minneapolis, Minnesota, *t* months after the start of the year. Temperature is measured in degrees Fahrenheit.

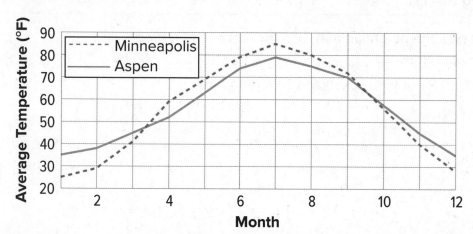

Which function had the higher average rate of change between the beginning of January and middle of March? What does this mean about the temperature in the two cities?

3. Here are two graphs representing functions *f* and *g*.

   Select **all** statements that are true about functions *f* and *g*.

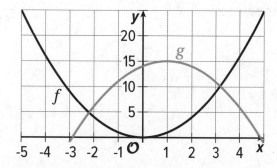

   (A.) $f(0) > g(0)$

   (B.) There are two values of *x* where $f(x) = g(x)$.

   (C.) $f(-1) < g(-1)$

   (D.) $f(-3) > g(4)$

NAME _____ DATE _____ PERIOD _____

4. The three graphs represent the progress of three runners in a 400-meter race.

The solid line represents runner A. The dotted line represents runner B. The dashed line represents runner C.

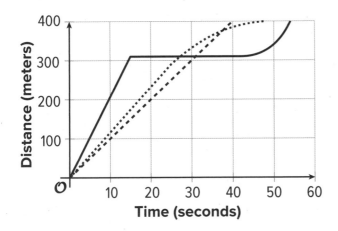

a. One runner ran at a constant rate throughout the race. Which one?

b. A second runner stopped running for a while. Which one? During which interval of time did that happen?

c. Describe the third runner's race. Be as specific as possible.

d. Who won the race? Explain how you know.

5. Function *A* gives the area, in square inches, of a square with side length *x* inches. **(Lesson 4-4)**

a. Complete the table.

x	0	1	2	3	4	5	6
A(x)							

b. Represent function *A* using an equation.

c. Sketch a graph of function *A*.

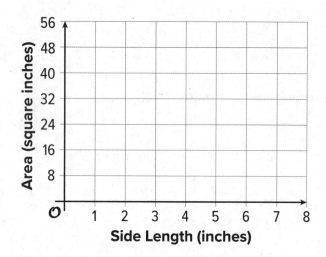

6. Function *f* is represented by $f(x) = 5(x + 11)$. **(Lesson 4-5)**

a. Find $f(-2)$.

b. Find the value of *x* such that $f(x) = 90$ is true.

Lesson 4-10

Domain and Range (Part 1)

NAME _____ DATE _____ PERIOD _____

Learning Goal Let's find all possible inputs and outputs for a function.

 ## Warm Up
10.1 Number of Barks

Earlier, you saw a situation where the total number of times a dog has barked was a function of the time, in seconds, after its owner tied its leash to a post and left. Less than 3 minutes after he left, the owner returned, untied the leash, and walked away with the dog.

1. Could each value be an input of the function? Be prepared to explain your reasoning.

   15                 $84\frac{1}{2}$ 300

2. Could each value be an output of the function? Be prepared to explain your reasoning.

   15                 $84\frac{1}{2}$ 300

Activity

10.2 Card Sort: Possible or Impossible?

Your teacher will give you a set of cards that each contain a number. Decide whether each number is a possible input for the functions described here. Sort the cards into two groups—possible inputs and impossible inputs. Record your sorting decisions.

1. The area of a square, in square centimeters, is a function of its side length, s, in centimeters. The equation $A(s) = s^2$ defines this function.

    a. Possible inputs:

    b. Impossible inputs:

2. A tennis camp charges $40 per student for a full-day camp. The camp runs only if at least 5 students sign up, and it limits the enrollment to 16 campers a day. The amount of revenue, in dollars, that the tennis camp collects is a function of the number of students that enroll.

    The equation $R(n) = 40n$ defines this function.

    a. Possible inputs:

    b. Impossible inputs:

3. The relationship between temperature in Celsius and the temperature in Kelvin can be represented by a function k. The equation $k(c) = c + 273.15$ defines this function, where c is the temperature in Celsius and $k(c)$ is the temperature in Kelvin.

    a. Possible inputs:

    b. Impossible inputs:

NAME _____ DATE _____ PERIOD _____

Activity

10.3 What about the Outputs?

In an earlier activity, you saw a function representing the area of a square (function A) and another representing the revenue of a tennis camp (function R). Refer to the descriptions of those functions to answer these questions.

1. Here is a graph that represents function A, defined by $A(s) = s^2$, where s is the side length of the square in centimeters.

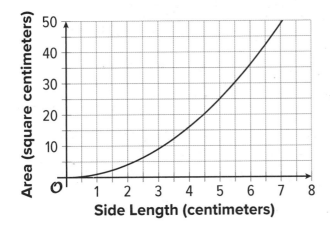

   a. Name three possible input-output pairs of this function.

   b. Earlier we describe the set of all possible input values of A as "any number greater than or equal to 0." How would you describe the set of all possible output values of A?

2. Function R is defined by $R(n) = 40n$, where n is the number of campers.

   a. Is 20 a possible output value in this situation? What about 100? Explain your reasoning.

   b. Here are two graphs that relate number of students and camp revenue in dollars. Which graph could represent function R? Explain why the other one could not represent the function.

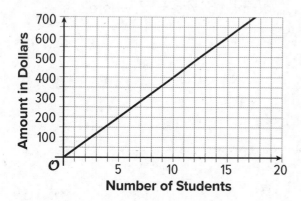 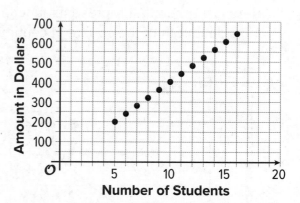

   c. Describe the set of all possible output values of R.

Are you ready for more?

If the camp wishes to collect at least $500 from the participants, how many students can they have? Explain how this information is shown on the graph.

NAME _____ DATE _____ PERIOD _____

Activity

10.4 What Could Be the Trouble?

Consider the function $f(x) = \dfrac{6}{x-2}$.

To find out the sets of possible input and output values of the function, Clare created a table and evaluated f at some values of x. Along the way, she ran into some trouble.

1. Find $f(x)$ for each x-value Clare listed. Describe what Clare's trouble might be.

x	-10	0	$\dfrac{1}{2}$	2	8
$f(x)$					

2. Use graphing technology to graph function f. What do you notice about the graph?

3. Use a calculator to compute the value you and Clare had trouble computing. What do you notice about the computation?

4. How would you describe the domain of function f?

Why do you think the graph of function *f* looks the way it does? Why are there two parts that split at $x = 2$, with one curving down as it approaches $x = 2$ from the left and the other curving up as it approaches $x = 2$ from the right?

Evaluate function *f* at different *x*-values that approach 2 but are not exactly 2, such as 1.8, 1.9, 1.95, 1.999, 2.2, 2.1, 2.05, 2.001, and so on. What do you notice about the values of $f(x)$ as the *x*-values get closer and closer to 2?

NAME _____ DATE _____ PERIOD _____

Summary
Domain and Range (Part 1)

The **domain** of a function is the set of all possible input values. Depending on the situation represented, a function may take all numbers as its input or only a limited set of numbers.

Function A gives the area of a square, in square centimeters, as a function of its side length, s, in centimeters.

- The input of A can be 0 or any positive number, such as 4, 7.5, or $\frac{19}{3}$. It cannot include negative numbers because lengths cannot be negative.

- The domain of A includes 0 and all positive numbers (or $s \geq 0$).

Function q gives the number of buses needed for a school field trip as a function of the number of people, n, going on the trip.

- The input of q can be 0 or positive whole numbers because a negative or fractional number of people doesn't make sense.

- The domain of q includes 0 and all positive whole numbers. If the number of people at a school is 120, then the domain is limited to all non-negative whole numbers up to 120 (or $0 \leq n \leq 120$).

Function v gives the total number of visitors to a theme park as a function of days, d, since a new attraction was open to the public.

- The input of v can be positive or negative. A positive input means days since the attraction was open, and a negative input means days before the attraction was open.

- The input can also be whole numbers or fractional. The statement $v(17.5)$ means 17.5 days after the attraction was open.

- The domain of v includes all numbers. If the theme park had been opened for exactly one year before the new attraction was open, then the domain would be all numbers greater than or equal to -365 (or $d \geq -365$).

The **range** of a function is the set of all possible output values. Once we know the domain of a function, we can determine the range that makes sense in the situation.

- The output of function A is the area of a square in square centimeters, which cannot be negative but can be 0 or greater, not limited to whole numbers. The range of A is 0 and all positive numbers.

- The output of q is the number of buses, which can only be 0 or positive whole numbers. If there are 120 people at the school, however, and if each bus could seat 30 people, then only up to 4 buses are needed. The range that makes sense in this situation would be any whole number that is at least 0 and at most 4.

- The output of function v is the number of visitors, which cannot be fractional or negative. The range of v therefore includes 0 and all positive whole numbers.

Glossary

domain

range

NAME _____ DATE _____ PERIOD _____

Practice
Domain and Range (Part 1)

1. The cost for an upcoming field trip is $30 per student. The cost of the field trip C, in dollars, is a function of the number of students x.

   Select **all** the possible outputs for the function defined by $C(x) = 30x$.

   (A.) 20

   (B.) 30

   (C.) 50

   (D.) 90

   (E.) 100

2. A rectangle has an area of 24 cm². Function f gives the length of the rectangle, in centimeters, when the width is w cm.

   Determine if each value, in centimeters, is a possible input of the function.

   3   0.5   48   -6   0

3. Select **all** the possible input-output pairs for the function $y = x^3$.

   (A.) (-1, -1)

   (B.) (-2, 8)

   (C.) (3, 9)

   (D.) $\left(\frac{1}{2}, \frac{1}{8}\right)$

   (E.) (4, 64)

   (F.) (1, -1)

4. A small bus charges $3.50 per person for a ride from the train station to a concert. The bus will run if at least 3 people take it, and it cannot fit more than 10 people.

   Function B gives the amount of money that the bus operator earns when n people ride the bus.

   a. Identify all numbers that make sense as inputs and outputs for this function.

   b. Sketch a graph of B.

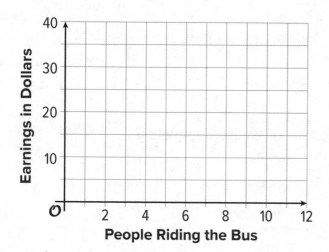

5. Two functions are defined by the equations $f(x) = 5 - 0.2x$ and $g(x) = 0.2(x + 5)$.

   Select **all** statements that are true about the functions. (Lesson 4-5)

   (A.) $f(3) > 0$

   (B.) $f(3) > 5$

   (C.) $g(-1) = 0.8$

   (D.) $g(-1) < f(-1)$

   (E.) $f(0) = g(0)$

NAME _____ DATE _____ PERIOD _____

6. The graph of function *f* passes through the coordinate points (0,3) and (4,6).

   Use function notation to write the information each point gives us about function *f*. (Lesson 4-3)

7. Match each feature of the graph with the corresponding coordinate point.

   If the feature does not exist, choose "none". (Lesson 4-6)

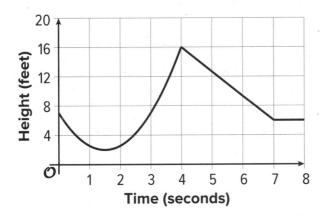

A. maximum   1. (0,7)

B. minimum   2. (1.5,2)

C. vertical intercept   3. (4,16)

D. horizontal intercept   4. none

8. The graphs show the audience, in millions, of two TV shows as a function of the episode number.

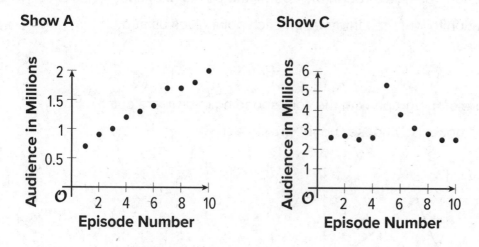

Show A

Show C

For each show, pick two episode numbers between which the function has a negative average rate of change, if possible. Estimate the average rate of change, or explain why it is not possible. **(Lesson 4-9)**

Lesson 4-11

Domain and Range (Part 2)

NAME _____ DATE _____ PERIOD _____

Learning Goal Let's analyze graphs of functions to learn about their domain and range.

Warm Up
11.1 Which One Doesn't Belong: Unlabeled Graphs

Which one doesn't belong?

A

B

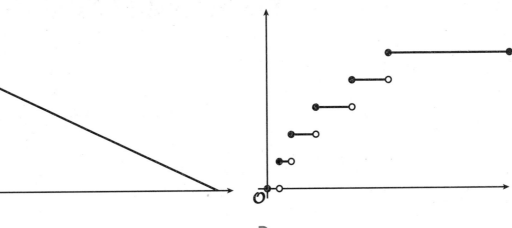

C

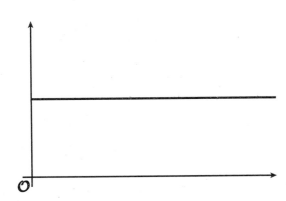

D

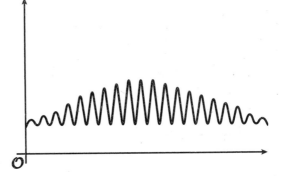

Activity

11.2 Time on the Swing

A child gets on a swing in a playground, swings for 30 seconds, and then gets off the swing.

1. Here are descriptions of four functions in the situation and four graphs representing them.

   The independent variable in each function is time, measured in seconds.

   Match each function with a graph that could represent it. Then, label the axes with the appropriate variables. Be prepared to explain how you make your matches.

A

B

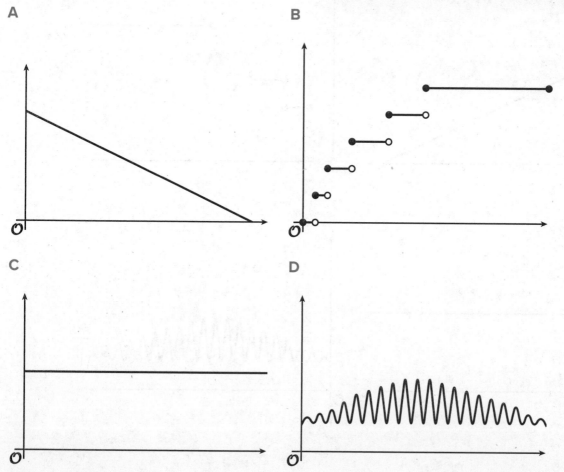

C

D

maximkabb © 123RF.com

- Function *h*: The height of the swing, in feet, as a function of time since the child gets on the swing

- Function *r*: The amount of time left on the swing as a function of time since the child gets on the swing

- Function *d*: The distance, in feet, of the swing from the top beam (from which the swing is suspended) as a function of time since the child gets on the swing

- Function *s*: The total number of times an adult pushes the swing as a function of time since the child gets on the swing

2. On each graph, mark one or two points that—if you have the coordinates—could help you determine the domain and range of the function. Be prepared to explain why you chose those points.

3. Once you receive the information you need from your teacher, describe the domain and range that would be reasonable for each function in this situation.

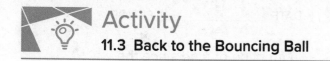
A tennis ball was dropped from a certain height. It bounced several times, rolled along for a short period, and then stopped. Function *H* gives its height over time.

Here is a partial graph of *H*. Height is measured in feet. Time is measured in seconds.

Use the graph to help you answer the questions.

Be prepared to explain what each value or set of values means in this situation.

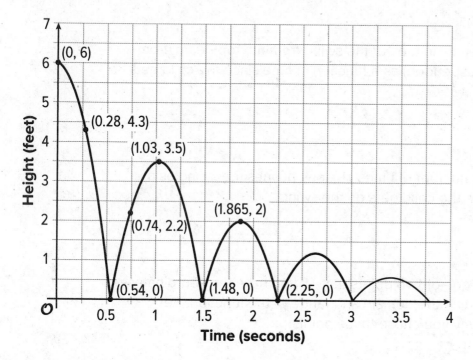

1. Find $H(0)$.

2. Solve $H(x) = 0$.

3. Describe the domain of the function.

4. Describe the range of the function.

NAME _____ DATE _____ PERIOD _____

In function *H*, the input was time in seconds and the output was height in feet.

Think about some other quantities that could be inputs or outputs in this situation.

1. Describe a function whose domain includes only integers. Be sure to specify the units.

2. Describe a function whose range includes only integers. Be sure to specify the units.

3. Sketch a graph of each function.

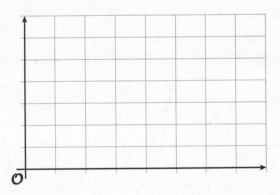
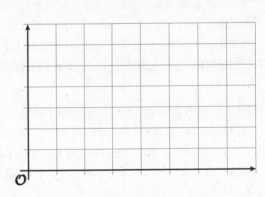

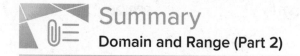

The graph of a function can sometimes give us information about its domain and range.

Here are graphs of two functions we saw earlier in the unit. The first graph represents the best price of bagels as a function of the number of bagels bought. The second graph represents the height of a bungee jumper as a function of seconds since the jump began.

What are the domain and range of each function?

The number of bagels cannot be negative but could include 0 (no bagels bought). The domain of the function therefore includes 0 and positive whole numbers, or $n \geq 0$.

The best price can be $0 (for buying 0 bagels), certain multiples of 1.25, certain multiples of 6, and so on. The range includes 0 and certain positive values.

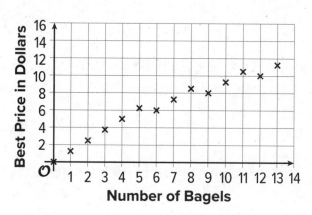

The domain of the height function would include any amount of time since the jump began, up until the jump is complete. From the graph, we can tell that this happened more than 70 seconds after the jump began, but we don't know the exact value of t.

The graph shows a maximum height of 80 meters and a minimum height of 10 meters. We can conclude that the range of this function includes all values that are at least 10 and at most 80.

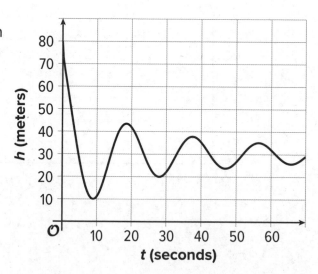

NAME _____ DATE _____ PERIOD _____

Practice
Domain and Range (Part 2)

1. A child tosses a baseball up into the air. On its way down, it gets caught in a tree for several seconds before falling back down to the ground.

   Select the best description of the range of this function.

   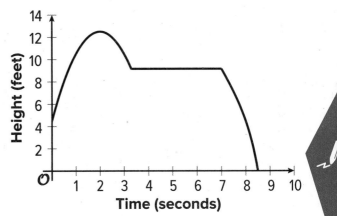

   A. The range includes all numbers from 5 to 12.5.

   B. The range includes all integers between 0 and 12.5.

   C. The range includes all numbers from 0 to 8.5.

   D. The range includes all numbers from 0 to 12.5.

2. To raise funds for a trip, members of a high school math club are holding a game night in the gym. They sell tickets at $5 per person. The gym holds a maximum of 250 people. The amount of money raised is a function of the number of tickets sold.

   Which statement accurately describes the domain of the function?

   A. all numbers less than 250

   B. all integers

   C. all positive integers

   D. all positive integers less than or equal to 250

3. C gives the cost, in dollars, of a cafeteria meal plan as a function of the number of meals purchased, n. The function is represented by the equation $C(n) = 4 + 3n$. **(Lesson 4-5)**

   a. Find a value of n such that $C(n) = 31$ is true.

   b. What does that value of n tell you about the cafeteria meal plan?

4. Lin completes a 5K using a combination of walking and running. Here are four graphs that represent four possible situations. Each graph shows the distance, in meters, as a function of time, in minutes.

Graph 1

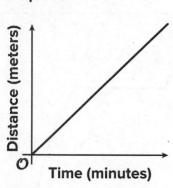

Graph 2

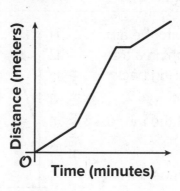

Graph 3

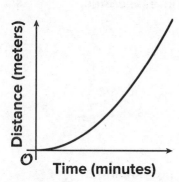

Graph 4

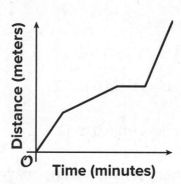

Match each description with a graph that could represent it.

A Lin starts out running, but then slows down to a jog. After 10 minutes, she stops for a water break. She then runs the rest of the way.

1 Graph 1

B Lin starts the race walking, gradually getting faster and faster.

2 Graph 2

C Lin jogs at a steady pace for the entire race.

3 Graph 3

D Lin starts out walking, then moves to a steady run. After 15 minutes, she stops to stretch a cramped leg. Then, she walks the rest of the way.

4 Graph 4

NAME _____ DATE _____ PERIOD _____

5. The graph of H shows the height, in meters, of a rocket t seconds after it was launched.

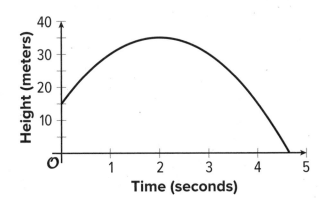

a. Find H(0). What does this value represent?

b. Describe the domain of this function.

c. Describe the range of this function.

d. Solve H(x) = 0. What does this value represent?

6. Mai has to decide between two cafeteria meal plans. Under plan A, each meal costs $2.50. Under plan B, one month of meals costs $30. (Lesson 4-5.)

a. Write an equation for function A, which gives the cost, in dollars, of buying n meals under plan A.

b. Write an equation for function B, which gives the cost, in dollars, of buying n meals under plan B.

c. Mai estimates that she'll buy 15 meals per month. Which meal plan should she choose? Explain your reasoning.

7. Kiran is playing a video game. He earns 3 stars for each easy level he completes and 5 stars for each difficult level he completes. He completes more than 20 levels total and earns 80 or more stars. (Lesson 2-25.)

a. Create a system of inequalities that describes the constraints in this situation. Be sure to specify what each variable represents.

b. Graph the inequalities and show the solution region.

c. Then, identify a point that represents a combination of stars and levels that is a solution to the system.

d. Interpret the point (5, 6) in the context of this situation and determine how many stars Kiran earns based on this point.

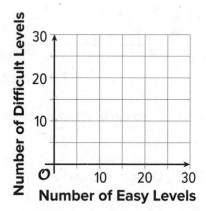

Lesson 4-12

Piecewise Functions

NAME _____ DATE _____ PERIOD _____

Learning Goal Let's look at functions that are defined in pieces.

Warm Up
12.1 Frozen Yogurt

A self-serve frozen yogurt store sells servings up to 12 ounces. It charges $0.50 per ounce for a serving between 0 and 8 ounces, and $4 for any serving greater than 8 ounces and up to 12 ounces. Choose the graph that represents the price as a function of the weight of a serving of yogurt. Be prepared to explain how you know.

A
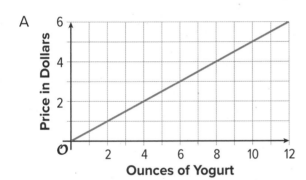

B
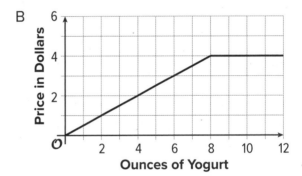

C
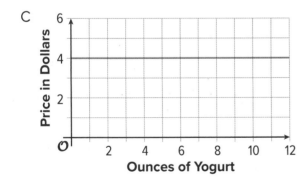

D
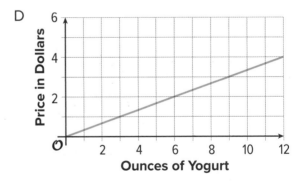

The relationship between the postage rate and the weight of a letter can be defined by a **piecewise function**.

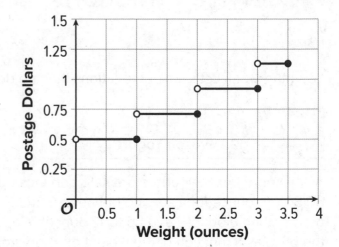

The graph shows the 2018 postage rates for using regular service to mail a letter.

1. What is the price of a letter that has the following weight?

   a. 1 ounce

   b. 1.1 ounces

   c. 0.9 ounce

2. A letter costs $0.92 to mail. How much did the letter weigh?

NAME _____ DATE _____ PERIOD _____

3. Kiran and Mai wrote some rules to represent the postage function, but each of them made some errors.

$$K(w) = \begin{cases} 0.50, & 0 \leq w \leq 1 \\ 0.71, & 1 \leq w \leq 2 \\ 0.92, & 2 \leq w \leq 3 \\ 1.13, & 3 \leq w \leq 3.5 \end{cases} \qquad M(w) = \begin{cases} 0.50, & 0 < w < 1 \\ 0.71, & 1 < w < 2 \\ 0.92, & 2 < w < 3 \\ 1.13, & 3 < w < 3.5 \end{cases}$$

Identify the error in each person's work and write a corrected set of rules.

Here is an image showing how the postal service specifies the different mailing rates.

Notice that it uses the language "weight not over (oz.)" to describe the different rates.

Explain or use a sketch to show how the graph would change if the postal service uses "under (oz.)" instead.

First Class Mail

Retail – Single Piece

Letters (Stamped)

Weight Not Over (oz.)	Rate
1	$0.50
2	$0.71
3	$0.92
3.5	$1.13

Activity

12.3 Bike Sharing

Function C represents the dollar cost of renting a bike from a bike-sharing service for t minutes. Here are the rules describing the function:

$$C(t) = \begin{cases} 2.50, & 0 < t \leq 30 \\ 5.00, & 30 < t \leq 60 \\ 7.50, & 60 < t \leq 90 \\ 10.00, & 90 < t \leq 120 \\ 12.50, & 120 < t \leq 150 \\ 15.00, & 150 < t \leq 720 \end{cases}$$

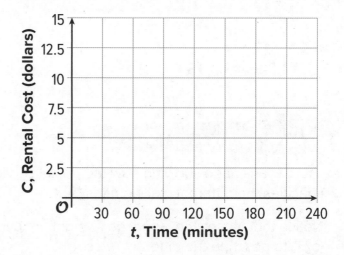

1. Complete the table with the costs for the given lengths of rental.

t (minutes)	C(t) (dollars)
0	
10	
25	
60	
75	
130	
180	

Sketch a graph of the function for all values of t that are at least 0 minutes and at most 240 minutes.

2. Describe in words the pricing rules for renting a bike from this bike sharing service.

3. Determine the domain and range of this function.

NAME _____ DATE _____ PERIOD _____

 Activity

12.4 Piecing It Together

Your teacher will give your group strips of paper with parts of a graph of a function. Gridlines are 1 unit apart. Arrange the strips of paper to create a graph for each of the following functions.

$$f(x) = \begin{cases} -5, & -10 < x < -5 \\ x, & -5 \le x < 0 \\ 1, & 0 \le x < 3 \\ x - 2, & 3 \le x < 8 \\ 6, & 8 \le x < 10 \end{cases} \qquad g(x) = \begin{cases} 5.5, & -10 < x \le -8 \\ 4, & -8 < x \le -3 \\ -x, & -3 < x \le 2 \\ -3.5, & 2 < x \le 5 \\ x - 5, & 5 < x \le 10 \end{cases}$$

To accurately represent each function, be sure to include a scale on each axis and add open and closed circles on the graph where appropriate.

A **piecewise function** has different descriptions or rules for different parts of its domain.

Function *f* gives the train fare, in dollars, for a child who is *t* years old based on these rules:

- Free for children under 5

- $5 for children who are at least 5 but younger than 11

- $7 for children who are at least 11 but younger than 16

The different prices for different ages tell us that function *f* is a piecewise function.

The graph of a piecewise function is often composed of pieces or segments. The pieces could be connected or disconnected. When disconnected, the graph appears to have breaks or steps.

Here is a graph that represents *f*.

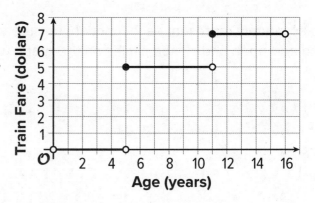

NAME _____ DATE _____ PERIOD _____

It is important to consider the value of the function at the points where the rule changes, or where the graph "breaks." For instance, when a child is exactly 5 years old, is the ride free, or does it cost $5?

On the graph, one segment ends at (5, 0) and another segment starts at (5, 5), but the function cannot have both 0 and 5 as outputs when the input is 5!

Based on the fare rules, the ride is free only if the child is under 5, which means:

- $f(5) = 0$ is false. On the graph, the point (5, 0) is marked with an open circle to indicate that it is *not* included in the first segment (which represents ages that qualify for a free ride).

- $f(5) = 5$ is true. The point (5, 5) has a solid circle to indicate that it is included in the middle segment (which represents ages that qualify for $5 fare).

The same reasoning applies when deciding how $f(11)$ and $f(16)$ should be shown on the graph.

- $f(11) = 7$ is true because 11-year-olds ride for $7. The point (11, 7) is a solid circle.

- $f(16) = 7$ is false because a 16-year-old no longer qualifies for a child's fare. The point (16, 7) is an open circle.

The fare rules can be expressed with function notation:

$$f(x) = \begin{cases} 0, & 0 < x < 5 \\ 5, & 5 \leq x < 11 \\ 7, & 11 \leq x < 16 \end{cases}$$

Glossary

piecewise function

1. A parking garage charges $5 for the first hour, $10 for up to two hours, and $12 for the entire day. Let G be the dollar cost of parking for t hours.

   a. Complete the table.

t (hours)	G (dollars)
0	
$\frac{1}{2}$	
1	
$1\frac{3}{4}$	
2	
5	

   b. Sketch a graph of G for $0 \le t \le 12$.

   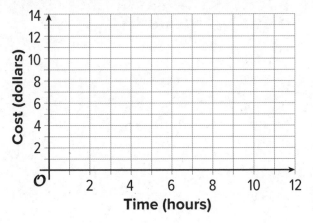

   c. Is G a function of t? Explain your reasoning.

   d. Is t a function of G? Explain your reasoning.

2. Is this a graph of a function? Explain your reasoning.

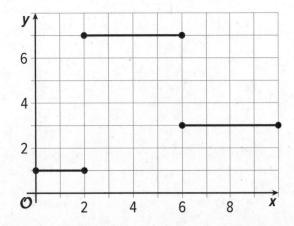

NAME _____ DATE _____ PERIOD _____

3. Use the graph of function *g* to answer these questions.

   a. What are the values of *g*(1), *g*(-12), and *g*(15)?

   b. For what *x*-values is *g*(*x*) = -6?

   c. Complete the rule for *g*(*x*) so that the graph represents it.

   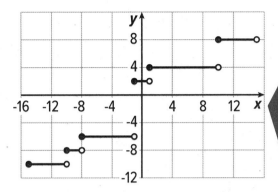

$$g(x) = \begin{cases} -10, & -15 \le x < -10 \\ \underline{\hspace{1cm}}, & -10 \le x < -8 \\ -6, & \underline{\hspace{1cm}} \le x < -1 \\ \underline{\hspace{1cm}}, & -1 \le x < 1 \\ 4, & \underline{\hspace{1cm}} \le x < \underline{\hspace{1cm}} \\ 8, & 10 \le x < 15 \end{cases}$$

4. This graph represents Andre's distance from his bicycle as he walks in a park. **(Lesson 4-6)**

   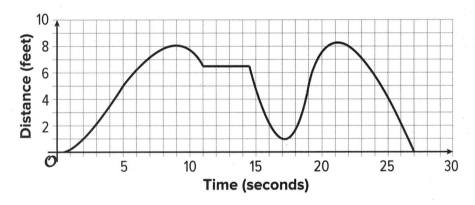

   a. For which intervals of time is the value of the function decreasing?

   b. For which intervals is it increasing?

   c. Describe what Andre is doing during the time when the value of the function is increasing.

5. The temperature was recorded at several times during the day. Function *T* gives the temperature in degrees Fahrenheit, *n* hours since midnight.

Here is a graph for this function. (Lesson 4-7)

    a. Describe the overall trend of temperature throughout the day.

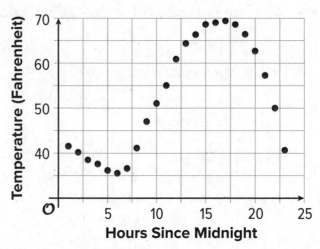

    b. Based on the graph, did the temperature change more quickly between 10:00 a.m. and noon, or between 8:00 p.m. and 10:00 p.m.? Explain how you know.

6. Explain why this graph does not represent a function. (Lesson 4-8).

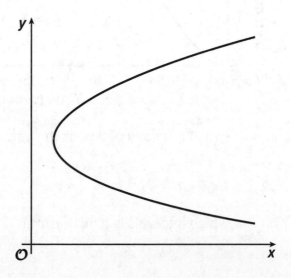

Lesson 4-13

Absolute Value Functions (Part 1)

NAME _____ DATE _____ PERIOD _____

Learning Goal Let's make some guesses and see how good they are.

Warm Up
13.1 How Good Were the Guesses?

Before this lesson, you were asked to guess the number of objects in the jar. The guesses of all students have been collected. Your teacher will share the data and reveal the actual number of objects in the jar.

Use that number to calculate the **absolute guessing error** of each guess, or how far the guess is from the actual number. Suppose the actual number of objects is 100.

- If your guess is 75, then the absolute guessing error is 25.

- If your guess is 110, then the absolute guessing error is 10.

Record the absolute guessing error of at least 12 guesses in Table A of the handout from your teacher (or elsewhere as directed).

Guess	Absolute Guessing Error

Guess	Absolute Guessing Error

Guess	Absolute Guessing Error

Activity

13.2 Plotting the Guesses

Refer to the table you completed in the warm-up, which shows your class' guesses and absolute guessing errors.

1. Plot at least 12 pairs of values from your table on the coordinate plane on the handout (or elsewhere as directed by your teacher).

2. Write down 1–2 other observations about the completed scatter plot.

3. Is the absolute guessing error a function of the guess? Explain how you know.

NAME _____ DATE _____ PERIOD _____

Are you ready for more?

Suppose there's another guessing contest that comes with a prize. Each class can submit one guess. It is up to the students to decide on the number to be submitted. Here are some ideas that have been proposed on how to decide on that number:

- Option A: Ask the person or persons who did really well in the previous guessing game to make a guess.

- Option B: Ask everyone to make a guess and have a discussion to narrow the list and then choose a number.

- Option C: Ask everyone to make a guess and find the mean of all the guesses.

- Option D: Ask everyone to make a guess and find the middle point between the largest number and the smallest number.

Which approach do you think would give your class the best chance of winning? Explain your reasoning.

Activity

13.3 Oops, Try Again!

Earlier, you guessed the number of objects in a container and then your teacher told you the actual number.

Suppose your teacher made a mistake about the number of objects in the jar and would like to correct it. The actual number of objects in the jar is _____.

1. Find the new absolute guessing errors based on this new information. Record the errors in Table B of the handout (or elsewhere as directed by your teacher).

2. Make 1–2 observations about the new set of absolute guessing errors.

3. Respond to each question.

   a. Predict how the scatter plot would change given the new actual number of objects. (Would it have the same shape as in the first scatter plot? If so, what would be different about it? If not, what would it look like?)

   b. Use technology to plot the points and test your prediction.

4. Can you write a rule for finding the output (absolute guessing error) given the input (a guess)?

NAME _____ DATE _____ PERIOD _____

Summary
Absolute Value Functions (Part 1)

Have you played a number guessing game where the guess that is closest to a target number wins?

In such a game, it doesn't matter if the guess is above or below the target number. What matters is how far off the guess is from the target number, or the *absolute guessing error*. The smaller the absolute guessing error, or the closer it is to 0, the better.

Suppose eight people made these guesses for the number of pretzels in a jar: 14, 15, 19, 21, 23, 24, 26, and 28. If the actual number of pretzels is 22, the absolute guessing error of each number is as shown in the table.

Guess	14	15	19	21	23	24	26	28
Absolute Guessing Error	8	7	3	1	1	2	4	6

In this case, 21 and 23 are both winning guesses. Even though one number is an underestimate and the other an overestimate, 21 and 23 are both 1 away from 22. Of all the absolute guessing errors, 1 is the smallest one.

If we plot the guesses and the guessing errors on a coordinate plane, the points would form a V shape. Notice that the V shape is above the horizontal axis, suggesting that all the vertical values are positive.

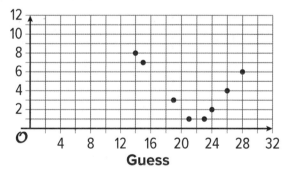

Suppose the actual number of pretzels is 19. The absolute guessing errors of the same eight guesses are shown in this table.

Guess	14	15	19	21	23	24	26	28
Absolute Guessing Error	5	4	0	2	4	5	7	9

Notice that all the errors are still positive. If we plot these points on a coordinate plane, they are also on or above the horizontal line and form a V shape.

Why does the relationship between guesses and absolute guessing errors always have this kind of graph? We will explore more in the next lesson!

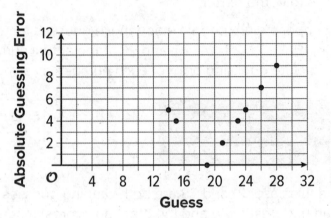

Glossary

absolute value

NAME _____ DATE _____ PERIOD _____

Practice
Absolute Value Functions (Part 1)

1. A group of ten friends played a number guessing game. They were asked to pick a number between 1 and 20. The person closest to the target number wins. The ten people made these guesses:

   a. The actual number was 14. Complete the table with the absolute guessing errors.

Guess	2	15	10	8	12	19	20	5	7	9
Absolute Guessing Error										

   b. Graph the guesses and absolute guessing errors.

   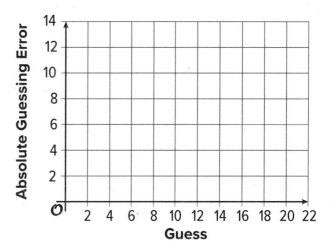

   c. Is the absolute guessing error a function of the guess? Explain how you know.

2. Bags of walnuts from a food producer are advertised to weigh 500 grams each. In a certain batch of 20 bags, most bags have an absolute error that is less than 4 grams.

Could this scatter plot represent those 20 bags and their absolute errors? Explain your reasoning.

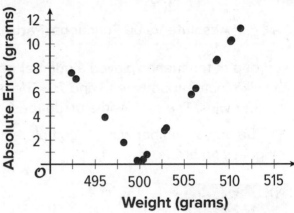

3. The class guessed how many objects were placed in a mason jar. The graph displays the class results, with an actual number of 47.

Suppose a mistake was made, and the actual number is 45.

Explain how the graph would change, given the new actual number.

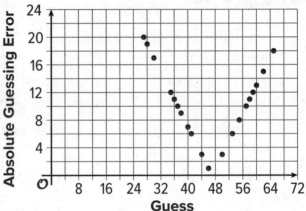

4. Function D gives the height of a drone t seconds after it lifts off. Sketch a possible graph for this function given that: (Lesson 4-3)

- $D(3) = 4$

- $D(10) = 0$

- $D(5) > D(3)$

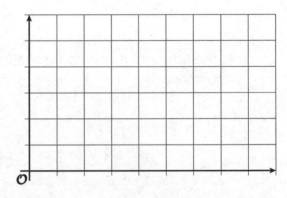

NAME _____ DATE _____ PERIOD _____

5. The population of a city grew from 23,000 in 2010 to 25,000 in 2015.
   (Lesson 4-7)
   a. What was the average rate of change during this time interval?

   b. What does the average rate of change tell us about the population growth?

6. Here is the graph of a function. (Lesson 4-7)

   Which time interval shows the largest rate of change?

   (A.) From 0 to 2 seconds

   (B.) From 0 to 3 seconds

   (C.) From 4 to 5 seconds

   (D.) From 6 to 8 seconds

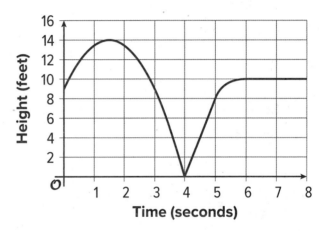

7. Here are the graphs of $L(x)$ and $R(x)$. (Lesson 4-12)

   $L(x)$ $R(x)$

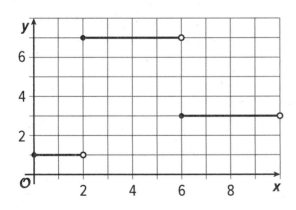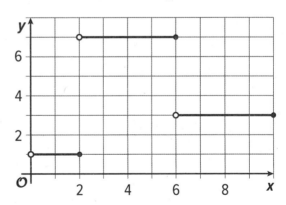

   a. What are the values of $L(0)$ and $R(0)$?

   b. What are the values of $L(2)$ and $R(2)$?

   c. For what x-values is $L(x) = 7$?

   d. For what x-values is $R(x) = 7$?

8. Select **all** systems that are equivalent to this system of

equations: $\begin{cases} 4x + 5y = 1 \\ x - y = \dfrac{3}{8} \end{cases}$ **(Lesson 2-16)**

(A.) $\begin{cases} 4x + 5y = 1 \\ 4x - 4y = \dfrac{3}{2} \end{cases}$

(B.) $\begin{cases} x + \dfrac{5}{4}y = \dfrac{1}{4} \\ x - y = \dfrac{3}{8} \end{cases}$

(C.) $\begin{cases} 4x + 5y = 1 \\ 5x - 5y = 3 \end{cases}$

(D.) $\begin{cases} 8x + 10y = 2 \\ 8x - 8y = 3 \end{cases}$

(E.) $\begin{cases} x + y = \dfrac{1}{5} \\ x - y = \dfrac{3}{8} \end{cases}$

Lesson 4-14

Absolute Value Functions (Part 2)

NAME _____ DATE _____ PERIOD _____

Learning Goal Let's investigate distance as a function.

Warm Up
14.1 Temperature in Toronto

Toronto is a city at the border of the United States and Canada, just north of Buffalo, New York. Here are twelve guesses of the average temperature of Toronto, in degrees Celsius, in February 2017.

5	2	-5	3	0	-1	1.5	4
-2.5	6	4	-0.5				

1. The actual average temperature of Toronto in February 2017 is 0 degrees Celsius.

   Use this information to sketch a scatter plot representing the guesses, x, and the corresponding absolute guessing errors, y.

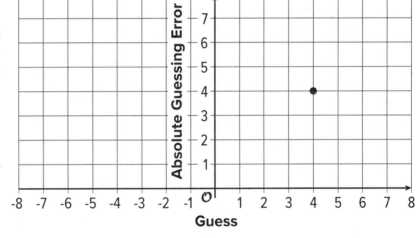

2. What rule can you write to find the output given the input?

The function A gives the distance of x from 0 on the number line.

1. Complete the table and sketch a graph of function A.

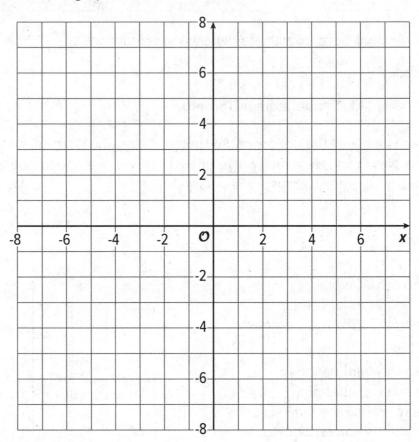

x	A(x)
8	
	5.6
π	
$\frac{1}{2}$	
	1
0	
$-\frac{1}{2}$	
-1	
-5.6	
	8

2. Andre and Elena are trying to write a rule for this function.

   - Andre writes: $A(x) = \begin{cases} x, & x \geq 0 \\ -x, & x < 0 \end{cases}$

   - Elena writes: $A(x) = |x|$

   Explain why both equations correctly represent the function A.

NAME _____ DATE _____ PERIOD _____

Activity

14.3 Moving Graphs Around

Here are equations and graphs that represent five absolute value functions.

$f(x) = |x|$ $g(x) = |x - 2|$ $h(x) = |x + 2|$

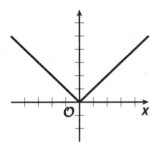   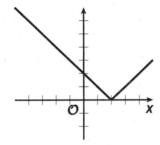   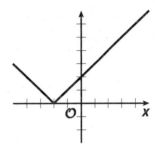

$j(x) = |x| - 2$ $k(x) = |x| + 2$

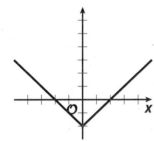   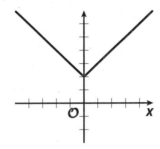

Notice that the number 2 appears in the equations for functions g, h, j, and k. Describe how the addition or subtraction of 2 affects the graph of each function.

Then, think about a possible explanation for the position of the graph. How can you show that it really belongs where it is on the coordinate plane?

1. Mark the minimum of each graph in the activity. Each point you marked represents the least output value of the function.

   In each function, what value of x gives that minimum output value?

2. Respond to each question.

   a. Another function is defined by $m(x) = |x + 11.5|$. What value of x produces the least output of function m? Be prepared to explain how you know.

   b. Describe or sketch the graph of m.

NAME _____ DATE _____ PERIOD _____

Activity

14.4 More Moving Graphs Around

1. Here are five equations and four graphs.

   Match each equation with a graph that represents it. One equation has no match.

   - Equation 1: $y = |x - 3|$

     A
     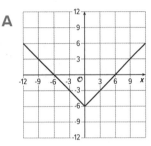

   - Equation 2:
     $y = |x - 9| + 3$

     B
     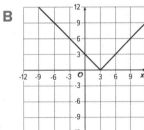

   - Equation 3:
     $y = |x| - 6$

     C
     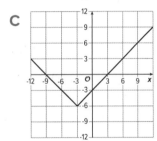

   - Equation 4: $y = |x + 3|$

     D
     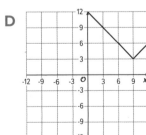

   - Equation 5:
     $y = |x + 3| - 6$

     E
     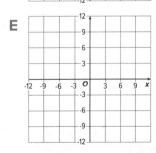

2. For the equation without a match, sketch a graph on the blank coordinate plane.

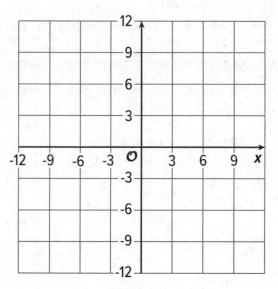

3. Use graphing technology to check your matches and your graph. Revise your matches and graphs as needed.

NAME _____ DATE _____ PERIOD _____

Summary
Absolute Value Functions (Part 2)

In a guessing game, each guess can be seen as an input of a function and each absolute guessing error as an output. Because absolute guessing error tells us how far a guess is from a target number, the output is distance.

Suppose the target number is 0.

- We can find the distance of a guess, x, from 0 by calculating $x - 0$. Because distance cannot be negative, what we want to find is $|x - 0|$, or simply $|x|$.

- If function f gives the distance of x from 0, we can define it with the equation:

$$f(x) = |x|$$

Function f is the **absolute value function**. It gives the distance of x from 0 by finding the absolute value of x.

The graph of function f is a V shape with the two lines converging at (0,0).

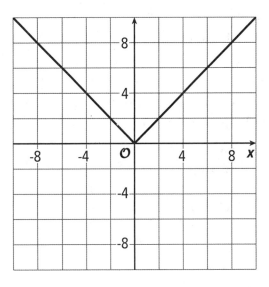

We call this point the **vertex** of the graph. It is the point where a graph changes direction, from going down to going up, or the other way around.

We can also think of a function like f as a *piecewise function* because different rules apply when x is less than 0 and when it is greater than 0.

Suppose we want to find the distance between x and 4.

- We can find the difference between x and 4 by calculating $x - 4$. Distance cannot be negative, so what we want is the absolute value of that difference: $|x - 4|$.

- If function p gives the distance of x from 4, we can define it with the equation:

$$p(x) = |x - 4|$$

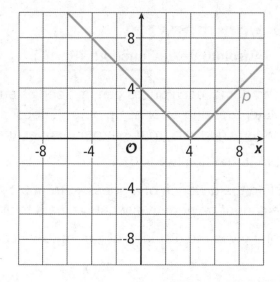

Now suppose we want to find the distance between x and -4.

- We can find the difference of x and -4 by calculating $x - (-4)$, which is equal to $x + 4$. Distance cannot be negative, so let's find the absolute value: $|x + 4|$.

- If function q gives the distance of x from -4, we can define it with the equation:

$$q(x) = |x + 4|$$

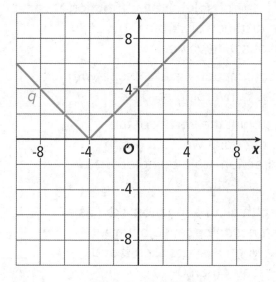

Notice that the graphs of p and q are like that of f, but they have shifted horizontally.

Glossary

absolute value function

NAME _____ DATE _____ PERIOD _____

Practice
Absolute Value Functions (Part 2)

1. The absolute value function can be defined using piecewise notation.

$$A(x) = \begin{cases} x, & x \geq 0 \\ -x, & x < 0 \end{cases}$$

Use this notation to find the following values:

a. $A(10)$

b. $A(0)$

c. $A(-3)$

d. $A(3.14159)$

e. $A(x) = 7$

f. $A(x) = -5$

2. Here are four equations of absolute value functions and three coordinate pairs. Each coordinate pair represents the vertex of the graph of an absolute value function.

Match the equation of each function with the coordinates of the vertex of its graph. The vertex coordinates of the graph of one equation are not shown.

A $p(x) = |x - 9|$ 1 (-9, 0)

B $q(x) = |x| + 9$ 2 (9, 0)

C $r(x) = |x + 9|$ 3 (0, -9)

D $t(x) = |x| - 9$

3. Function G is defined by the equation $G(x) = |x|$.

   Function R is defined by the equation $R(x) = |x| + 2$.

   Describe how the graph of function R relates to the graph of G, or sketch the graphs of the two functions to show their relationship.

4. Here is the graph of a function.

   Select the equation for the function represented by the graph.

   (A.) $y = |x| - 5$

   (B.) $y = |x| + 5$

   (C.) $y = |x - 5|$

   (D.) $y = |x + 5|$

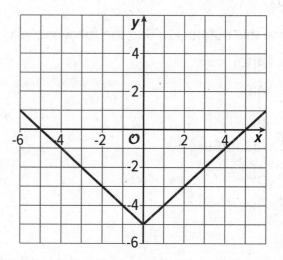

NAME _____ DATE _____ PERIOD _____

5. The temperature was recorded at several times during the day. Function *T* gives the temperature in degrees Fahrenheit, *n* hours since midnight. **(Lesson 4-7)**

Here is a graph for this function.

a. Pick two consecutive points and connect them with a line segment. Estimate the slope of that line. Explain what that estimated value means in this situation.

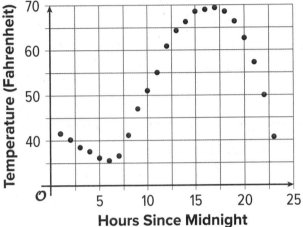

b. Pick two non-consecutive points and connect them with a line segment. Estimate the slope of that line. Explain what that estimated value means in this situation.

6. A tennis ball is dropped from an initial height of 30 feet. It bounces 5 times, with each bounce height being about $\frac{2}{3}$ of the height of the previous bounce. (Lesson 4-8)

Sketch a graph that models the height of the ball over time. Be sure to label the axes.

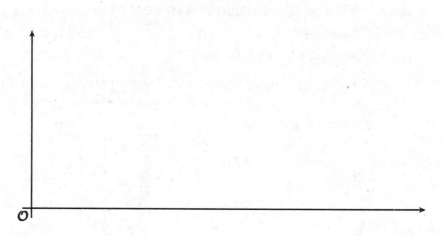

7. Here are two graphs representing functions f and g.

Identify at least two values of x at which the inequality $g(x) > f(x)$ is true. (Lesson 4-9)

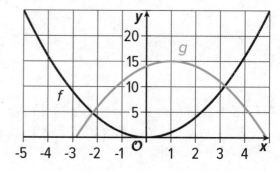

Lesson 4-15

Inverse Functions

NAME _____ DATE _____ PERIOD _____

Learning Goal Let's define functions forward and backward.

Warm Up

15.1 What Does It Say?

Here is an *encoded* message, a message that has been converted into a code.

WRGDB LV D JRRG GDB.

Can you figure out what it says in English? How was the original message encoded?

Activity

15.2 Caesar Says Shift

1. Now it's your turn to write a secret code!

   a. Write a short and friendly message with 3–4 words.

   b. Pick a number from 1 to 10. Then, encode your message by shifting each letter that many steps forward or backward in the alphabet, wrapping around from Z to A as needed.

   Consider using this table to create a key for your cipher.

Plain Text	A	B	C	D	E	F	G	H	I	J	K	L	M	N	O	P	Q	R	S	T	U	V	W	X	Y	Z
Cipher Text																										

   c. Give your encoded message to a partner to decode. If requested, give the number you used.

   d. Decode the message from your partner. Ask for their number, if needed.

NAME _____ DATE _____ PERIOD _____

2. Suppose *m* and *c* each represent the position number of a letter in the alphabet, but *m* represents the letters in the original message and *c* the letters in your secret code.

   a. Complete the table.

Letter in Message				
m	6	9	19	8
c				
Letter in Code				

   b. Use *m* and *c* to write an equation that can be used to *encode* an original message into your secret code.

   c. Use *m* and *c* write an equation that can be used to *decode* your secret code into the original message.

Are you ready for more?

There are 26 letters in the alphabet, so only the numbers 1–26 make sense for *m* and *c*.

1. Try using the equation that you wrote to encode the letters A, B, Y, and Z. Did you end up with position numbers or *c* values that are less than 1 or greater than 26? For which letters?

2. Use your encoding equation to plot the (m, c) pairs for all the letters in the alphabet.

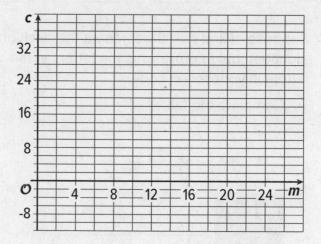

3. Look for the points whose c value is less than 1 or greater than 26. What letters should they be in the code? Plot the points where they should be according to the rule of your cipher.

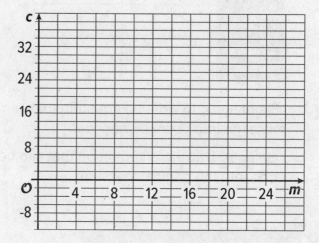

4. Did you end up with a graph of a piecewise function? If so, can you describe the different rules that apply to different domains of the function?

NAME _____ DATE _____ PERIOD _____

Activity

15.3 U.S. Dollars and Mexican Pesos

An American traveler who is heading to Mexico exchanges some U.S. dollars for Mexican pesos. At the time of his travel, 1 dollar can be exchanged for 19.32 pesos.

At the same time, a Mexican businesswoman who is in the United States is exchanging some Mexican pesos for U.S. dollars at the same exchange rate.

1. Find the amount of money in pesos that the American traveler would get if he exchanged:

   a. 100 dollars

   b. 500 dollars

2. Write an equation that gives the amount of money in pesos, p, as a function of the dollar amount, d, being exchanged.

3. Find the amount that the Mexican businesswoman would get if she exchanged:

   a. 1,000 pesos

   b. 5,000 pesos

4. Explain why it might be helpful to write the inverse of the function you wrote earlier. Then, write an equation that defines the inverse function.

Summary
Inverse Functions

Sometimes it is useful to reverse a function so that the original output is now the input.

Suppose Han lives 400 meters from school and walks to school. A linear function gives Han's distance to school, D, in meters, after he has walked w meters from home, and is defined by:

$$D = 400 - w$$

With this equation, if we know how far Han has walked from home, w, we can easily find his remaining distance to school, D. Here, w is the input and D is the output.

What if we know Han's remaining distance to school, D, and want to know how far he has walked, w?

We can find out by solving for w:

$$D = 400 - w$$
$$D + w = 400$$
$$w = 400 - D$$

The equation $w = 400 - D$ represents the *inverse* of the original function.

With this equation, we can easily find how far Han has walked from home if we know his remaining distance to school. Here, w and D have switched roles: w is now the output and D the input.

In general, if a function takes a as its input and gives b as its output, its **inverse function** takes b as the input and gives a as the output.

Glossary

inverse (function)

NAME _____ DATE _____ PERIOD _____

Practice
Inverse Functions

1. Noah's cousin is exactly 7 years younger than Noah. Let C represent Noah's cousin's age and N represent Noah's age. Ages are measured in years.

   a. Write a function that defines the cousin's age as a function of Noah's age. What are the input and output of this function?

   b. Write the inverse of the function you wrote. What are the input and output of this inverse function?

2. Noah's cousin is exactly 7 years younger than Noah. Let M represent Noah's cousin's age in months and N represent Noah's age in years.

   a. If Noah is 15 years old, how old is his cousin, in months?

   b. When Noah's cousin is 132 months old, how old is Noah, in years?

   c. Write a function that gives the age of Noah's cousin in months, as a function of Noah's age in years.

   d. Write the inverse of the function you wrote. What are the input and the output of this inverse function?

3. Each equation represents a function. For each, find the inverse function.

   a. $c = w + 3$

   b. $y = x - 2$

   c. $y = 5x$

   d. $w = \dfrac{d}{7}$

4. The number of years, y, is a function of the number of months, m. The number of months, m, is also a function of the number of years, y.

    a. Write two equations, one to represent each function.

    b. Explain why the two functions are inverses.

5. Sketch a graph to represent each quantity described as a function of time. Be sure to label the vertical axis. (Lesson 4-8)

Swing: The height of your feet above ground while swinging on a swing at a playground

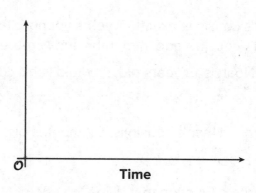

Slide: The height of your shoes above ground as you walk to a slide, go up a ladder, and then go down a slide

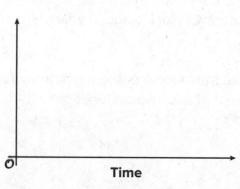

Merry-go-round: Your distance from the center of a merry-go-round as you ride the merry-go-round

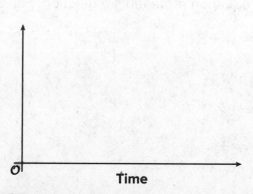

NAME _____ DATE _____ PERIOD _____

Merry-go-round, again: Your distance from your friend, who is standing next to the merry-go-round as you go around

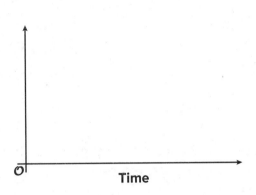

6. Lin charges $5.50 per hour to babysit. The amount of money earned, in dollars, is a function of the number of hours that she babysits.

Which of the following inputs is impossible for this function?

(A.) -1

(C.) 5

(B.) 2

(D.) 8

7. The instructions for cooking a steak with a pressure cooker can be represented with this set of rules, where x represents the weight of a steak in ounces and $f(x)$ the cooking time in minutes. (Lesson 4-12)

$$f(x) = \begin{cases} 7, & 8 \leq x \leq 12 \\ 8, & 12 < x \leq 13 \\ 9, & 13 < x \leq 14 \\ 10, & 14 < x \leq 15 \\ 11, & 15 < x \leq 16 \end{cases}$$

a. Describe the instructions in words so that they can be followed by someone using the pressure cooker.

b. Graph function f.

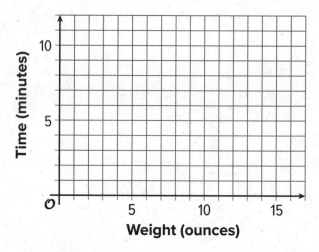

8. The absolute value function $Q(x) = |x|$ gives the distance from 0 of the point x on the number line. (Lesson 4-14)

Q can also be defined using piecewise notation: $Q(x) = \begin{cases} x, & x \geq 0 \\ -x, & x < 0 \end{cases}$

Determine if each point is on the graph of Q. For each point that you believe is *not* on the graph of Q, change the output coordinate so that the point is on the graph of Q.

a. (-3, 3)

b. (0, 0)

c. (-5, -5)

d. (-72, 72)

e. $\left(\frac{4}{5}, -\frac{4}{5}\right)$

Lesson 4-16

Finding and Interpreting Inverse Functions

NAME _____ DATE _____ PERIOD _____

Learning Goal Let's find the inverse of linear functions.

Warm Up
16.1 Shopping for Cookbooks

Lin is comparing the cost of buying cookbooks at different online stores.

- Store A sells them at $9 each and offers free shipping.

- Store B sells them at $9 each and charges $5 for shipping.

- Store C sells them at p dollars and charges $5 for shipping.

- Store D sells them at p dollars and charges f dollars for shipping.

1. Write an equation to represent the total cost, T, in dollars as a function of n cookbooks bought at each store.

2. Write an equation to find the number of books, n, that Lin could buy if she spent T dollars at each store.

If we know the temperature in degrees Celsius, C, we can find the temperature in degrees Fahrenheit, F, using the equation:

$$F = \frac{9}{5}C + 32$$

1. Complete the table with temperatures in degrees Fahrenheit or degrees Celsius.

C	0	100	25			
F				104	50	62.6

2. The equation $F = \frac{9}{5}C + 32$ represents a function. Write an equation to represent the inverse function. Be prepared to explain your reasoning.

3. The equation $R = \frac{9}{5}(C + 273.15)$ defines the temperature in degrees Rankine as a function of the temperature in degrees Celsius.

   Show that the equation $C = (R - 491.67) \cdot \frac{5}{9}$ defines the inverse of that function.

NAME _____ DATE _____ PERIOD _____

Are you ready for more?

It was cold enough in Alaska one day so that the temperature was the same in degrees Fahrenheit and degrees Celsius. How cold was it? Explain or show how you know.

Activity

16.3 Info Gap: Custom Mugs

Your teacher will give you either a problem card or a data card. Do not show or read your card to your partner.

If your teacher gives you the *data card*:

1. Silently read the information on your card.

2. Ask your partner "What specific information do you need?" and wait for your partner to *ask* for information. *Only* give information that is on your card. (Do not figure out anything for your partner!)

3. Before telling your partner the information, ask "Why do you need that information?"

4. After your partner solves the problem, ask them to explain their reasoning and listen to their explanation.

If your teacher gives you the *problem card*:

1. Silently read your card and think about what information you need to answer the question.

2. Ask your partner for the specific information that you need.

3. Explain to your partner how you are using the information to solve the problem.

4. Solve the problem and explain your reasoning to your partner.

Pause here so your teacher can review your work. Ask your teacher for a new set of cards and repeat the activity, trading roles with your partner.

NAME _____ DATE _____ PERIOD _____

Activity

16.4 Tables and Seats

At a party, hexagonal tables are placed side by side along one side, as shown here.

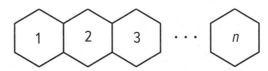

1. Explain why the equation $S = 4n + 2$ represents the number of seats, S, as a function of the number of tables, n.

2. What domain and range make sense for this function?

3. Write an equation to represent the inverse of the given function. Explain what this inverse function tells us.

4. How many tables are needed if the following number of people are attending the party? Be prepared to explain your reasoning.

   a. 94 people

   b. 95 people

5. What domain makes sense for the inverse function? Is it the same set of values as the range of the original function? Explain your reasoning.

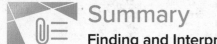

It is helpful to interpret the inverse of a function in terms of a situation and the quantities it represents.

Suppose a linear function gives the dollar cost, C, of renting some equipment for n hours. The function is defined by this equation:

$$C = 8.25n + 30$$

If we know the number of hours of rental, n, we can substitute it into the expression $8.25n + 30$ and evaluate it to find the cost, C.

What is the inverse of this function, and what does it tell us about the length and cost of rental?

To find the inverse, let's solve for n:

$$8.25n + 30 = C$$
$$8.25n = C - 30$$
$$n = \frac{C - 30}{8.25}$$

If we know the cost of rental, C, we can substitute it into the expression $\frac{C - 30}{8.25}$ and evaluate it to find the hours of rental, n.

Notice that the equation defining the inverse function is found by reversing the process that defines the original linear function.

- The original rule, $C = 8.25n + 30$, tells us to multiply the input, n, by 8.25 and add 30 to the result to get the output, C.

- The rule of the inverse function, $\frac{C - 30}{8.25}$, suggests that we subtract 30 from the input and then divide the result by 8.25 to get the output n.

NAME _____ DATE _____ PERIOD _____

Practice

Finding and Interpreting Inverse Functions

1. Tickets to a family concert cost $10 for adults and $3 for children. The concert organizers collected a total of $900 from ticket sales.

   a. In this situation, what is the meaning of each variable in the equation $10A + 3C = 900$?

   b. If 42 adults were at the concert, how many children attended?

   c. If 140 children were at the concert, how many adults attended?

   d. Write an equation to represent C as a function of A. Explain what this function tells us about the situation.

   e. Write an equation to represent A as a function of C. Explain what this function tells us about the situation.

2. A school group has $600 to spend on T-shirts. The group is buying from a store that gives them a $5 discount off the regular price per shirt.
   $n = \dfrac{600}{p - 5}$ gives the number of shirts, n, that can be purchased at a regular price, p.
   $p = \dfrac{600}{n} + 5$ gives the regular price, p, of a shirt when n shirts are bought.

   a. What is n when p is 20?

   b. What is p when n is 40?

   c. Is one function an inverse of the other? Explain how you know.

3. Functions f and g are inverses, and $f(-2) = 3$. Is the point $(3, -2)$ on the graph of f, on the graph of g, or neither?

4. Here are two equations that relate two quantities, p and Q:

$$Q = 7p + 1,999 \qquad\qquad p = \frac{Q - 1,999}{7}$$

Select **all** statements that are true about p and Q.

A. $Q = 7p + 1,999$ could represent a function, but $p = \frac{Q - 1,999}{7}$ could not.

B. Each equation could represent a function.

C. $p = \frac{Q - 1,999}{7}$ could represent a function, but $Q = 7p + 1,999$ could not.

D. The two equations represent two functions that are inverses of one another.

E. If $Q = 7p + 1,999$ represents a function, then the inverse function can be defined by $p = 7Q - 1,999$.

5. Elena plays the piano for 30 minutes each practice day. The total number of minutes p that Elena practiced last week is a function of n, the number of practice days. (Lesson 4-10)

Find the domain and range for this function.

NAME _____ DATE _____ PERIOD _____

6. The graph shows the attendance at a sports game as a function of time in minutes. **(Lesson 4-11)**

   a. Describe how attendance changed over time.

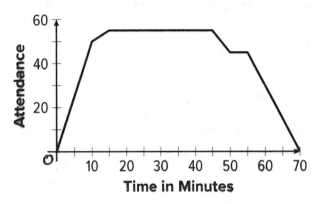

   b. Describe the domain.

   c. Describe the range.

7. Two children set up a lemonade stand in their front yard. They charge $1 for every cup. They sell a total of 15 cups of lemonade. The amount of money the children earned, *R* dollars, is a function of the number of cups of lemonade they sold, *n*. **(Lesson 4-11)**

   a. Is 20 part of the domain of this function? Explain your reasoning.

   b. What does the range of this function represent?

   c. Describe the set of values in the range of *R*.

   d. Is the graph of this function discrete or continuous? Explain your reasoning.

8. Here is the graph of function *f*, which represents Andre's distance from his bicycle as he walked in a park. (Lesson 4-6)

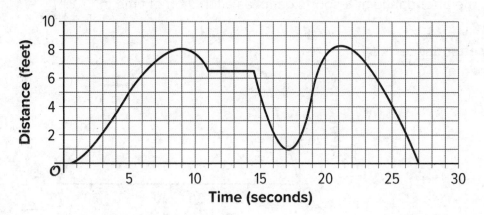

a. Estimate *f*(5).

b. Estimate *f*(17).

c. For what values of *t* does *f*(t) = 8?

d. For what values of *t* does *f*(t) = 6.5?

e. For what values of *t* does *f*(t) = 10?

Lesson 4-17

Writing Inverse Functions to Solve Problems

NAME _____ DATE _____ PERIOD _____

Learning Goal Let's use inverse functions to solve problems.

 **Warm Up**
17.1 Water in a Tank

A tank contained some water. The function w represents the relationship between t, time in minutes, and the amount of water in the tank in liters. The equation $w(t) = 80 - 2.5t$ defines this function.

1. Discuss with a partner:

    a. How is the water in the tank changing? Be as specific as possible.

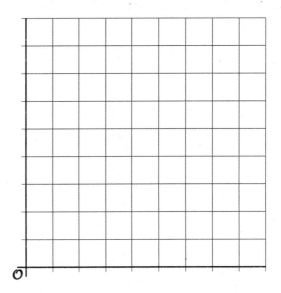

    b. What does $w(t)$ represent? Is $w(t)$ the input or the output of this function?

2. Sketch a graph of the function. Be sure to label the axes.

Activity

17.2 Another Look at the Tank

A tank contained 80 liters of water. The function w represents the relationship between t, time in minutes, and the amount of water in the tank in liters. The equation $w(t) = 80 - 2.5t$ defines this function.

1. How much water will be in the tank after 13 minutes?

2. How many minutes will it take until the tank has 5 liters of water?

3. In this situation, what information can we gain from the inverse of function w?

4. Find the inverse of function w. Be prepared to explain or show your reasoning.

5. How would the graph of the inverse function of w compare to the graph of w? Describe or sketch your prediction.

NAME _____ DATE _____ PERIOD _____

Activity
17.3 Phones in Homes

In 2004, less than 5% of the homes in the U.S. relied only on a cell phone. Since then, the percentage of homes that use only cell phones has increased.

Here are the percentages of homes with only cell phones from 2004 to 2009.

Years Since 2004	Percentage
0	4.4
1	6.7
2	9.6
3	13.6
4	17.5
5	22.7

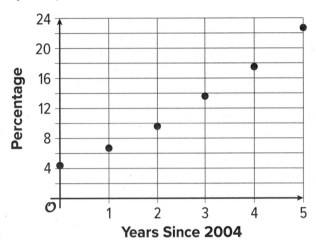

1. Suppose a linear function, P, gives us the percentage of homes with only cell phones as a function of years since 2004, t.

   Fit a line on the scatter plot to represent this function and write an equation that could define the function. Use function notation.

2. Use your equation to find the value of $P(6)$. Then, explain what it means in this situation.

3. Use your equation to solve $P(t) = 30$ for t. What does the solution represent?

4. Suppose we want to know when the percentage of homes with only cell phones would reach 50%, 75%, or 100% (assuming that the trend continues and the function stays valid). What equation could be written to help us find the years that correspond to those percentages? Show your reasoning.

How well do you think your model will work to predict the percentage of homes with only a cell phone in future years, for example, a decade or two decades from now? Explain your reasoning.

NAME _____ DATE _____ PERIOD _____

 ## Summary

Writing Inverse Functions to Solve Problems

The water in a rain barrel is being drained and used to water a garden. Function v gives the volume of water remaining in the barrel, in gallons, t minutes after it started being drained. This equation represents the function:

$$v(t) = 60 - 2.25t$$

From the equation and description, we can reason that there were 60 gallons of water in the rain barrel, and that it was being drained at a constant rate of 2.25 gallons per minute.

This equation is handy for finding out the amount of water left in the barrel after some number of minutes. In other words, it helps us find the output, $v(t)$, when we know the input, t.

Suppose we want to know how long it would take before the barrel has 20 gallons of water remaining, or how long it would take to empty the barrel. Let's find the inverse of function v so that the volume of water is the input and time is the output.

Even though the equation is in function notation, we can still solve for t as we had done before:

$$v(t) = 60 - 2.25t$$

$$v(t) + 2.25t = 60$$

$$2.25t = 60 - v(t)$$

$$t = \frac{60 - v(t)}{2.25}$$

This equation now shows t as the output and $v(t)$ as the input. We can easily find or estimate the time when the barrel will have 20 gallons remaining or when it will be empty by substituting 20 or 0 for $v(t)$, and then evaluating $\frac{60 - 20}{2.25}$ or $\frac{60 - 0}{2.25}$, respectively.

Practice

Writing Inverse Functions to Solve Problems

1. Respond to each question.

   a. The table shows the value of a car, in thousands of dollars, each year after it was purchased. Plot the data values, and find a line that fits the data.

Age (years)	Value (thousands of dollars)
0	30.0
1	22.5
2	19.0
3	16.0
4	13.5
5	11.4

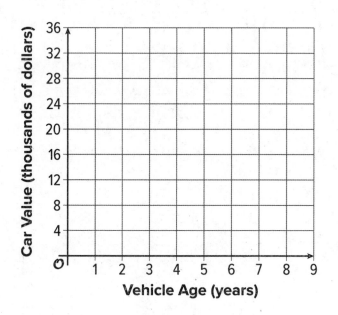

   b. Write an equation for the linear function, C, that gives the value of the car, in thousands of dollars, when its age is t years.

   c. What does C(6) mean in this situation? What is the value of C(6)?

   d. In this situation, what does the solution to the equation C(t) = 2 tell us? Find that solution.

   e. Write an equation that would allow us to find the age of the car when we know C(t).

NAME _____ DATE _____ PERIOD _____

f. Use your equation to estimate the vehicle age when the value of the car will be $500.

2. The distance d, in kilometers, that a car travels at a speed of 80 km per hour, for t hours, is given by the equation $d = 80t$.

 a. If the car has gone 120 kilometers, how long has it been traveling?

 b. Rewrite the equation to represent time, t, as a function of distance, d.

3. Match each function to its inverse.

 A. $y = 2x - 3$ **1.** $x = \dfrac{y + 2}{3}$

 B. $y = 3x$ **2.** $x = \dfrac{y + 3}{2}$

 C. $y = 3x - 2$ **3.** $x = 3y + 2$

 D. $y = x - 2$ **4.** $x = y + 2$

 E. $y = x + 2$ **5.** $x = \dfrac{y}{3}$

 F. $y = \dfrac{x - 2}{3}$ **6.** $x = y - 2$

4. Functions h and j are inverses. When x is -10, the value of $h(x)$ is 7, or $h(-10) = 7$.

 a. What is the value of $j(7)$?

 b. Determine if each point is on the graph of h, on the graph of j, or neither. Explain your reasoning.

   i. (-10, 7)

   ii. (7, -10)

5. Crickets make chirping sounds by rubbing their wings together. The number of chirps they make is closely related to the temperature of their environment. When the temperature is between 55 and 100 degrees Fahrenheit, we can tell the temperature by counting the number of chirps!

A formula that is commonly used to find the temperature in degrees Fahrenheit is:

Count the number of chirps in 14 seconds, and then add 40 to get the temperature.

Let n be the number of chirps that crickets make in 14 seconds and F be the temperature in degrees Fahrenheit. (Lesson 4-16)

a. What is the temperature when a cricket chirps 52 times in 14 seconds?

b. Write an equation that defines F as a function of n.

c. How many chirps would we expect to hear in 14 seconds when it is 60 degrees Fahrenheit?

d. Write an equation that defines n as a function of F.

NAME _____ DATE _____ PERIOD _____

6. Describe the domain and range of the function this graph represents. **(Lesson 4-12)**

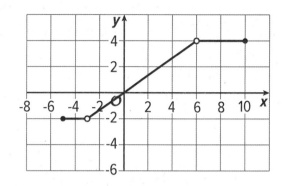

7. The parking rate R for a car in a garage is a function of t, the hours it is parked. **(Lesson 4-12)**

   a. Find $R(1)$.

   b. Find $R(4.5)$.

   c. Find $R(8)$.

STARTING NOVEMBER 1
DAILY PARKING RATES
COLLEGE & LAKEVIEW PARKING GARAGE

HOURS	RATE
0 - 2	FREE
2 - 2.5	$2.00
2.5 - 3	$3.00
3 - 3.5	$4.00
3.5 - 4	$5.00
4 - 5	$6.00
5 - 6	$7.00
6 - 7	$8.00
>7	$8.00
LOST TICKET	$8.00

Illustrative Math

8. Here are rules that define function f. (Lesson 4-12)

$$f(x) = \begin{cases} 2, & -5 \leq x \leq 1 \\ x, & 1 < x < 5 \\ 7, & 5 \leq x \leq 7 \end{cases}$$

Draw the graph of f.

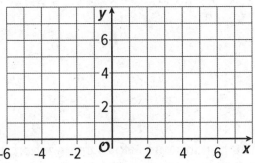

Lesson 4-18

Using Functions to Model Battery Power

NAME _____ DATE _____ PERIOD _____

Learning Goal Let's use functions to model data and make predictions.

 Warm Up

18.1 Devices

Think about an electronic device with a battery that you have to charge on a regular basis.

1. What device is it?

2. When you are using the device, about how long does it take the battery to go from 100% charged until the time you plug it in again to recharge?

3. About how long does it take to charge to 100% starting from 0% or nearly 0%?

4. Suppose you plugged in your device when the battery was 50% charged.

   How long do you think it would take to recharge the device to 100% compared to the time it would take if the device was at 0%? Would it be exactly half the time, more than half the time, or less than half the time it would take if starting from 0%?

Activity

18.2 Charging a Phone

A cell phone is plugged in to be charged. The table shows the percent of battery power at some times after it was plugged in.

Time	Percent Charged
11:00 a.m.	6%
11:10 a.m.	15%
11:30 a.m.	35%
11:40 a.m.	43%

At what time will the battery be 100% charged? Use the data to find out and explain or show your reasoning.

Activity

18.3 How Long Will It Last?

1. The image shows the battery usage of a cell phone 9 hours since it was fully charged.

   It also shows a prediction that the battery would last 8 more hours.

Illustrative Math

NAME _____ DATE _____ PERIOD _____

a. Write an equation for a model that fits the data in the image and gives the percent of battery power, p, as a function of time since the phone was fully charged, t. Show your reasoning.

If you get stuck, consider creating a table of values or a scatter plot of the data.

b. Based on your function, what percentage of power would the battery have 4 hours after this image was taken? What about 5 hours after the image was taken? Show your reasoning.

2. Here are two more images showing the battery usage at two later times, before the battery was charged again.

a. How well did the function you wrote predict the battery power 4 and 5 hours since the first image was taken (that is, 13 and 14 hours after the battery was fully charged)? Explain or show your reasoning.

b. What do you notice from the images about the change in the prediction between $t = 13$ and $t = 14$?

c. Write a new equation for a function that would better fit the data shown in the last image.

NAME _____ DATE _____ PERIOD _____

Are you ready for more?

Would a piecewise function be a better model for capturing the data shown in all three images? If so, what might the rules of that function be?

1. Two cyclists, A and B, are going on a bike ride and are meeting at an orchard. They left home at the same time.

   Functions A and B give their distance from the orchard, in miles, after riding for x hours. The functions are defined by these equations:

   $$A(x) = 48.5 - 21x \qquad\qquad B(x) = 42 - 16.8x$$

   For each question, explain or show your reasoning.

   a. Which cyclist lives farther away from the orchard?

   b. Who will get to the orchard first? How much earlier will that cyclist arrive?

   c. Is there a time when both cyclists are the same distance from the orchard?

NAME _____ DATE _____ PERIOD _____

2. Each equation defines a function. Write an equation for the inverse function. **(Lesson 4-17)**

   a. $y(x) = 65 + 5x$

   b. $f(t) = 3.5 - 0.5t$

   c. $P(n) = \dfrac{n}{3} - 1.2$

3. The number of chirps that crickets make is closely related to the temperature of their environment. When the temperature is between 12 and 38 degrees Celsius, we can tell the temperature by counting the number of chirps!

   A formula that is commonly used to find the temperature in degrees Celsius is:

   Count the number of chirps in 25 seconds, divide by 3, then add 4 to get the temperature.

   Let m be the number of chirps that crickets make in 25 seconds and C be the temperature in degrees Celsius. **(Lesson 4-16)**

   a. What is the temperature when 84 chirps are heard in 25 seconds?

   b. Write an equation that defines C as a function of m.

   c. How many chirps would we expect to hear in 25 seconds when it is 14 degrees Celsius?

   d. Write an equation that defines the inverse of the function you wrote. Explain what the inverse function tells us about the situation.

4. A college student borrows $360 from his cousin to repair his car. He agrees to pay $15 per week until the loan is paid off. (Lesson 4-17)

a. Function L represents the amount owed, w weeks after the student borrows the money. Write an equation to represent this function. Use function notation.

b. Write an equation to represent the inverse of function L. Explain what information it tells us about the situation.

c. How many weeks will it take the student to pay off the loan?

5. A family bought a used car that had been driven 12,000 miles.

The table shows the total distance, in miles, that the car has traveled each year since the purchase. (Lesson 4-17)

Years Since Purchased	Total Miles Traveled
0	12,000
1	15,140
2	18,525
3	21,750

a. On average, how many miles does the family drive each year? Explain or show your reasoning.

b. Write an equation that could define function M, which gives the total miles traveled, t years since the purchase. Use function notation.

c. Write an equation that is the inverse of function M. Explain what information it tells us about the situation.

d. If the family's driving trend continues, when will the car have traveled 50,000 miles? Explain or show your reasoning.

Learning Targets

Lesson	Learning Target(s)
4-1 Describing and Graphing Situations	• I can explain when a relationship between two quantities is a function. • I can identify independent and dependent variables in a function, and use words and graphs to represent the function. • I can make sense of descriptions and graphs of functions and explain what they tell us about situations.
4-2 Function Notation	• I can use function notation to express functions that have specific inputs and outputs. • I understand what function notation is and why it exists. • When given a statement written in function notation, I can explain what it means in terms of a situation.

(continued on the next page)

(continued from the previous page)

Lesson	Learning Target(s)
4-3 Interpreting & Using Function Notation	• I can describe the connections between a statement in function notation and the graph of the function. • I can use function notation to efficiently represent a relationship between two quantities in a situation. • I can use statements in function notation to sketch a graph of a function.
4-4 Using Function Notation to Describe Rules (Part 1)	• I can make sense of rules of functions when they are written in function notation, and create tables and graphs to represent the functions. • I can write equations that represent the rules of functions.
4-5 Using Function Notation to Describe Rules (Part 2)	• I can use technology to graph a function given in function notation, and use the graph to find the values of the function. • I know different ways to find the value of a function and to solve equations written in function notation. • I know what makes a function a linear function.

(continued on the next page)

(continued from the previous page)

Lesson	Learning Target(s)
4-6 Features of Graphs	• I can identify important features of graphs of functions and explain what they mean in the situations represented.
	• I understand and can use the terms "horizontal intercept," "vertical intercept," "maximum," and "minimum" when talking about functions and their graphs.
4-7 Using Graphs to Find Average Rate of Change	• I understand the meaning of the term "average rate of change."
	• When given a graph of a function, I can estimate or calculate the average rate of change between two points.
4-8 Interpreting and Creating Graphs	• I can explain the average rate of change of a function in terms of a situation.
	• I can make sense of important features of a graph and explain what they mean in a situation.
	• When given a description or a visual representation of a situation, I can sketch a graph that shows important features of the situation.

(continued on the next page)

(continued from the previous page)

Lesson	Learning Target(s)
4-9 Comparing Graphs	• I can compare the features of graphs of functions and explain what they mean in the situations represented.
	• I can make sense of an equation of the form $f(x) = g(x)$ in terms of a situation and a graph, and know how to find the solutions.
	• I can make sense of statements about two or more functions when they are written in function notation.
4-10 Domain and Range (Part 1)	• I know what is meant by the "domain" and "range" of a function.
	• When given a description of a function in a situation, I can determine reasonable domain and range for the function.
4-11 Domain and Range (Part 2)	• When given a description of a function in a situation, I can determine reasonable domain and range for the function.

(continued on the next page)

(continued from the previous page)

Lesson	Learning Target(s)
4-12 Piecewise Functions	• I can make sense of a graph of a piecewise function in terms of a situation, and sketch a graph of the function when the rules are given.
	• I can make sense of the rules of a piecewise function when they are written in function notation and explain what they mean in the situation represented.
	• I understand what makes a function a piecewise function.
4-13 Absolute Value Functions (Part 1)	• Given a set of numerical guesses and a target number, I can calculate absolute errors and create a scatter plot of the data.
	• I can analyze and describe features of a scatter plot that shows absolute error data.
	• I can describe the general relationship between guesses and absolute errors using words or equations.

(continued on the next page)

(continued from the previous page)

Lesson	Learning Target(s)
4-14 Absolute Value Functions (Part 2)	• I can describe the effects of adding a number to the expression that defines an absolute value function.
	• I can explain the meaning of absolute value function in terms of distance.
	• When given an absolute value function in words or in function notation, I can make sense of it, and can create a table of values and a graph to represent it.
4-15 Inverse Functions	• I understand the meaning of "inverse function" and how it could be found.
	• When given a linear function that represents a situation, I can use words and equations to describe the inverse function.
4-16 Finding and Interpreting Inverse Functions	• I can explain the meaning of an inverse function in terms of a situation.
	• When I have an equation that defines a linear function, I know how to find its inverse.

(continued on the next page)

(continued from the previous page)

Lesson	Learning Target(s)
4-17 Writing Inverse Functions to Solve Problems	• I can write a linear function to model given data and find the inverse of the function. • When given a linear function defined using function notation, I know how to find its inverse.

Notes

Glossary

A

absolute value The absolute value of a number is its distance from 0 on the number line.

absolute value function The function f given by $f(x) = |x|$.

association In statistics we say that there is an association between two variables if the two variables are statistically related to each other; if the value of one of the variables can be used to estimate the value of the other.

average rate of change The average rate of change of a function f between inputs a and b is the change in the outputs divided by the change in the inputs: $\dfrac{f(b) - f(a)}{b - a}$. It is the slope of the line joining $(a, f(a))$ and $(b, f(b))$ on the graph.

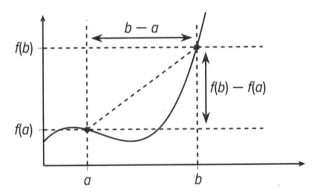

B

bell-shaped distribution A distribution whose dot plot or histogram takes the form of a bell with most of the data clustered near the center and fewer points farther from the center.

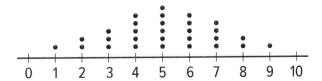

bimodal distribution A distribution with two very common data values seen in a dot plot or histogram as distinct peaks. In the dot plot shown, the two common data values are 2 and 7.

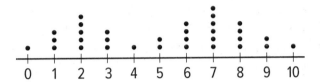

C

categorical data Categorical data are data where the values are categories. For example, the breeds of 10 different dogs are categorical data. Another example is the colors of 100 different flowers.

categorical variable A variable that takes on values which can be divided into groups or categories. For example, color is a categorical variable which can take on the values, red, blue, green, etc.

causal relationship A causal relationship is one in which a change in one of the variables causes a change in the other variable.

coefficient In an algebraic expression, the coefficient of a variable is the constant the variable is multiplied by. If the variable appears by itself then it is regarded as being multiplied by 1 and the coefficient is 1.

The coefficient of x in the expression $3x + 2$ is 3. The coefficient of p in the expression $5 + p$ is 1.

completing the square Completing the square in a quadratic expression means transforming it into the form $a(x + p)^2 - q$, where a, p, and q are constants.

Completing the square in a quadratic equation means transforming into the form $a(x + p)^2 = q$.

constant term In an expression like $5x + 2$ the number 2 is called the constant term because it doesn't change when x changes. In the expression $5x - 8$ the constant term is -8, because we think of the expression as $5x + (-8)$. In the expression $12x - 4$ the constant term is -4.

constraint A limitation on the possible values of variables in a model, often expressed by an equation or inequality or by specifying that the value must be an integer. For example, distance above the ground d, in meters, might be constrained to be non-negative, expressed by $d \geq 0$.

correlation coefficient A number between -1 and 1 that describes the strength and direction of a linear association between two numerical variables. The sign of the correlation coefficient is the same as the sign of the slope of the best fit line. The closer the correlation coefficient is to 0, the weaker the linear relationship. When the correlation coefficient is closer to 1 or -1, the linear model fits the data better. The first figure shows a correlation coefficient which is close to 1, the second a correlation coefficient which is positive but closer to 0, and the third a correlation coefficient which is close to -1.

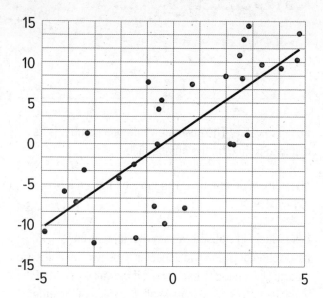

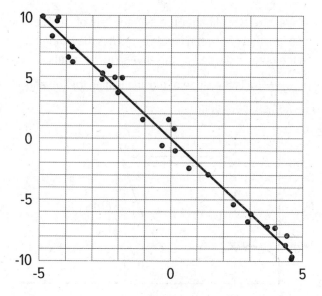

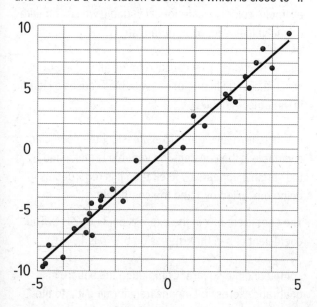

D

decreasing (function) A function is decreasing if its outputs get smaller as the inputs get larger, resulting in a downward sloping graph as you move from left to right.

A function can also be decreasing just for a restricted range of inputs. For example the function f given by $f(x) = 3 - x^2$, whose graph is shown, is decreasing for $x \geq 0$ because the graph slopes downward to the right of the vertical axis.

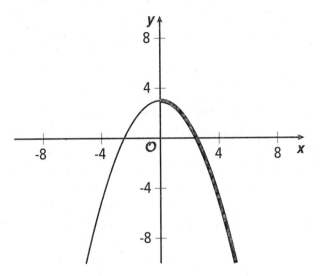

dependent variable A variable representing the output of a function.

The equation $y = 6 - x$ defines y as a function of x. The variable x is the independent variable, because you can choose any value for it. The variable y is called the dependent variable, because it depends on x. Once you have chosen a value for x, the value of y is determined.

distribution For a numerical or categorical data set, the distribution tells you how many of each value or each category there are in the data set.

domain The domain of a function is the set of all of its possible input values.

E

elimination A method of solving a system of two equations in two variables where you add or subtract a multiple of one equation to another in order to get an equation with only one of the variables (thus eliminating the other variable).

equivalent equations Equations that have the exact same solutions are equivalent equations.

equivalent systems Two systems are equivalent if they share the exact same solution set.

exponential function An exponential function is a function that has a constant growth factor. Another way to say this is that it grows by equal factors over equal intervals. For example, $f(x) = 2 \cdot 3^x$ defines an exponential function. Any time x increases by 1, $f(x)$ increases by a factor of 3.

F

factored form (of a quadratic expression) A quadratic expression that is written as the product of a constant times two linear factors is said to be in factored form. For example, $2(x - 1)(x + 3)$ and $(5x + 2)(3x - 1)$ are both in factored form.

five-number summary The five-number summary of a data set consists of the minimum, the three quartiles, and the maximum. It is often indicated by a box plot like the one shown, where the minimum is 2, the three quartiles are 4, 4.5, and 6.5, and the maximum is 9.

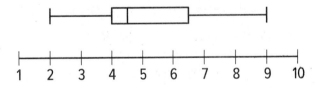

function A function takes inputs from one set and assigns them to outputs from another set, assigning exactly one output to each input.

function notation Function notation is a way of writing the outputs of a function that you have given a name to. If the function is named f and x is an input, then $f(x)$ denotes the corresponding output.

G

graph of a function The graph of a function is the set of all of its input-output pairs in the coordinate plane.

growth factor In an exponential function, the output is multiplied by the same factor every time the input increases by one. The multiplier is called the growth factor.

growth rate In an exponential function, the growth rate is the fraction or percentage of the output that gets added every time the input is increased by one. If the growth rate is 20% or 0.2, then the growth factor is 1.2.

H

horizontal intercept The horizontal intercept of a graph is the point where the graph crosses the horizontal axis. If the axis is labeled with the variable x, the horizontal intercept is also called the x-intercept. The horizontal intercept of the graph of $2x + 4y = 12$ is $(6,0)$.

The term is sometimes used to refer only to the x-coordinate of the point where the graph crosses the horizontal axis.

I

increasing (function) A function is increasing if its outputs get larger as the inputs get larger, resulting in an upward sloping graph as you move from left to right.

A function can also be increasing just for a restricted range of inputs. For example the function f given by $f(x) = 3 - x^2$, whose graph is shown, is increasing for $x \leq 0$ because the graph slopes upward to the left of the vertical axis.

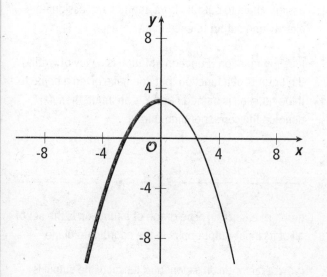

independent variable A variable representing the input of a function.

The equation $y = 6 - x$ defines y as a function of x. The variable x is the independent variable, because you can choose any value for it. The variable y is called the dependent variable, because it depends on x. Once you have chosen a value for x, the value of y is determined.

intercept A point on the graph of a function which is also on one of the axes.

inverse (function) Two functions are inverses to each other if their input-output pairs are reversed, so that if one function takes a as input and gives b as an output, then the other function takes b as an input and gives a as an output. You can sometimes find an inverse function by reversing the processes that define the first function in order to define the second function.

irrational number An irrational number is a number that is not rational. That is, it cannot be expressed as a positive or negative fraction, or zero.

L

linear function A linear function is a function that has a constant rate of change. Another way to say this is that it grows by equal differences over equal intervals. For example, $f(x) = 4x - 3$ defines a linear function. Any time x increases by 1, $f(x)$ increases by 4.

linear term The linear term in a quadratic expression (In standard form) $ax^2 + bx + c$, where a, b, and c are constants, is the term bx. (If the expression is not in standard form, it may need to be rewritten in standard form first.)

M

maximum A value of a function that is greater than or equal to all the other values, corresponding to the highest point on the graph of the function.

minimum A value of a function that is less than or equal to all the other values, corresponding to the lowest point on the graph of the function.

model A mathematical or statistical representation of a problem from science, technology, engineering, work, or everyday life, used to solve problems and make decisions.

N

negative relationship A relationship between two numerical variables is negative if an increase in the data for one variable tends to be paired with a decrease in the data for the other variable.

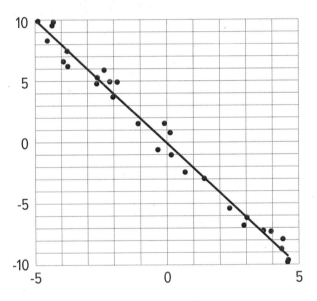

non-statistical question A non-statistical question is a question which can be answered by a specific measurement or procedure where no variability is anticipated, for example:

- How high is that building?
- If I run at 2 meters per second, how long will it take me to run 100 meters?

numerical data Numerical data, also called measurement or quantitative data, are data where the values are numbers, measurements, or quantities. For example, the weights of 10 different dogs are numerical data.

O

outlier A data value that is unusual in that it differs quite a bit from the other values in the data set. In the box plot shown, the minimum, 0, and the maximum, 44, are both outliers.

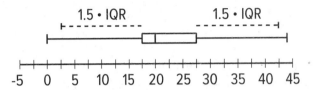

P

perfect square A perfect square is an expression that is something times itself. Usually we are interested in situations where the something is a rational number or an expression with rational coefficients.

piecewise function A piecewise function is a function defined using different expressions for different intervals in its domain.

positive relationship A relationship between two numerical variables is positive if an increase in the data for one variable tends to be paired with an increase in the data for the other variable.

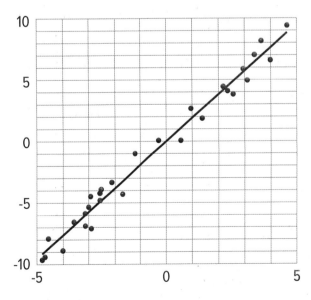

Q

quadratic equation An equation that is equivalent to one of the form $ax^2 + bx + c = 0$, where a, b, and c are constants and $a \neq 0$.

quadratic expression A quadratic expression in x is one that is equivalent to an expression of the form $ax^2 + bx + c$, where a, b, and c are constants and $a \neq 0$.

quadratic formula The formula $x = \dfrac{-b \pm \sqrt{b^2 - 4ac}}{2a}$ that gives the solutions of the quadratic equation $ax^2 + bx + c = 0$, where a is not 0.

quadratic function A function where the output is given by a quadratic expression in the input.

R

range The range of a function is the set of all of its possible output values.

rational number A rational number is a fraction or the opposite of a fraction. Remember that a fraction is a point on the number line that you get by dividing the unit interval into b equal parts and finding the point that is a of them from 0. We can always write a fraction in the form $\dfrac{a}{b}$ where a and b are whole numbers, with b not equal to 0, but there are other ways to write them. For example, 0.7 is a fraction because it is the point on the number line you get by dividing the unit interval into 10 equal parts and finding the point that is 7 of those parts away from 0. We can also write this number as $\dfrac{7}{10}$.

The numbers 3, $-\dfrac{3}{4}$, and 6.7 are all rational numbers. The numbers π and $-\sqrt{2}$ are not rational numbers, because they cannot be written as fractions or their opposites.

relative frequency table A version of a two-way table in which the value in each cell is divided by the total number of responses in the entire table or by the total number of responses in a row or a column. The table illustrates the first type for the relationship between the condition of a textbook and its price for 120 of the books at a college bookstore.

	$10 or Less	More than $10 but Less than $30	$30 or More
new	0.025	0.075	0.225
used	0.275	0.300	0.100

residual The difference between the y-value for a point in a scatter plot and the value predicted by a linear model. The lengths of the dashed lines in the figure are the residuals for each data point.

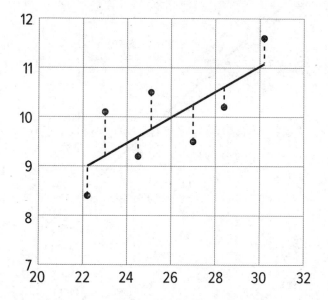

S

skewed distribution A distribution where one side of the distribution has more values farther from the bulk of the data than the other side, so that the mean is not equal to the median. In the dot plot shown, the data values on the left, such as 1, 2, and 3, are further from the bulk of the data than the data values on the right.

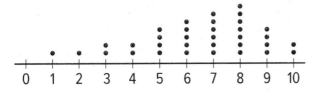

solution to a system of equations A coordinate pair that makes both equations in the system true.

On the graph shown of the equations in a system, the solution is the point where the graphs intersect.

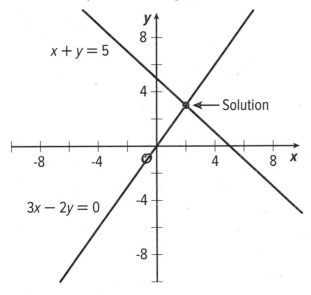

solutions to a system of inequalities All pairs of values that make the inequalities in a system true are solutions to the system. The solutions to a system of inequalities can be represented by the points in the region where the graphs of the two inequalities overlap.

standard deviation A measure of the variability, or spread, of a distribution, calculated by a method similar to the method for calculating the MAD (mean absolute deviation). The exact method is studied in more advanced courses.

standard form (of a quadratic expression) The standard form of a quadratic expression in x is $ax^2 + bx + c$, where a, b, and c are constants, and a is not 0.

statistic A quantity that is calculated from sample data, such as mean, median, or MAD (mean absolute deviation).

statistical question A statistical question is a question that can only be answered by using data and where we expect the data to have variability, for example:

- Who is the most popular musical artist at your school?
- When do students in your class typically eat dinner?
- Which classroom in your school has the most books?

strong relationship A relationship between two numerical variables is strong if the data is tightly clustered around the best fit line.

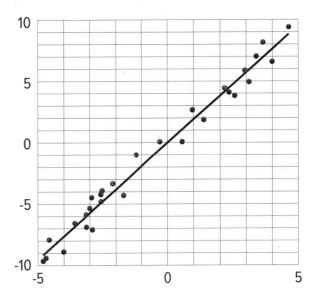

substitution Substitution is replacing a variable with an expression it is equal to.

symmetric distribution A distribution with a vertical line of symmetry in the center of the graphical representation, so that the mean is equal to the median. In the dot plot shown, the distribution is symmetric about the data value 5.

system of equations Two or more equations that represent the constraints in the same situation form a system of equations.

system of inequalities Two or more inequalities that represent the constraints in the same situation form a system of inequalities.

two-way table A way of organizing data from two categorical variables in order to investigate the association between them.

	has a cell phone	does not have a cell phone
10–12 years old	25	35
13–15 years old	38	12
16–18 years old	52	8

uniform distribution A distribution which has the data values evenly distributed throughout the range of the data.

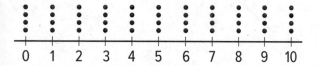

variable (statistics) A characteristic of individuals in a population that can take on different values

vertex (of a graph) The vertex of the graph of a quadratic function or of an absolute value function is the point where the graph changes from increasing to decreasing or vice versa. It is the highest or lowest point on the graph.

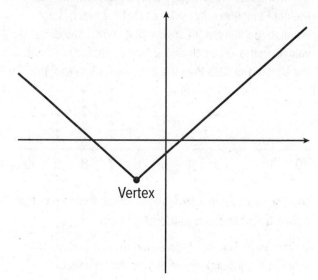

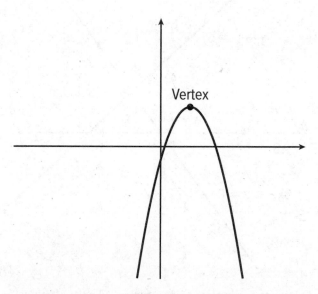

vertex form (of a quadratic expression) The vertex form of a quadratic expression in x is $a(x - h)^2 + k$, where a, h, and k are constants, and a is not 0.

vertical intercept The vertical intercept of a graph is the point where the graph crosses the vertical axis. If the axis is labeled with the variable y, the vertical intercept is also called the y-intercept.

Also, the term is sometimes used to mean just the y-coordinate of the point where the graph crosses the vertical axis. The vertical intercept of the graph of $y = 3x - 5$ is (0, -5), or just -5.

W

weak relationship A relationship between two numerical variables is weak if the data is loosely spread around the best fit line.

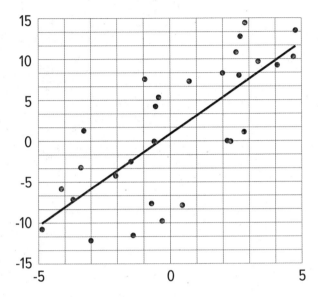

Z

zero (of a function) A zero of a function is an input that yields an output of zero. If other words, if $f(a) = 0$ then a is a zero of f.

zero product property The zero product property says that if the product of two numbers is 0, then one of the numbers must be 0.

Index